Comprehensive College Algebra

Building Mathematical Insights Through Logic and Exercises

First Edition

By Xiang Ji and Ge Mu
Pennsylvania State University

cognella
academic publishing

Bassim Hamadeh, CEO and Publisher
Christopher Foster, General Vice President
Michael Simpson, Vice President of Acquisitions
Jessica Knott, Managing Editor
Kevin Fahey, Marketing Manager
Jess Busch, Senior Graphic Designer
Melissa Barcomb, Acquisitions Editor
Sarah Wheeler, Senior Project Editor
Stephanie Sandler, Licensing Associate

First published in the United States of America in 2013 by Cognella, Inc.

File licensed by www.depositphotos.com

Printed in the United States of America

ISBN: 978-1-62131-010-5

cognella
academic publishing

www.cognella.com   800-200-3908

Contents

Preface

We believe everybody can learn mathematics well through good instructions and enough training. Based on the past four years' teaching experience of mathematics in universities, together with our study experience as undergraduates, the authors of this book developed their own ideas on how to teach and learn college algebra.

A better understanding of mathematics depends on a correct and thorough understanding of concepts. Mathematics does not become simpler when students avoid accurate definitions but concentrate on problem-solving only. The basic concepts are like foundations of the building of mathematics. Only with rigorous definitions and theorems, can one have a clear picture of the materials; one can even develop the method to solve problems by himself and avoid the awkward situation of reciting the methods without understanding, like floating in a balloon without root.

Mathematics skills are developed by plenty of thinking and training. The ability of critical thinking is a major goal of mathematics learning. It grows with mathematics skills through practice. After diving into the detailed exercises, we not only get a betting understanding of materials, but also learn to think following logic.

With these ideas in mind, we designed the book into the current format: a concise but rigorous and complete introduction of college algebra with typical examples followed by adequate and diverse exercises of different levels.

College algebra is a subject that combines the contents of elementary algebra (opposite to modern/abstract algebra) and coordinate geometry (or analytic geometry) of 2-dimension, aiming to provide solid backgrounds and sufficient trainings for other courses, not limited to mathematics classes. Elementary algebra starts from the discovery of numbers with their operations. The key idea of elementary algebra is to bring in variables, which allows us to build algebraic expressions (for example, polynomials, rational expressions, radical expressions) as well as functions. The major task of elementary algebra is to solve algebraic equations and inequalities, or intrinsically to solve polynomial equations. Coordinate geometry was born with the invention of coordinate systems, which makes it possible to algebraize geometric objects, such as points, lines and circles. The philosophy of coordinate geometry is to use algebra as tools to study geometry.

All materials are presented in a mathematical way, following the logic of the subjects. The book starts from the development of the real number system, but we add an introduction of sets before it, which allows us to build everything on a firm foundation. The readers are assumed to be familiar with natural numbers, integers, rational numbers as well as their operations, although we provide an systematical introduction of number systems in the book.

The exercises are generally from easy to hard. The 'true or false' and 'multiple choice' problems mainly test the understanding of concepts and theorems. This is the part which embeds our philosophy of emphasizing definitions, propositions, theorems as well as the conditions for them to hold. There are also many routine exercises for training purpose. Some comprehensive problems are arranged at the end. We encourage the reader to solve some of these problems independently. This would be a great opportunity for thinking mathematically, and you may get the sense of achievement after working them out.

To improve the skill of calculations, we strongly discourage the use of calculators. All problems are within the ability of hand calculations. Because some of the exercises are of high difficulty level, we do recommend using the book under instruction. Moreover, we encourage the reader to keep all numbers in their accurate form, not to replace any fractions, roots and other irrational numbers such as $\frac{5}{7}$, $\sqrt{3}$ and π by finite decimal approximations.

We use some abbreviations in the book. 'i.e.' stands for 'that is'; 'e.g.' stands for 'for example'; 'D.N.E' stands for 'does not exist'; '$\Rightarrow$' stands for 'implies'; '$\Leftrightarrow$' stands for 'be equivalent to'. There are also some other conventions. All of them are explained in context. In our book, 0 is considered as a natural number. This convention is different from most college algebra textbooks, but consistent with the recognition of most mathematicians.

The book is for college students but can also be used by high school students who are willing to complete their knowledge of algebra, to seek more training, or to work on some challenging and interesting problems. The book can serve either as a textbook, as a workbook, or just as a reference.

It is a happy and fulfilling experience to write the book, full of fruitful discussions as well as efficient cooperation. We would endlessly thank our parents and family for their love, support and encouragement. We are also grateful to University Readers Inc. which finally made this book come into being; and to the editor, Sarah Wheeler, for her passionate help and assistance.

December 21, 2012

Sets

In order to have a convenient language to present the subject of college algebra, as well as to build the subject on a firm foundation, we start the book with an introduction of sets. Sets are one of the most fundamental concepts in mathematics. However, as a metaconcept, there is no rigorous mathematical definition, except verbal descriptions together with some requirements.

Sets have great power in describing collections of objects. Besides the basic concepts, several important operations, which will be used later in the book, together with the propositions they satisfy are discussed as well.

1.1 Sets and Subsets

A set is a well-defined collection of objects. An object belonging to the collection is called an element of the set. By convention we use capital letters to denote sets and small letters to denote objects. An object a is an element of a set A is denoted by $a \in A$, reading as 'a belongs to A'; otherwise we write $a \notin A$ (a does not belong to A). For example,

the collection $\mathbb{N}$ of all natural numbers

the collection L of all small English letters

the collection of all universities in the US in 2012

are all sets. $1 \in \mathbb{N}$ but $1 \notin L$.

We can define a set either intensionally (by specifying the exact properties of its elements) or extensionally (by listing all its elements). For example, the set of all planets in the solar system can either be defined intensionally as

$$X = \{\text{planets in the solar system}\}$$

or extensionally as

$$Y = \{\text{Mercury, Venus, Earth, Mars, Jupiter, Saturn, Uranus, Neptune}\}.$$

Here the objects are enclosed by braces '{' and '}' to indicate the set made of them. Some sets, especially those consisting of infinitely many elements, can only be defined intensionally, e.g.

the set P of all prime numbers. We also use set-builder notation to define sets intensionaly. The set-builder notation has the form

$$\{x|\text{properties of } x\},$$

which contains two parts separated by '|'. The letter x before the separator stands for a general element in the set; the part after the separator describes the properties of x. For example, using set-builder notation, the set P above can be written as $\{a|a \text{ is a prime number}\}$; similarly $X = \{x|x \text{ is a planet in the solar system}\}$.

For a set A to be well-defined we require the following:

- Given any object, we should be able to determine if it is an element of A;

- All elements of A are distinct, i.e. any repeating element is considered as a single element, e.g. $\{1, 2\}$ and $\{1, 1, 2, 1\}$ are same sets;

- The elements of A are not in order, e.g. $\{1, 2, 3\}$ and $\{3, 2, 1\}$ are the same sets.

The empty set is the set containing no elements, denoted by $\emptyset$. The empty set is unique.

Given a set A, a set S is called a subset of A, denoted by $S \subseteq A$ or $A \supseteq S$, if any element of S is also an element of A, i.e. $a \in S$ implies $a \in A$. For example $\{1, 2\} \subseteq \{1, 2, 3\}$ and $P \subseteq \mathbb{N}$. A is not a subset of B can be denoted by $A \nsubseteq B$. For example, $\{1, 4\} \nsubseteq \{1, 2, 3\}$, $\mathbb{N} \nsubseteq P$(not all natural numbers are prime).

The empty set $\emptyset$ is a subset of any set, i.e. $\emptyset \subseteq A$ for all A. Any set A is always a subset of itself, i.e. $A \subseteq A$.

Two sets A and B are equal, denoted by $A = B$, if they contain exactly the same elements. Mathematically $A = B$ if and only if $A \subseteq B$ and $B \subseteq A$. In the previous example of planets in solar system, X and Y are equal.

S is a proper subset of A if $S \subseteq A$ but $S \neq A$, denoted by $S \subset A$. For example, $\emptyset \subset \{1, 3\}$, $\{3\} \subset \{1, 3\}$, but $\{1, 3\} \not\subset \{1, 3\}$. In fact, any subset of A except A itself is a proper subset of A.

We need to distinguish the two notations $3 \in \{1, 3\}$ and $\{3\} \subset \{1, 3\}$, both of which are legal. $3 \in \{1, 3\}$ means 3 is an element of the set $\{1, 3\}$; $\{3\} \subset \{1, 3\}$ means the set containing only one number 3 is a proper subset of $\{1, 3\}$, which is true as well.

Exercises.

1. True or false.

(1) $\emptyset \subset \emptyset$

(2) $\emptyset \subseteq \emptyset$

(3) $\emptyset \in \emptyset$

(4) $\emptyset \subset \{\emptyset\}$

(5) $\emptyset \in \{\emptyset\}$

(6) $\{a\} = \{a, \{a\}\}$

(7) $\{a, b\} \subseteq \{a, b, c, \{a\}\}$

(8) $\{a, b\} \in \{a, b, \{a, b, c\}\}$

(9) $\{a,b\} \in \{a,b,c,\{a,b\}\}$

(10) $\{a,b\} \in \{a,b,c,\{\{a,b\}\}\}$

2. Find sets using set-builder notation.

(1) the collection of nonnegative integers less than 6

(2) the collection of all odd numbers

(3) the set of all rational numbers

3. Rewrite the set extensionally.

(1) $A = \{x|x \text{ is a digit}\}$

(2) $B = \{x|x = 2 \text{ or } x = 5\}$

(3) $C = \{x|x \text{ is an integer greater than 3 but less than 13}\}$

(4) $D = \{x|x^2 - 1 = 0, x > 3\}$

(5) $E = \{(x,y)|x,y \text{ are integers and } -1 \le x < 2, -1 < y \le 1\}$

4. Find all subsets of $\{1,2,3\}$.

1.2 Operations on Sets

There are several operations on sets. In this section we are going to introduce intersection, union, difference, complement and Cartesian product.

The intersection of two sets A and B is the set $A \cap B = \{x|x \in A \text{ and } x \in B\}$, i.e. the collection of all common elements of A and B. A element x belongs to $A \cap B$ if and only if x belongs to both A and B. Two sets A,B are disjoint if $A \cap B = \emptyset$, i.e A and B has no common elements.

The union of two sets A and B is the set $A \cup B = \{x | x \in A \text{ or } x \in B\}$, i.e. the collection of all elements of A and all elements of B. A element x belongs to $A \cup B$ if and only if x belongs to at least on of A and B.

The difference of two sets A and B is the set $A - B = \{x \in A | x \notin B\}$, i.e. the collection of all elements belonging to A but not belonging to B.

Given a set E and a subset S, the complement of S in E is the subset $S^c = E - S$. In this case we treat E as the universal set. For two subsets $A, B \subset E$, $A - B = A \cap B^c$.

The Cartesian product of two sets A and B is the set $A \times B = \{(a, b) | a \in A \text{ and } b \in B\}$, i.e. the collection of all ordered pairs (a, b) with the first component a belonging to A and the second component b belonging to B. Similarly, $A \times B \times C = \{(a, b, c) | a \in A, b \in B, c \in C\}$ is the set of all ordered triples.

Suppose E is the universal set and A, B are two subsets of E. We can use Venn diagrams to visualize all above operations on A, B except the Cartesian product.

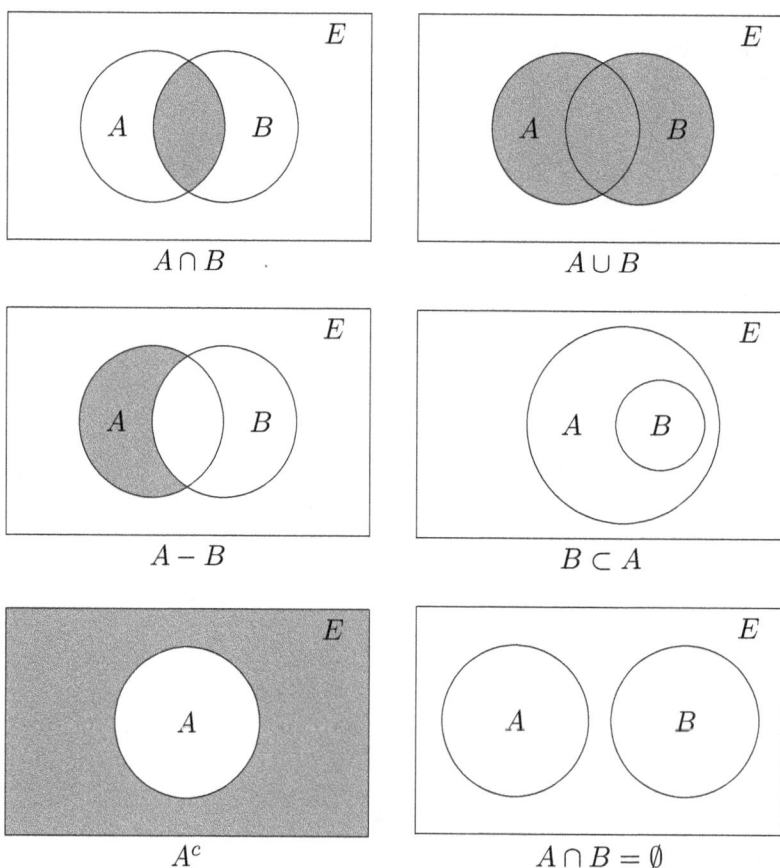

$A \cap B$

$A \cup B$

$A - B$

$B \subset A$

A^c

$A \cap B = \emptyset$

Example. If $E = \{1, 2, 3, 4, 5\}$ is the universal set and $A = \{2, 5\}$, $B = \{2, 3, 4\}$ are two subsets, then

$A \cap B = \{2\}$, $A \cup B = \{2, 3, 4, 5\}$,

$A^c = \{1, 3, 4\}$, $B^c = \{1, 5\}$,

$A - B = \{5\}$, $B - A = \{3, 4\}$

$A \times B = \{(2, 2), (2, 3), (2, 4), (5, 2), (5, 3), (5, 4)\}$.

Exercises.

1. If $E = \{1, 2, 3, 4, 5, 6\}$ is the universal set, $A = \{1, 4\}$, $B = \{1, 2, 5\}$, $C = \{2, 4\}$, find

(1) B^c

(2) $A - B$

(3) $(A \cap B) \cup C^c$

(4) $(A \cap B)^c$

(5) $A \cup C$

(6) $B - (A \cup C)$

2. If A, B, C, D are subsets of $\mathbb{Z}$(the set of integers), where

$A = \{1, 2, 7, 8\}$

$B = \{x^2 | x \in \mathbb{Z}, x^2 < 50\}$

$C = \{x \in \mathbb{Z} | 0 \leq x \leq 20, \text{the remainder of } x \text{ dividing 3 is 1}\}$

$D = \{2^k | k = 0, 1, 2, 3, 4, 5, 6\}$

Find the following sets by listing their elements.

(1) $A \cup B \cup C \cup D$

(2) $A \cap B \cap C \cap D$

(3) $B - (A \cup C)$

(4) $(A^c \cap C) \cup D$

3. Suppose A, B, C are all subsets of E. Find the following subsets by Venn diagrams.

(1) $A^c \cap B^c$ (2) $A^c \cup B^c$

(3) $(A - (B \cup C)) \cup ((B \cup C) - A)$ (4) $A \cap (B^c \cup C)$

4. If $A \subseteq B$, find

(1) $A \cup B$ (2) $A \cap B$

(3) $A - B$ (4) $B^c \cap A$

1.3 Propositions of Operations on Sets

Suppose A, B, C are subsets of E. The operations of sets satisfy the following propositions.

- $\emptyset^c = E$, $E^c = \emptyset$

- $A \cap E = A$, $A \cup E = E$

- $A \cap A = A$, $A \cup A = A$

- $A \cap \emptyset = \emptyset$, $A \cup \emptyset = A$

- $A \cap A^c = \emptyset$, $A \cup A^c = E$

- (Double Complement) $(A^c)^c = A$

- (Commutative Law) $A \cap B = B \cap A$, $A \cup B = B \cup A$

- (Associative Law) $(A \cap B) \cap C = A \cap (B \cap C)$, $(A \cup B) \cup C = A \cup (B \cup C)$

- (Distributive Law) $A \cap (B \cup C) = (A \cap B) \cup (A \cap C)$, $A \cup (B \cap C) = (A \cup B) \cap (A \cup C)$

- (Difference Law) $A - B = A \cap B^c$

- (De Morgan's Law) $(B \cup C)^c = B^c \cap C^c$, $(B \cap C)^c = B^c \cup C^c$

- $A \times \emptyset = \emptyset \times A = \emptyset$

- $A \times (B \cap C) = (A \times B) \cap (A \times C)$, $A \times (B \cup C) = (A \times B) \cup (A \times C)$

Exercises.

1. Suppose $E = \{1, 2, 3, 4, 5, 6, 7\}$ is the universal set, and $A = \{2, 4, 6\}$, $B = \{2, 3\}$, $C = \{2, 5, 6\}$ are three subsets. Verify the associative, distributive, difference and De Morgan's law.

2. Define $A \triangle B = (A - B) \cup (B - A)$, called the symmetric difference of A and B. Prove $A \triangle B = (A \cup B) - (A \cap B)$. Draw the Venn diagram of $A \triangle B$.

Number Systems

Numbers together with their operations form the foundation of algebra as well as mathematics. It is a long history for our ancestors to discover and create all different types of numbers. Numbers are classified into sets, which are called number systems. There are five important number systems: the system of natural numbers, the system of integers, the system of rational numbers, the system of real numbers and the system of complex numbers.

Each number system, besides the set of specified numbers, are also characterized by the operations defined over it — addition and multiplication. Different number systems have different properties on these two operations. The two operations together with the properties they satisfy, are considered as part of the number system.

In this chapter, we are going to develop the first four number systems by order, leaving the complex numbers to be introduced later in Chapter 10. By default, a number means a real number in this book.

2.1   Natural Numbers

The numbers $0, 1, 2, 3, 4, 5, \cdots$ are called natural numbers, which are abstracted from the counting of objects. The set of all natural numbers is denoted by $\mathbb{N}$, i.e. $\mathbb{N} = \{0, 1, 2, 3, 4, 5, \cdots\}$ In this book, we adopt the convention that 0 is a natural number for convenience and simplicity.

We are already familiar with the operations of addition and multiplication on natural numbers. Let us take a brief review. Let m, n, p, q denote natural numbers in the following.

Addition. The notation is $m + n$. Both m and n are called summands, and the result is called the sum. For example, $2 + 3 = 5$, $15 + 7 = 22$.

Multiplication. The notation is $m \times n$. Both m and n are called multiplicands, and the result is called the product. For example, $3 \times 5 = 15$, $14 \times 6 = 84$. The notation $m \times n$ sometimes is abbreviated to $m \cdot n$ or mn. For example, $5 \times m = 5 \cdot m = 5m$. However, the '·' of multiplication is easily confused with the '.' of decimals in handwriting. We do not recommend the dot notation when both operands are numbers.

These operations satisfy the following properties:

- Commutative property: $m + n = n + m$, $m \times n = n \times m$;
- Associative property: $(m + n) + p = m + (n + p)$, $(m \times n) \times p = m \times (n \times p)$;
- Distributive property: $m \times (n + p) = m \times n + m \times p$;
- Special numbers: $m + 0 = m = 0 + m$, $m \times 1 = m = 1 \times m$, $m \times 0 = 0 = 0 \times m$.

Sometimes, the operations of subtraction and division of natural numbers can be applied. If $m + n = p$, define $p - m = n$. Here p is called the minuend; m is called the subtrahend; and n is called the difference. For example, $10 - 4 = 6$, $31 - 17 = 14$. Some subtractions, e.g. $3 - 5$, are not well-defined in the set of natural numbers ($3 - 5$ is not a natural number!). The non-closedness of $\mathbb{N}$ under subtraction motivates the discovery of negative integers.

If $m \times n = q$, define $q \div m = n$. Here q is called the dividend; m is called the divisor; and n is called the quotient. In this case, we say q can be divided by m. For example, $12 \div 4 = 3$, $45 \div 5 = 9$. Not all divisions can be done. For example, $19 \div 5 \notin \mathbb{N}$. This leads to, in one direction, the discovery of rational numbers (fractions); in the other direction, the development of the number theory (by division with remainders, $19 = 5 \times 3 + 4$). Pay attention, 0 cannot be the divisor, i.e. for any $m \in \mathbb{N}$, $m \div 0$ is not well-defined or D.N.E.

Besides, the exponentiations (or powers) of natural numbers can be broadly defined. If $m, n \neq 0$, m to the n-th power (or the n-th power of m, m to the power n), denoted by m^n, is the natural number

$$\overbrace{m \times \cdots \times m}^{n \text{ times}}.$$

In the notation m^n, m is called the base, and n is called the exponent. When $m \neq 0$, m^0 is defined to be 1. 0^0 is not well-defined. In particular, $m^2 = m \cdot m$ is called the square of m, which reads 'm squared'; similarly, $m^3 = m \cdot m \cdot m$ is called the cube of m, which reads 'm cubed'. For example, $4^2 = 4 \times 4 = 16$, $3^3 = 3 \times 3 \times 3 = 27$, $5^1 = 5$, $4^0 = 1$.

When combining these operations together, we apply the standard operator precedence.

$$\text{power} \succ \text{(parentheses)} \succ \text{[brackets]} \succ \{\text{braces}\} \succ (\times, \div) \succ (+, -)$$

Here '$\succ$' stands for 'prior to'. The following example demonstrates how the precedence works.

$$
\begin{aligned}
&[11 + (5 + 7) \times 3 - 23] \div 2^2 + 1 \\
=&[11 + (5 + 7) \times 3 - 23] \div 4 + 1 \\
=&[11 + 12 \times 3 - 23] \div 4 + 1 \\
=&[11 + 36 - 23] \div 4 + 1 \\
=&[11 + 13] \div 4 + 1 \\
=&24 \div 4 + 1 \\
=&6 + 1 \\
=&7.
\end{aligned}
$$

There is a partial order on the set of natural numbers. Define

- $m \geq n$ (m is no less than n), if $m - n \in \mathbb{N}$ exists. For example, $1 \geq 0$, $5 \geq 5$, $10 \geq 7$.
- $m > n$ (m is greater than n), if $m \geq n$ but $m \neq n$. For example, $1 > 0$, $9 > 7$, but $5 \not> 5$.

- $m \leq n$ (m is no greater than n), if $n \geq m$. For example, $0 \leq 1, 5 \leq 5$.

- $m < n$ (m is less than n), if $n > m$. For example, $4 < 5, 7 < 10$.

Here '$\geq$', '$\leq$', '$>$', '$<$' are all inequality signs. Besides, we also have '$\neq$' (not equal to), e.g. $2 \neq 3$. It is legal to combine the order relation in the same direction. For example, $0 < 1, 1 \leq 2, 2 < 3$ can be abbreviated to $0 < 1 \leq 2 < 3$. However, $5 < 7 > 3$ is illegal, since the direction of inequality signs contradict. From the definition, $m \geq n$ is equivalent to $m > n$ or $m = n$.

Exercises.

1. True or false.

(1) 0 is not a natural number.

(2) $(27 + 32) - 13 = 32 + (27 - 13)$.

(3) 0 multiplied by any natural number is 0.

(4) $(2 + 3) \times 5 = 2 + 3 \times 5$.

(5) 0 divided by any natural number is 0.

(6) $84 \div (3 + 4) = 84 \div 3 + 84 \div 4$.

(7) Any natural number divided by itself is 1.

(8) $3 \times 7 - 2 \times 7 = (3 - 2) \times 7$.

(9) The sum and product of two natural numbers are natural numbers.

(10) $6 \times (4 + 3) \times 7 = 6 \times 4 + 3 \times 7$.

(11) The difference of any two natural numbers is a natural number.

(12) $0 \div 1 = 1 \div 0 = 0$.

(13) Every natural number can be divide by 1.

(14) $m - 0 = m$ for any $m \in \mathbb{N}$.

(15) For any $n \in \mathbb{N}$, $n^2 \div n = n$.

(16) $23 \times 32 = 726$.

(17) If $0 \neq p \in \mathbb{N}$ can be divided by $m \in \mathbb{N}$, then $p \geq m$.

(18) Given $m, n, p \in \mathbb{N}$, if $m \geq n, n \geq p$, then $m \geq p$.

(19) $9 > 8 > 7 > 6 > 5 > 4 > 3 > 2 > 1 > 0 < 10 < 11 < 12$ is legal.

(20) No natural number is less than 0.

2. Multiple choice.

(1) Which of the following is correct?

    A. $13 \times 12 = 146$ B. $15 \times 13 = 225$ C. $16 \times 14 = 204$ D. $17 \times 19 = 323$

(2) Which of the following calculation is correct?

    A. $36 + 10 \div 2 = 46 \div 2 = 23$

    B. $(87 - 15) \div 6 \times 3 = (87 - 15) \div 18 = 72 \div 18 = 4$

    C. $91 \times 99 = 91 \times (100 - 1) = 9100 - 91 = 9009$

D. $35 - (9 + 16) = 35 - 9 + 16 = 26 + 16 = 42$

(3) If $m = 3, n = 2$, then $(m + 2n) \times (2m - n) \div (2n - m)$ equals

    A. 20   B. 28   C. 21   D. 27

(4) $19, 999, 0, 100, 66, 2, 37$ should be ordered as

    A. $19 < 999 > 0 < 100 > 66 > 2 < 37$ B. $19 \leq 999 \geq 0 \leq 100 \geq 66 \geq 2 \leq 37$

    C. $999 \geq 100 \geq 66 \geq 37 \geq 19 \geq 2 \geq 0$ D. $999 > 100 \geq 0 \leq 2 < 19 < 37$

(5) Which of the following is not true for all $q \in \mathbb{N}$?

    A. $q + 0 = q$ B. $0 \div q = 0$ C. $q - q = 0$ D. $q \div 1 = q$

(6) Which of the following is true for all $m \in \mathbb{N}$?

    A. $0 \times m = m$ B. $m \div 0 = 0$ C. $1 \times m = m$ D. $m \div m = 1$

(7) In the set of natural numbers, which of the following is true for all $m, n, p \in \mathbb{N}$?

    A. $(m + n)p = np + mp$ B. $p \div (m + n) = p \div m + p \div n$

    C. $(m + n) \div p = m \div p + n \div p$ D. $(m \cdot n) \div p = m \cdot (n \div p)$

(8) In the set of natural numbers, given $m, n \in \mathbb{N}$ and $m > n$, we cannot derive

    A. $m \geq n + 1$ B. $m + 1 > n + 1$ C. $m - 1 \geq n$ D. $m + 1 \geq n + 1$

(9) Given $m, n, p \in \mathbb{N}$, which of the following is not correct?

    A. If $m > n, n \geq p$, then $m > p$.   B. If $m < n, n \geq p$, then $m \geq p$.

    C. If $m \leq n, n < p$, then $m < p$.   D. If $m > n, n > p$, then $m \geq p$.

(10) Which group gives all digits that cannot be the ones digit of the square of a natural number?

    A. $3, 4, 9, 0$ B. $4, 5, 6, 7$ C. $1, 5, 6, 9$ D. $2, 3, 7, 8$

3. Calculate.

(1) 8×7 (2) $15 + 27$

(3) $100 - 37$ (4) 11×12

(5) $84 \div 7$ (6) $128 - 36$

(7) $29 + 75$ (8) 13×6

(9) 4×15 (10) 17×3

(11) $85 \div 5$ (12) $17 + 58$

(13) $74 - 39$ (14) 12×17

(15) $56 \div 4$ (16) $64 \div 16$

(17) 23×4 (18) 15×13

(19) $91 \div 13$ (20) 16×18

(21) $108 \div 18$ (22) $123 - 68$

(23) 14×16

(24) $47 + 85$

(25) 96×7

(26) 13×21

(27) 19×11

(28) $252 \div 14$

(29) $328 - 149$

(30) $962 \div 37$

4. Calculate.

(1) 7^2

(2) 2^4

(3) 3^3

(4) 5^3

(5) 9^2

(6) 10^3

(7) 2^6

(8) 4^5

(9) 3^4

(10) 8^3

(11) 6^3

(12) 5^4

(13) 11^2

(14) 12^2

(15) 13^2

(16) 14^2

(17) 15^2

(18) 16^2

(19) 17^2

(20) 18^2

(21) 19^2

(22) 21^2

(23) 25^2

(24) 37^2

(25) 42^2

(26) 65^2

(27) 53^2

(28) 101^2

(29) 77^2

(30) 89^2

5. Calculate.

(1) $5 \times (9 - 3) - 14$

(2) $(107 - 39) \div 17$

(3) $128 + 35 \times 3$

(4) $700 - 125 \times 3$

(5) $330 \div 5 + 46 \times 7$

(6) $104 \times 9 - 256 \div 8$

(7) $145 - 150 \div 2 + 23$

(8) $18 \times 6 + 522 \div 3$

(9) $89 \times 2 + 86$

(10) $450 \div 6 + 29 \times 6$

(11) $784 \div 8 + 105 \times 4$

(12) $252 \div 9 \div (11 - 4)$

(13) $522 \div (328 - 319) + 42$

(14) $(42 + 18) \times (56 - 27)$

(15) $162 \div 6 - 96 \div 8$

(16) $305 \times (400 - 395) - 278$

(17) $520 \times 2 - 149 \times 5$

(18) $3 + (289 - 198) \times 2$

(19) $7362 \div 9 \times 7$

(20) $64 \times 8 + 78 \times 2$

(21) $(439 + 717) \div 68$

(22) $668 \div 4 - 387 \div 9$

(23) $156 + 187 \div 17 \times 9$

(24) $(488 + 32 \times 5) \div 12$

(25) $325 \div 13 \times (275 - 258)$

(26) $(17 - 5) \times (8 - 6) \times (22 - 17) - 78$

(27) $(5 + 23 \times 3 - 2 \times 7) \div 6$

(28) $(99 - 23 \times 3 + 3 \times 7) \div (54 \div 18)$

(29) $21 - 2[(3 + 9 \times 4) \div (15 - 2)] - 3[(47 - 3 \times 4) \div (33 - 26)]$

(30) $[(21 \times 2 - 17) \times 6 + 19] \div [2 \times (8 - 2) + 1]$

6. Find the characteristics of the natural number m.

(1) m can be divided by 1

(2) m can be divided by 2

(3) m can be divided by 3

(4) m can be divided by 4

(5) m can be divided by 5

(6) m can be divided by 6

(7) m can be divided by 8

(8) m can be divided by 9

(9) m can be divided by 10

(10) m can be divided by 11

2.2 Integers

The set of natural numbers is not closed under subtraction, which indicates that it can be extended. For any $m \in \mathbb{N}$, define

$$0 - m = -m.$$

Here $-m$ reads 'negative m'. Immediately, we get the new numbers, $-0, -1, -2, -3, -4, -5, \cdots$ From the definition, it is easy to see $-0 = 0 - 0 = 0$. For a natural number m, to emphasize that it has a negative correspondence, we also write it as $+m$ (positive m).

The numbers $\cdots, -3, -2, -1, 0, 1, 2, 3, \cdots$ are called integers, among which $-1, -2, -3, \cdots$ are negative integers; $1, 2, 3, \cdots$ are positive integers; 0 is neither positive nor negative. The set of all integers is denoted by

$$\mathbb{Z} = \{0, \pm 1, \pm 2, \pm 3, \cdots\}.$$

Besides, $\mathbb{Z}^+ = \{1, 2, 3, \cdots\}$ denotes the set of all positive integers, and $\mathbb{Z}^- = \{-1, -2, -3, \cdots\}$ denotes the set of all negative integers. In the set notation, we have $\mathbb{N} = \mathbb{Z}^+ \cup \{0\}$.

In the following of this section, let m, n denote natural numbers, while p, q denote integers.

The characteristic operations of integers are also addition and multiplication, which are defined by extending the operations on natural numbers.

Addition.

- $m + n$ is defined as before;

- $m + (-n) = \begin{cases} m - n & \text{if } m \geq n \\ -(n - m) & \text{if } n < m; \end{cases}$

- $(-m) + n$ is defined similarly;

- $(-m) + (-n) = -(m + n)$.

For example, $5 + 6 = 11$, $-6 + 4 = -(6 - 4) = 2$, $7 + (-2) = 7 - 2 = 5$, $(-2) + (-3) = -(2 + 3) = -5$.

Multiplication.

- $m \times n$ is defined as before;

- $(-m) \times n = m \times (-n) = -(m \times n)$;

- $(-m)(-n) = mn$.

For example, $2 \times 3 = 6$, $(-2) \times 3 = 2 \times (-3) = -6$, $(-2) \times (-3) = 6$.

The operations of addition and multiplication over $\mathbb{Z}$ satisfy ① the commutative property (both addition and multiplication), ② the associative property (both addition and multiplication) and ③ the distribute property. Besides, for any integer p,

④ $p + 0 = p = 0 + p$, $p \times 1 = 1 \times p = p$;

⑤ there is an integer p', such that $p + p' = 0$.

The integer p' is unique, called the opposite of p. For example, the opposite of 2 is -2; the opposite of -5 is 5; the opposite of 0 is 0. p' is the opposite of p, is equivalent to, p is the opposite of p'. For convenience, we also say p and p' are opposites. We need a notation for the opposite. When $p > 0$, its opposite is just $-p$, so we can abuse the notation to denote the opposite of p constantly by $-p$. The opposite of the opposite is the number itself. Immediately,

$$-(-p) = p.$$

For example, $-(-5) = 5$, which also states that the opposite of -5 is 5.

By convention, the positive sign '+' can be placed in front of any numbers without changing anything. For example, $+3 = 3$, $+(-2) = -2$, $+0 = 0$. It is possible that more than one signs appear in front of a number. The above proposition also tells us how to simplify: the final sign is determined by the number of negative signs.

- Odd number of '−' ⇒ the final sign is '−'. For example, $-[-(-4)] = -4$ as there are 3 (odd) negative signs.

- Even number of '−' ⇒ the final sign is '+'. For example, $-[+(-5)] = 5$ as there are 2 (even) negative signs.

At this point, the signs '+, −' have two meanings: as positive/negative signs and as operators of addition and subtraction. However, the two meanings won't be mixed: as signs, '$\pm$' only apply to single numbers ('+' reads 'positive' and '−' reads negative), but as operators they need two operands ('+' reads 'plus' and '−' reads 'minus'). When combining all possible operations, the rule of operator precedence adds the signs '$\pm$' in. Again, '$\succ$' stands for 'prior to'.

$$\text{power} \succ \pm(\text{sign}) \succ (\,) \succ [\,] \succ \{\,\} \succ (\times, \div) \succ (+, -)(\text{addition,subtraction})$$

From the rule, $-p \times q$ should be understood as $(-p) \times q$, although the result does equal $-(p \times q)$. However, $-p + q$ should not be confused with $-(p + q)$; $-p^m = -(p^m)$ is different from $(-p)^m$ (see below for the definition).

The above listed five properties of the addition and multiplication operations have several well-known corollaries:

- $0 \times p = 0 = p \times 0$;

- $-1 \times p = -p$, $-1 \times (-p) = p$; e.g. $-1 \times 2 = -2$, $-1 \times (-2) = 2$;

- $-p \times q = -(pq)$, $(-p)(-q) = pq$; e.g. $-2 \times 3 = -(2 \times 3) = -6$, $(-2)(-3) = 2 \times 3 = 6$.

For integers, the subtraction can always be done, with the result again integers. Using the notation of opposites, subtractions can be defined elegantly by additions.

Subtraction. $p - q = p + (-q)$. For example, $2 - 5 = 2 + (-5) = -(5 - 2) = -3$, $2 - (-1) = 2 + 1 = 3$, $(-9) - (-6) = (-9) + [-(-6)] = (-9) + 6 = -3$.

Unfortunately the division of integers can not be carried out inclusively. We say p divides q, or q can be divided by p, denoted by $p|q$, if there exists $r \in \mathbb{Z}$, such that $q = pr$. In this case, we write $q \div p = r$. For example $-6 \div 3 = -2$, $6 \div (-3) = -2$, $-6 \div (-3) = 2$. To make the quotient unique, we require that 0 can never be the divisor.

Exponentiations (powers) of integers are integers when the exponents are natural numbers.

$$p^m = \overbrace{p \cdots p}^{m \text{ times}} \quad \text{if } m > 0;$$
$$p^0 = 1 \quad \text{if } p \neq 0.$$

For example, $(-2)^0 = 1$, $(-2)^1 = -2$, $(-2)^2 = (-2) \times (-2) = 4$, $(-2)^3 = (-2) \times (-2) \times (-2) = -8$. Again, 0^0 is not well-defined.

The partial order on integers is defined similarly. $p \geq q$ if $p - q \in \mathbb{N}$. The other three inequality relations ($>, \leq, <$) are derived as in Section 2.1. It is obvious that positive integers $> 0 >$ negative integers. It follows that $p \geq q$ if and only if $p - q \geq 0$. Similar results hold for the other three inequality relations. As a result, when comparing numbers with same signs, we can check the sign of their differences. Compare $-3, -4$ as an example: since

$-3 - (-4) = -3 + 4 = 1 > 0$, immediately $-3 > -4$. From $p > 0$, we easily get $-p < 0$. Based on this, we have the equivalence

$$p > q \Longleftrightarrow -p < -q.$$

As an application, $-3 > -4$ can be derived directly from $3 < 4$.

We also need to know how to find the greatest common divisor (GCD) and the least common multiples (LCM) of two or more integers. Let $p_1, \cdots, p_m$ be m integers. The GCD of $p_1, \cdots, p_m$ is the greatest (positive) integer d such that d divides each p_i. The GCD of $p_1, \cdots, p_m$ always exists, and it is unique. We denote it by $(p_1, \cdots, p_m)$. The LCM of $p_1, \cdots, p_m$ is the least nonnegative integer m that can be divided by each p_i. The LCM of $p_1, \cdots, p_m$ always exists, and it is unique. We denote it by $[p_1, \cdots, p_m]$.

In order to find GCDs and LCMs, we need the concept of primes. An integer $p \neq \pm 1$ is called a prime or a prime number, if it has exactly four divisors, ± 1 and $\pm p$. For example, $\pm 2, \pm 3, \pm 5, \pm 7, \pm 11$ are all primes. Integers other than $0, \pm 1$ that are not primes are called composite. ± 1 are neither prime not composite. The fundamental theorem of arithmetic asserts that any integer $p \neq 0, \pm 1$ can be decomposed as a finite product of primes. For example, $4 = 2^2$, $-6 = -2 \times 3$, $8 = 2^3$, $-12 = -2^2 \times 3$.

The GCD and LCM of a group of integers can be found by the following procedure: (1) decompose all numbers in the group as products of primes; (2) the product of the common primes with the smallest exponents is the GCD; (3) the product of all primes with the biggest exponents is the LCM. For example, consider 60 and -36. $60 = 2^2 \times 3 \times 5$, and $-36 = 2^2 \times 3^2$. 2 and 3 are the common primes. The lowest power of 2 is 2^2, while the lowest power of 3 is 3^1, as a result the GCD $(60, -36) = 2^2 \times 3 = 12$. 2^2, 3^2 and 5^1 are the highest powers of all prime factors 2, 3 and 5 separately, so the LCM $[60, -36] = 2^2 \times 3^2 \times 5 = 180$.

Two integers p, q are relatively prime if $(p, q) = 1$. For example, 20 and 9 are relatively prime, so are -120 and 77.

Exercises.

    1. True or false.

(1) 0 is an integer as well as a natural number.

(2) The sum of two integers is bigger than their difference.

(3) 0 is a prime.

(4) The difference of two integers is smaller than the minuend.

(5) 0 is neither positive nor negative.

(6) -77 is a prime.

(7) The set of integers contains positive integers and negative integers.

(8) An integer is either positive or negative.

(9) $m^0 = 0$ when $m \in \mathbb{Z}$.

(10) Even numbers are not primes.

(11) 0 is the smallest integer.

(12) Two different odd numbers must be relatively prime.

(13) 0 has no opposite.

(14) In $\mathbb{Z}$, if the product is odd, both multiplicands are odd.

(15) 1 is the smallest positive prime.

(16) If m, n are opposite integers, then $mn < 0$.

(17) If the product of two integers are negative, both of them are negative.

(18) If an integer can be divided by both 4 and 6, it can be divided by 24.

(19) Positive integers are natural numbers, i.e $\mathbb{Z}^+ \subseteq \mathbb{N}$.

(20) Two integers are relatively prime means they have no common divisors.

2. Multiple Choice.

(1) Which of the following is correct?

    A. $(2-1)^2 = 2^2 - 1^2$ B. $144 \div 12 \times 12 = 1$

    C. $(-3-2)^3 = 3^3 \times 2^3$ D. $(3^2)^4 - (3^4)^2 = 0$

(2) $(10^3 - 20^3) \div 1000$ equals

    A. -3 B. -30 C. -7 D. -70

(3) Which of the following is incorrect?

    A. $2^5 - 2^4 = 2^4$ B. $(-3)^3(-2)^3 = 6^3$

    C. $(2^3)^2 = 2^6$ D. $(-5)^4 = -5^4$

(4) Which of the following is not correct?

    A. The sum of a positive integer and a negative integer is zero.

    B. The difference of two negative integers can be positive.

    C. The product of two negative integers can not be negative.

    D. Only the product of a positive integer and a negative integers is negative.

(5) Which of the following is correct?

    A. $+(-5) - \{-[+(-5)]\} = 0$ B. $+[-(+5)] - \{-[-(-5)] = 0\}$

    C. $-(-5) - \{-[-(-5)]\} = 0$ D. $-[-(-5)] - \{+[-(-5)] = 0\}$

(6) A composite should have at least _____ positive divisors.

    A. 1   B. 2   C. 3   D. 4

(7) Which of the following is correct?

    A. $-2 \times (2-3) = -2 \times 2 - 2 \times 3$

    B. $(-2-3) \times (-1) = (-2) \times (-1) - (-3) \times (-1)$

    C. $(-7+2) \times (-3) = 7 \times 3 - 2 \times 3$

    D. $(2+3) \times (-5) = -2 \times 5 - 3 \times (-5)$

(8) The biggest number among the following is

    A. $3 \times 3^3 - 2 \times 2^2$ B. $(3 \times 3)^2 - (2 \times 2)^2$

    C. $(3^3)^2 - (2^3)^2$ D. $33^2 - 22^2$

(9) The smallest number among the following is

   A. $(-2 \times 3)^3$ B. $(-3 - 2)^3$ C. $(-3)(-2)^3$ D. $(-3)^3(-2)^3$

(10) If the product of two integers is even, then

   A. both of them are even.

   B. both of them are odd.

   C. one of them is even and the other is odd.

   D. at least one of them is even.

(11) Which of the following is a decomposition of 252 as a product of primes?

   A. $4 \times 7 \times 27$ B. $2 \times 14 \times 3^3$ C. $2^2 \times 3^3 \times 7$ D. $6 \times 9 \times 14$

(12) Which of the following is correct?

   A. $(12, 18, 20) = 4$ B. $(0, 1) = 0$

   C. $[0, 1] = 0$ D. $[-20, 30, 50] = 600$

(13) How many different positive factors does -12 have?

   A. 4   B. 5   C. 6   D. 8

(14) Given $m, n, p \in \mathbb{Z}$, which of the following is not true?

   A. If $(m, n) = 1$, then $(m, n, p) = 1$.   B. If $[m, n] = mn$, then $(m, n) = 1$.

   C. If $5|m$, $7|n$, then $35|mn$.   D. If $4|p$, $15|p$, then $60|p$.

(15) Which of the following is correct?

   A. Two composites are never relatively prime.

   B. An integer is either prime or composite.

   C. Two different prime numbers must be relatively prime.

   D. Odd number and even number must be relatively prime.

(16) Given $m, n \in \mathbb{Z}$, which of the following condition allows us to derive $m > n$?

   A. $-m \geq -n$ B. $-m \leq -n$ C. $-m < -n$ D. $2 - m > 2 - n$

3. Fill in the blanks.

(1) The smallest positive integers is _____; the biggest negative integer is _____; the smallest positive prime is _____; the smallest positive composite is _____.

(2) The product of a positive integer and a negative integer is _____. If the product of two integers is positive, the two multiplicands are _____.

(3) The GCD of all odd numbers are _____; the GCD of all even numbers are _____.

(4) The composites between 1 and 10 are _____.

(5) If $m = 2 \times 3 \times 5$, $n = 2 \times 5 \times 7$, then $(m, n) = $ _____, $[m, n] = $ _____.

(6) If two integers m, n are relatively prime, then $(m, n) = $ _____, $[m, n] = $ _____.

(7) If two integers a, b satisfies $a \div b = 10$, then $(a, b) = $ _____, $[a, b] = $ _____.

(8) Given $n \in \mathbb{Z}^+$, $(1, n) = $ _____, $[1, n] = $ _____, $(0, n) = $ _____, $[0, n] = $ _____.

(9) Suppose $p, q \in \mathbb{Z}^+$ and $(p, q) = k$. The GCD of $(m \div k)$ and $(n \div k)$ is _____.

(10) The numbers $17, -21, -45, 63, -11, 0$ should be ordered as _____.

4. Simplify the signs.

(1) $-(+2)$

(2) $-(-3)$

(3) $+(-5)$

(4) $-[+(-3)]$

(5) $-[-(+8)]$

(6) $-[-(-16)]$

(7) $-\{+[-(-19)]\}$

(8) $+\{-[-(-61)]\}$

5. Find all primes and composites in the following integers.

$$0, 1, -2, 12, -13, 19, 26, -33, 47, 54, -57, -61, 75, -91, 97, 99, -143, -159, -221$$

Primes:

Composites:

6. Calculate.

(1) $(-2) + 3$

(2) $-15 - 9$

(3) $-63 \div (-7)$

(4) $25 - 74$

(5) $-35 - (-83)$

(6) $-47 - (-13)$

(7) $(-3) \times (-2) - 3 - 2$

(8) $-7 \times 5 + (-4) \times (-6)$

(9) $-3^2 \times 2^3 + (-3)^3 \times 2^2$

(10) $-6^3 \div (34 - 37)^2$

(11) $365 \times [74 - 2^3 \times 7 - (-2)^4]$

(12) $21 + (23 - 327) \div 19$

(13) $-24 - (-45) - 14 \times (-12)$

(14) $[53 + (-588) \div 21] \times (33 - 47)$

(15) $(264 - 1000) \div (29 - 52)$

(16) $(-357) \div (-21) \times (-13) - (-213)$

(17) $-9^2 - (45 - 29) \times (-14)$

(18) $36 - (-19) \times (-14) - (+23)$

7. Order the numbers.

(1) $4, -5, 0$

(2) $-100, -101, -99$

(3) $-73, 121, -22, 69, -53$

(4) $1^2 - 2^2, 1^3 - 2^3, 2^3 - 3^3, 2^2 - 3^2, 3^2 - 5^2$

(5) $(-2)^3, -3^2, (-2)^1, -(-3)^0, (-2)^2, -(-3)^1, 2^0$

8. Find the GCDs and LCMs.

(1) $4, 6$

(2) $-5, 9$

(3) $-26, 39$

(4) $13, -6$

(5) $-45, -60$

(6) $36, 60$

(7) $27, 72$

(8) $-58, 87$

(9) $0, -2, -3, 7$

(10) $1, 2, 3, 5$

(11) $18, -24, 60$

(12) $-24, 36, -48$

9. (1) Find all divisors of 24.

(2) Find all common divisors of 84 and 105.

10. If $a - 6$ and a are opposite numbers, find a.

11. Lucy's home is 3km away from school. She rides a bicycle to school every day at the speed of 15km/h. If one day she rode against wind at the speed of 10km/h for the first 1km, how fast should she ride for the rest to reach school on time?

2.3 Rational Numbers

The quotient of two integers may not be an integer, which requires extension of the numbers system. Given two integers p, q where $q \neq 0$, denote the quotient $p \div q$ by the common fraction

$$\frac{p}{q}.$$

A number of this form is called a rational number, in which p is called the denominator and $q \neq 0$ is called the numerator. For example, $\frac{1}{2}, \frac{-6}{2}, \frac{-9}{-7}, \frac{19}{-11}$ are all rational numbers. Integers are always rational numbers, as any integer $n = n \div 1$ can always be written as a fraction $\frac{n}{1}$. The set of all rational numbers are denoted by $\mathbb{Q}$, i.e.

$$\mathbb{Q} = \{\frac{p}{q} | p, q \in \mathbb{Z}, q \neq 0\}.$$

Fractions have the following fundamental principle: given $0 \neq r \in \mathbb{Z}$,

$$\frac{p}{q} = \frac{pr}{qr},$$

i.e. the result won't change if multiplying the denominator and numerator by the same nonzero integer. Exchanging the two sides of the identity, a fraction also keeps unchanged if canceling a common factor between the denominator and numerator. By the fundamental principle, a rational number may have a lot of (infinitely many) representations. For example, $\frac{1}{2} = \frac{-1}{-2} = \frac{2}{4} = \frac{-3}{-6} = \cdots$.

A rational number $\frac{p}{q}$ is in lowest terms, if $(p, q) = 1$, i.e. the denominator and the numerator are relative prime. By the fundamental principle of fractions, every rational number can be simplified into its lowest terms (or reduced form) by canceling all common factors between the denominator and the numerator. For example, $\frac{-30}{24} = \frac{-5 \times 6}{4 \times 6} = \frac{-5}{4}$.

All five operations $(+, -, \times, \div, \text{power})$ can be defined for rational numbers. Suppose $p, q, r, s \in \mathbb{Z}, m, n \in \mathbb{N}$ in the following.

Addition. If $q, s \neq 0$,

$$\frac{p}{q} + \frac{r}{s} = \frac{ps + rq}{qs}.$$

The idea is to use the fundamental principle of fractions to make the summands have common denominators and then add the numerators. In practice, we use the least common denominators (LCD), i.e. the LCM of the denominators. For example, the LCD of $\dfrac{3}{-2}$ and $\dfrac{2}{3}$ is $[-2,3] = 6$, and $\dfrac{3}{-2} = \dfrac{3 \times (-3)}{(-2) \times (-3)} = \dfrac{-9}{6}$, $\dfrac{2}{3} = \dfrac{2 \times 2}{3 \times 2} = \dfrac{4}{6}$, as a result

$$\frac{3}{-2} + \frac{2}{3} = \frac{-9}{6} + \frac{4}{6} = \frac{-9+4}{6} = \frac{-5}{6}.$$

Multiplication. If $q, s \neq 0$,

$$\frac{p}{q} \cdot \frac{r}{s} = \frac{pr}{qs},$$

i.e. multiplying the top together and the bottom together. In practice, we cancel common factors between the top and the bottom before multiplying the numerators and denominators. For example, the numerator of $\dfrac{-4}{7}$ and the denominator of $\dfrac{3}{2}$ have a common factor 2, so we can cancel them in the multiplication:

$$\frac{-4}{7} \cdot \frac{3}{2} = \frac{-4 \times 3}{7 \times 2} = \frac{-2 \times 3}{7 \times 1} = \frac{-6}{7}.$$

All properties of addition and multiplication on integers keep valid for these operations on rational numbers. In particular, any rational number $\dfrac{p}{q}$ has a unique opposite. For example, the opposite of $\dfrac{-5}{6}$ is $\dfrac{5}{6}$; the opposite of $\dfrac{9}{7}$ is $\dfrac{-9}{7}$. Since $\dfrac{-p}{q} + \dfrac{p}{q} = \dfrac{-p+p}{q} = 0$, $\dfrac{-p}{q}$ is the opposite of $\dfrac{p}{q}$. To keep the consistency of notations, denote the opposite of $\dfrac{p}{q}$ by $-\dfrac{p}{q}$. Immediately

$$-\frac{p}{q} = \frac{-p}{q} = \frac{p}{-q}.$$

The last equality holds since $\dfrac{-a}{b} = \dfrac{(-1)(-a)}{(-1)b} = \dfrac{a}{-b}$. As a result, any rational number can be written in the form $\pm\dfrac{m}{n}$ with $m, n \in \mathbb{N}$ and $n \neq 0$, i.e. taking out all negative signs to the front of the fraction. For example $\dfrac{3}{-2} = \dfrac{-3}{2} = -\dfrac{3}{2}$, $\dfrac{-3}{-2} = +\dfrac{3}{2}$. The rule in the last example is commonly used, which is summarized as

$$\frac{-p}{-q} = \frac{p}{q}.$$

With opposites, the subtraction of rational numbers can be defined.

Subtraction. If $q, s \neq 0$,

$$\frac{p}{q} - \frac{r}{s} = \frac{p}{q} + (-\frac{r}{s}) = \frac{ps - qr}{qs}.$$

For example, $-2 - \dfrac{1}{2} = -2 + (-\dfrac{1}{2}) = -(\dfrac{2}{1} + \dfrac{1}{2}) = -(\dfrac{4}{2} + \dfrac{1}{2}) = -\dfrac{5}{2}$.

Besides the properties of operations shared by rational numbers and integers, there is one more for rational numbers only. Any $0 \neq a \in \mathbb{Q}$ has a reciprocal $b \in \mathbb{Q}$, such that $ab = 1$. For example, the reciprocal of 2 is $\dfrac{1}{2}$, the reciprocal of $-\dfrac{5}{4}$ is $-\dfrac{4}{5}$. 0 has no reciprocal. The reciprocal of $a \neq 0$ is unique, denoted by $\dfrac{1}{a}$. It is easy to see, if $p, q \neq 0$, $\dfrac{q}{p}$ is the reciprocal of $\dfrac{p}{q}$, thus

$$\frac{1}{\dfrac{p}{q}} = \frac{q}{p}.$$

This means the fundamental principle of fractions keeps valid in $\mathbb{Q}$: given $a, b, c \in \mathbb{Q}$ and $b, c \neq 0$, $\dfrac{a}{b} = \dfrac{ac}{bc}$. With reciprocals, the operation of divisions can be defined inclusively in $\mathbb{Q}$.

Division. If $q, r, s \neq 0$

$$\frac{p}{q} \div \frac{r}{s} = \frac{p}{q} \times \frac{s}{r}.$$

For example, $\dfrac{3}{5} \div (-\dfrac{6}{25}) = \dfrac{3}{5} \cdot (-\dfrac{25}{6}) = -\dfrac{3}{5} \cdot \dfrac{25}{6} = -\dfrac{5}{2}$.

Exponentiation(Power). Given $n \in \mathbb{N}$,

$$\left(\frac{p}{q}\right)^n = \overbrace{\frac{p}{q} \times \cdots \times \frac{p}{q}}^{n \text{ times}} = \frac{p^n}{q^n} \quad \text{if } m > 0;$$

$$\left(\frac{p}{q}\right)^0 = 1 \quad \text{if } p \neq 0;$$

$$0^0 \text{ D.N.E.}$$

For example, $\left(\dfrac{2}{3}\right)^3 = \dfrac{2}{3} \cdot \dfrac{2}{3} \cdot \dfrac{2}{3} = \dfrac{2^3}{3^3} = \dfrac{8}{27}$, $\left(\dfrac{-5}{9}\right)^2 = \dfrac{(-5)^2}{9^2} = \dfrac{25}{81}$.

The rule of operator precedence keeps the same.

Any rational number can be represented in the form of a common fraction $\pm\dfrac{m}{n}$ with $m, n \in \mathbb{N}, n \neq 0$. Such a fraction is called proper if $m < n$, and improper if $m \geq n$. For example, $-\dfrac{4}{7}$ is proper, but $\dfrac{10}{9}$ and $-\dfrac{2}{2}$ are improper. Improper fractions can be written as integers or mixed numbers, i.e. sums of nonzero integers and proper fractions. For example, $2\dfrac{3}{4}$ denotes the number $2 + \dfrac{3}{4} = \dfrac{8}{4} + \dfrac{3}{4} = \dfrac{11}{4}$; $-1\dfrac{1}{3}$ equals $-(1 + \dfrac{1}{3}) = -\dfrac{4}{3}$. To write $-\dfrac{25}{7}$ as a mixed number, since $25 = 3 \times 7 + 4$, $-\dfrac{25}{7} = -\dfrac{3 \times 7 + 4}{7} = -(3 + \dfrac{4}{7}) = -3\dfrac{4}{7}$.

A rational number $\dfrac{p}{q}$ ($q \neq 0$) can also be represented in the form of a decimal, by carrying out the division $p \div q$. The decimal determined by a rational number is either terminating or repeating. For example, $\dfrac{2}{5} = 0.4$ is terminating, while $-\dfrac{1}{3} = -0.333\cdots = -0.\bar{3}$ and $\dfrac{71}{55} = 1.290909090\cdots = 1.2\overline{90}$ are repeating. Here, the short line notation means the beneath pattern of digits repeat infinitely many times. Conversely, all terminating and repeating decimals can be written as fractions. As a result, there is a one-one correspondence between rational numbers and decimals that are terminating or repeating. Given a finite decimal, we can change it into an equal common fraction by the fundamental principal of fractions. For example, $1.2 = \dfrac{1.2}{1} = \dfrac{1.2 \times 10}{1 \times 10} = \dfrac{12}{10} = \dfrac{6}{5}$, $-0.25 = -\dfrac{0.25}{1} = -\dfrac{0.25 \times 100}{1 \times 100} = -\dfrac{25}{100} = -\dfrac{1}{4}$. The idea is to treat the decimal a as a fraction $\dfrac{a}{1}$ and multiply the denominator and numerator by some power of 10 to clear the decimal point.

When adding or multiplying two finite decimals, we can either do it directly or change them into fractions. For example, $1.5 \times (-2.4)$ can be simplified either by $1.5 \times (-2.4) = -1.5 \times 2.4 = 3.6$, or by $1.5 \times (-2.4) = \dfrac{3}{2} \times \left(-\dfrac{12}{5}\right) = -\dfrac{18}{5}$. If only one operand is a finite decimal, depending on whether the other operand can be written as a terminating decimal or not, we can choose to convert both operands into decimals or both into fractions. For example, $\dfrac{4}{3} + (-0.4)$ can only be calculated by $\dfrac{4}{3} + (-0.4) = \dfrac{4}{3} - \dfrac{2}{5} = \dfrac{14}{15}$, while $-\dfrac{1}{8} - 0.25$ can be either $-0.125 - 0.25 = -0.375$ or $-\dfrac{1}{8} - \dfrac{1}{4} = -\dfrac{3}{8}$.

By the fundamental principle of fractions, we can simplify complex fractions. For example, $\dfrac{1.2}{1.8} = \dfrac{1.2 \times 10}{1.8 \times 10} = \dfrac{12}{18}$. Notice that 12 and 18 has a common factor 6. Canceling this factor we get $\dfrac{1.2}{1.8} = \dfrac{12}{18} = \dfrac{2}{3}$. To simplify $\dfrac{\frac{4}{7}}{\frac{2}{3}}$, multiply the top and bottom by the LCD 21 of the top and bottom fractions.

$$\frac{\dfrac{4}{7}}{\dfrac{2}{3}} = \frac{\dfrac{4}{7} \times 21}{\dfrac{2}{3} \times 21} = \frac{4 \times 3}{2 \times 7}.$$

Canceling the common factor 2, the result is $\dfrac{6}{7}$.

At last, let us define the partial order on $\mathbb{Z}$. First, a rational number $\dfrac{m}{n}$ ($m, n \in \mathbb{N}, n \neq 0$, e.g. $\dfrac{3}{5}$) is positive, and its opposite $-\dfrac{m}{n}$ (e.g. $-\dfrac{3}{5}$) is negative. Denote the set of positive rational numbers by $\mathbb{Q}^+$ and the set of negative rational numbers by $\mathbb{Q}^-$. Given two rational numbers $\dfrac{p}{q}$ and $\dfrac{r}{s}$, define $\dfrac{p}{q} > \dfrac{r}{s}$ if $\dfrac{p}{q} - \dfrac{r}{s} > 0$. The other three inequality relations are defined accordingly.

From the above definition, when comparing rational numbers, we may make them in their LCD format and compare the numerators. For example, the LCD of $-\dfrac{2}{3}$ and $-\dfrac{4}{5}$ is 15. $-\dfrac{2}{3} = \dfrac{-10}{15}$, $-\dfrac{4}{5} = \dfrac{-12}{15}$, as $-10 > -12$ (since $10 < 12$), we have $-\dfrac{2}{3} > -\dfrac{4}{5}$. Another way to compare rational numbers is to compare their decimal forms. As $-\dfrac{2}{3} = -0.\bar{6}$, $-\dfrac{4}{5} = -0.8$, it is easy to get $-0.\bar{6} > -0.8$.

When working with rational numbers, we always simplify the result in simplest forms. For fractions, keep in mind that the denominator should never be zero.

Exercises.

    1. True or false.

(1) 0 is not a rational number.

(2) Integers are rational numbers, i.e. $\mathbb{Z} \subseteq \mathbb{Q}$.

(3) $\mathbb{Z} \cap \mathbb{Q}^+ = \mathbb{N}$.

(4) Multiplying the denominator and numerator by the same number, the fraction does not change.

(5) Adding the same number to both the denominator and numerator, the fraction does not change.

(6) Dividing the denominator by 3, to keep the fraction unchanged, the numerator must be multiplied by $\frac{1}{3}$.

(7) Dividing the denominator by 4 and keeping the numerator, the fraction becomes 4 times of the original one.

(8) Given $a, b, c \in \mathbb{Q}$ and $b, c \neq 0$, we have $\dfrac{a}{b} = \dfrac{\frac{a}{c}}{\frac{b}{c}}$.

(9) For any common fraction $\dfrac{m}{n}$ $(m, n \in \mathbb{Z}, n \neq 0)$, we have $\dfrac{m}{n} = \dfrac{1}{\frac{n}{m}}$.

(10) $\dfrac{-16}{20} = \dfrac{4}{-5}$.

(11) $\dfrac{-39}{91}$ is in lowest terms.

(12) All rational numbers have opposites as well as reciprocals.

(13) The reciprocal of a negative rational number is always greater than itself.

(14) The reciprocal of a positive integer is always less than itself.

(15) The reciprocal of a nonzero integer is no greater than 1.

(16) The opposite of the reciprocal of a nonzero rational number equals the reciprocal of the opposite of it.

(17) The reciprocal of the opposite of 0.3 is -3.

(18) $-0.3 > -\dfrac{1}{3}$.

(19) $-\dfrac{2}{3} > -\dfrac{3}{4}$.

(20) $-\dfrac{1}{4} < -\dfrac{1}{3}$.

(21) $-100 > \dfrac{1}{100}$.

(22) Integers must be greater than proper fractions.

(23) There is no positive integers between $+5.01$ and $+5.99$.

(24) There is no positive common fractions between -2 and 0.

2. Multiple choice.

(1) -5 is not a

    A. rational number   B. integer

    C. natural number   D. negative rational number

(2) -1.25 is a

    A. negative non-rational number   B. rational number

    C. decimal but not a rational number   D. positive rational number

(3) Multiplying the denominator by 3 and dividing the numerator by 6, the fraction becomes ———— times of the original one.

    A. $\dfrac{1}{9}$ B. $\dfrac{1}{2}$ C. $\dfrac{1}{18}$ D. $\dfrac{1}{2}$

(4) Which of the following is in lowest terms?

    A. $\dfrac{-15}{-21}$ B. $\dfrac{17}{51}$ C. $-\dfrac{26}{39}$ D. $\dfrac{2}{-3}$

(5) If $\dfrac{3}{5} = \dfrac{3+6}{5+x}$, then

    A. $x = 5$ B. $x = 6$ C. $x = 15$ D. $x = 10$

(6) Which of the following is wrong?

    A. $\dfrac{0}{2}$ is in lowest terms.

    B. Any terminating decimal can be written as a common fraction in lowest terms.

    C. The lowest-term form of an integer n is $\dfrac{n}{1}$.

    D. The denominator of a fraction can not be 0.

(7) Given the fraction $\dfrac{5-a}{5-a}$ with $a \in \mathbb{Q}$, which of the following is correct?

    A. It equals 1 for all a.   B. The value depends on a.

    C. It equals 1 when $a \neq 5$.   D. It equals 0 when $a = 5$.

(8) How many numbers are there whose reciprocal is the same as itself?

    A. 0   B. 1   C. 2   D. 4

(9) If the opposite of the reciprocal of a number is in $\mathbb{Z}^+$, then this number could be

    A. 2   B. -3 C. $\dfrac{1}{2}$ D. $-\dfrac{1}{7}$

(10) In the set of rational numbers, which of the following is correct?

    A. The opposite of a number must be negative.

    B. The opposite of the opposite of a number must be positive.

    C. The opposite of a number must have a reciprocal.

    D. If the opposite has a reciprocal, then the number itself has a reciprocal.

(11) Which of the following is correct?

A. Nonnegative rational numbers are just positive rational numbers.

B. $1 \div \dfrac{1}{0} = 1 \times \dfrac{0}{1} = 0$.

C. The set of rational numbers consists of positive and negative rational numbers.

D. $\mathbb{Q}$ consists of integers and non-integer common fractions.

(12) Which of the following is not correct?

A. Rational numbers are either terminating decimals or repeating decimals.

B. $\mathbb{Q} = \mathbb{Q}^+ \cup \{0\} \cup \mathbb{Q}^-$.

C. Integers are not common fractions.

D. A number of the form $\dfrac{m}{n}$ $(m, n \in \mathbb{Z}, n \neq 0)$ is called a rational number.

3. Find opposites and reciprocals of the following numbers.

$$-1, \quad 7, \quad -\frac{1}{11}, \quad \frac{2}{3}, \quad \frac{9}{-4}, \quad \frac{-1}{2}, \quad 1\frac{2}{7}, \quad -3\frac{11}{13}, \quad 0, \quad -2\frac{35}{44}, \quad -2\frac{62}{53}$$

Opposites:

Reciprocals:

4. Write the following common fractions in lowest terms.

(1) $\dfrac{48}{64}$

(2) $\dfrac{-12}{15}$

(3) $\dfrac{-60}{-144}$

(4) $\dfrac{44}{-28}$

(5) $-\dfrac{-20}{-45}$

(6) $\dfrac{45}{-105}$

(7) $-\dfrac{36}{-72}$

(8) $\dfrac{-84}{-35}$

(9) $\dfrac{-360}{315}$

(10) $-\dfrac{18}{75}$

(11) $\dfrac{-121}{-187}$

(12) $-\dfrac{-65}{91}$

5. Rewrite the fractions into decimals.

(1) $\dfrac{9}{2}$

(2) $-\dfrac{5}{4}$

(3) $-\dfrac{7}{8}$

(4) $\dfrac{3}{5}$

(5) $\dfrac{13}{20}$

(6) $-\dfrac{6}{25}$

(7) $-\dfrac{25}{16}$

(8) $\dfrac{121}{40}$

(9) $\dfrac{27}{45}$ (10) $-\dfrac{169}{104}$

(11) $\dfrac{-13}{9}$ (12) $\dfrac{16}{11}$

6. Rewrite the decimals into fractions.

(1) 0.3 (2) -1.7

(3) -5.6 (4) 0.64

(5) 1.42 (6) -2.125

(7) 2.25 (8) -4.375

(9) -1.55 (10) $+3.16$

7. Simplify. Write the answer in lowest terms.

(1) $\dfrac{0.05}{0.3}$ (2) $\dfrac{-2.34}{0.9}$

(3) $\dfrac{-0.33}{1.21}$ (4) $\dfrac{-9.6}{-1.44}$

(5) $\dfrac{0.75}{2.75}$ (6) $\dfrac{-0.375}{1.75}$

(7) $\dfrac{\dfrac{7}{-6}}{\dfrac{-14}{5}}$ (8) $\dfrac{\dfrac{1}{13}}{-\dfrac{13}{15}}$

(9) $\dfrac{\dfrac{2}{16}}{18}$ (10) $\dfrac{2\dfrac{15}{21}}{-3}$

(11) $-\dfrac{-\dfrac{2}{3}}{\dfrac{5}{6}}$ (12) $-\dfrac{-1\dfrac{1}{27}}{-1\dfrac{2}{33}}$

(13) $\dfrac{\dfrac{30}{69}}{\dfrac{7}{-23}}$ (14) $\dfrac{\dfrac{-25}{78}}{\dfrac{-55}{-52}}$

(15) $\dfrac{\dfrac{6}{-7}}{\dfrac{-7}{6}}$ (16) $\dfrac{5}{-2\dfrac{2}{5}}$

(17) $\dfrac{-4.5}{5\dfrac{1}{2}}$ (18) $\dfrac{\dfrac{-5}{7}}{-7.5}$

(19) $\dfrac{\dfrac{1.2}{0.32}}{\dfrac{-45}{9.6}}$ (20) $\dfrac{\dfrac{1}{-1.2}}{\dfrac{3}{4}}$

(21) $\dfrac{\dfrac{-125}{21}}{\dfrac{1}{0.49}}$ (22) $\dfrac{\dfrac{39}{85}}{-0.36}$

8. Write fractions in their LCD forms.

(1) $\dfrac{3}{2}, -\dfrac{5}{6}$ (2) $\dfrac{3}{14}, \dfrac{2}{21}$

(3) $-\dfrac{3}{8}, -\dfrac{7}{6}, \dfrac{5}{12}$ (4) $\dfrac{17}{12}, -\dfrac{7}{4}, \dfrac{-5}{9}, \dfrac{2}{-3}$

9. Simplify.

(1) $(-2.7) + (-4.5)$ (2) $(-3.6) + (+2.4)$

(3) $-\dfrac{15}{4} + \dfrac{4}{3}$ (4) $16 \div (-6)$

(5) $\dfrac{2}{3} - (-\dfrac{1}{6})$ (6) $-4.5 - (-3.2)$

(7) $-16\dfrac{1}{3} + 29\dfrac{1}{6}$ (8) $-2\dfrac{1}{4} + 3\dfrac{5}{6}$

(9) $-1.2 + \dfrac{2}{3}$

(10) $-\dfrac{8}{5} - 3.7$

(11) $-\dfrac{5}{6} \times \dfrac{11}{10}$

(12) $\dfrac{1}{2} \times (-\dfrac{2}{3})$

(13) $-8\dfrac{3}{4} \times (-4\dfrac{4}{7})$

(14) $-5\dfrac{2}{5} \times 1\dfrac{1}{9}$

(15) -9.84×5

(16) $\dfrac{16}{25} \times (-0.625)$

(17) $(-6.5) \div 0.13$

(18) $375 \div (-\dfrac{3}{2})$

(19) $(-\dfrac{4}{9}) \div (-\dfrac{8}{15})$

(20) $-3.3 \div \dfrac{9}{8}$

(21) $\dfrac{\dfrac{5}{7} \cdot (-\dfrac{8}{3})}{\dfrac{55}{6}}$

(22) $-\dfrac{-\dfrac{10}{3}}{\dfrac{2}{7} \cdot (-\dfrac{28}{15})}$

(23) $\dfrac{-2}{\dfrac{80}{36} \cdot \dfrac{9}{25}}$

(24) $\dfrac{(-\dfrac{2}{5}) \cdot (-\dfrac{3}{4})}{\dfrac{3}{5} \cdot (-\dfrac{1}{2})}$

10. Calculate.

(1) $(-1.5)^2$

(2) $(-\dfrac{1}{7})^2$

(3) $(-\dfrac{3}{4})^3$

(4) $(-\dfrac{5}{3})^2$

(5) $-(-\dfrac{2}{5})^3$

(6) $(-1\dfrac{5}{11})^2$

(7) $-(2\frac{2}{5})^2$

(8) $-(-3\frac{1}{3})^3$

11. Calculate.

(1) $-3\frac{6}{7} \div (-0.3)^2 \times (-0.1)$

(2) $\frac{2}{3} \times (-\frac{4}{5}) \times (-\frac{25}{8}) \times \frac{18}{5}$

(3) $-0.25^2 \div (-\frac{1}{2})^4 \times (-1)^{21} \times (-5)^3$

(4) $-[-2.5^2 \times (-1)^{23}] \times \frac{1}{8} \div (0.5)^3$

(5) $[\frac{1}{2} - \frac{1}{3} \times (-\frac{1}{2}) - \frac{1}{6}]^2$

(6) $1 + \frac{1}{2} \cdot (-3)^2 + \frac{3}{2} \cdot 3^2 + (-3)^3$

(7) $(1 - \frac{1}{2^2})(1 - \frac{1}{3^2})(1 - \frac{1}{4^2})(1 - \frac{1}{5^2})$

(8) $(-2^4 - 5.1 \times 5 + 3 \times 5 + 3^3)^2$

(9) $\dfrac{(2 - \frac{5}{2})^2 + 4 \times \frac{1}{2} - 3}{-\dfrac{4 + (-1)^5}{-5}}$

(10) $[\frac{5}{18} - \frac{7}{12} + 1 - (-\frac{2^2}{9})] \div (-\frac{5}{36})$

12. Order the numbers.

(1) $1, -0.01, -1, 0, 0.5$

(2) $-\frac{1}{3}, -\frac{1}{2}, \frac{1}{3}, \frac{1}{2}$

(3) $-1, -\frac{6}{7}, -\frac{7}{6}$

(4) $-0.3, -\frac{1}{3}, -\frac{1}{4}$

(5) $\frac{4}{15}, \frac{3}{11}$

(6) $-\frac{2}{3}, -\frac{5}{8}$

(7) $-\frac{18}{17}, -\frac{21}{20}, -\frac{24}{23}$

(8) $\frac{17}{69}, \frac{15}{67}$

(9) $-\dfrac{7}{9}, -\dfrac{11}{13}, -\dfrac{13}{15}$ $\qquad\qquad$ (10) $-\dfrac{32}{66}, -\dfrac{36}{75}, -\dfrac{46}{94}$

13. Find all integers between $-4\dfrac{3}{7}$ and $2\dfrac{1}{5}$.

14. If the reciprocal of a number P is 3, find the opposite of P.

15. Find the reciprocal of the opposite of 0.224.

16. Find the 4th power of the reciprocal of the opposite of $\dfrac{3}{5}$.

2.4 Real Numbers

Besides terminating and repeating decimals ,there are also non-repeating (infinite) decimals, which are called irrational numbers. They can never be written as common fractions. Examples of irrational numbers are π, $\sqrt{2}$, $-\sqrt{3}$, etc.

Rational numbers and irrational numbers together are called real numbers. The set of real numbers is denoted by $\mathbb{R}$. We have the following inclusion relation

$$\mathbb{N} \subset \mathbb{Z} \subset \mathbb{Q} \subset \mathbb{R}.$$

$$\text{Real Numbers}\begin{cases}\text{Rational Numbers}\begin{cases}\text{Integers}\begin{cases}\text{Positive Integers}\\0\\\text{Negative Integers}\end{cases}\!\!\!\!\!\!\Big\}\text{Natural Numbers}\\\text{Nonintegers — noninteger fractions}\end{cases}\\\text{Irrational Numbers}\end{cases}$$

Real numbers can also be classified as positive real numbers and negative real numbers.

$$\text{Real Numbers}\begin{cases}\text{Positive Real Numbers}\\0\\\text{Negative Real Numbers}\end{cases}$$

The set of all positive real numbers is denoted by $\mathbb{R}^+$, while the set of all negative real numbers is denoted by $\mathbb{R}^-$. 0 is neither positive nor negative. In this book, a number means a real number by default.

All operations $(+, -, \times, \div)$ are defined over $\mathbb{R}$. These operations for rational numbers are already defined. We only need to specify how to handle irrational numbers. As non-repeating decimals, irrational numbers can be approximated to any precision by terminating decimals. For example, $\sqrt{2} \doteq 1.41421356237$, $\pi \doteq 3.1415927$. In calculations, we just replace these irrational numbers by their approximations. The results won't be accurate, but we have to compromise at this moment. For example, $-\sqrt{2} + 3 \doteq -1.414 + 3 = 1.586$, $\frac{1}{3} \times \sqrt{3} \doteq 1.732 \div 3 = 0.577\overline{3}$.

According to the rule of multiplications, the sign of a multiple product depends only on the number of negative multiplicands.

- Odd number of negative multiplicands $\Longrightarrow$ the product is negative. For example, $(-2) \times (-7) \times (-11) = -2 \times 7 \times 11$;

- Even number of negative multiplicands $\Longrightarrow$ the product is positive. For example, $3 \times (-9) \times (-7) \times 6 = 3 \times 9 \times 7 \times 6$.

Exponentiations (powers) of real numbers are again repeated multiplications. Given $a \in \mathbb{R}$, $n \in \mathbb{Z}^+$, $a^n = a \cdots a$ (n times). $a^0 = 1$ if $a \neq 0$; 0^0 is not well-defined (D.N.E). Applying the above result about the signs of multiple products, immediately, we get

$$(-1)^n = \begin{cases} 1 & \text{if } n \text{ is even} \\ -1 & \text{if } n \text{ is odd} \end{cases} \qquad (-a)^n = \begin{cases} a^n & \text{if } n \text{ is even} \\ -a^n & \text{if } n \text{ is odd} \end{cases}$$

Particularly, $(-a)^2 = a^2$, $(-a)^3 = -a^3$. For example, $(-3)^4 = 3^4 = 81$, $(-1.5)^3 = -1.5^3 = -\left(\frac{3}{2}\right)^3 = -\frac{27}{8}$.

An important observation is that the square of a number can not be negative, i.e.

$$a^2 \geq 0 \text{ for any } a \in \mathbb{R}.$$

As a result, $a^2 = 0$ if and only if $a = 0$.

The operations of addition and multiplication over $\mathbb{R}$ satisfy the same properties of these operations over $\mathbb{Q}$. In the following, let a, b, c denote real numbers.

Properties of Addition:

- Commutative Property: $a + b = b + a$;

- Associative Property: $(a + b) + c = a + (b + c)$;

- Identity Property: $a + 0 = 0 + a = a$;

- Inverse Property: any $a \in \mathbb{R}$ has a unique opposite, denoted by $-a$, which satisfies $a + (-a) = 0$.

Properties of Multiplication:

- Commutative Property: $a \cdot b = b \cdot a$;

- Associative Property: $(a \cdot b) \cdot c = a \cdot (b \cdot c)$;

- Identity Property: $a \cdot 1 = 1 \cdot a = a$;

- Inverse Property: any $0 \neq a \in \mathbb{R}$ has a reciprocal, denoted by $\frac{1}{a}$, which satisfies $a \cdot \frac{1}{a} = 1$.

Compatible Property of Addition and Multiplication:

- Distributive Property: $a(b + c) = ab + ac$.

From these properties, each real number has an opposite, and every nonzero number has a reciprocal. For example, the opposite of $\sqrt{2}$ is $-\sqrt{2}$; the reciprocal of π is $\frac{1}{\pi}$. With opposites and reciprocals, the operations of subtraction and division of real numbers can be defined (same as in Section 2.3). The notations, as well as the related concepts of opposites, reciprocals, powers, additions, subtractions, multiplications and divisions keep consistent for all the number systems ($\mathbb{N}$, $\mathbb{Z}$, $\mathbb{Q}$, $\mathbb{R}$). The rule of operator precedence applies to $\mathbb{R}$ as well. The fundamental principle of fractions keeps valid in the set of real numbers, i.e. $\frac{a}{b} = \frac{ac}{bc}$ for any $a, b, c \in \mathbb{R}$ and $b, c \neq 0$. Besides, we can also derive all other commonly used properties:

- Multiplication Property of Zero: $a \cdot 0 = 0 = 0 \cdot a$;

- Multiplication Property of Opposites: $(-a)b = a(-b) = -ab$, $(-a)(-b) = ab$;

- Multiplication Property of Negative One: $(-1) \cdot a = -a$.

One important application of the Multiplication Property of Negative One is to distribute the negative sign:

$$-(a + b) = -a - b.$$

Then it is natural to have $-(-a+b) = a-b$ and so on. For example, $-(99-128) = -99+128$, $-[-5 + (-7)] = 5 + 7$.

The distributive property also includes the following special cases.

$$a(b - c) = ab - ac, \quad a(-b - c) = -ab - ac, \quad a(-b + c) = -ab + ac.$$

For example, $-3(5 - 7) = -3 \times 5 + (-3) \times (-7) = -3 \times 5 + 3 \times 7$.

All properties listed here can be used to simplify numerical expressions.

Examples. Simplify.

(1) $2.7 \times (-123) + 2.7 \times 23$

(2) $20 \times (-23) \times (-5)$

(3) $-\dfrac{4}{3} + \dfrac{15}{7} - \dfrac{5}{3} - (-\dfrac{6}{7})$

Solutions.

(1) By the distributive property

$$2.7 \times (-123) + 2.7 \times 23 = 2.7 \times (-123 + 23) = 2.7 \times (-100) = -270.$$

(2) First, $20 \times (-23) \times (-5) = 20 \times 23 \times 5$ after simplifying the negative signs. By the commutative and associative properties of multiplication,

$$20 \times 23 \times 5 = (20 \times 5) \times 23 = 100 \times 23 = 2300.$$

(3) By the commutative and associative properties of addition,

$$-\frac{4}{3} + \frac{15}{7} - \frac{5}{3} - (-\frac{6}{7}) = (-\frac{4}{3} - \frac{5}{3}) + [\frac{15}{7} - (-\frac{6}{7})] = -\frac{9}{3} + \frac{21}{7} = -3 + 3 = 0.$$

There is a geometric model of the set of real numbers, i.e. the real number line. A line provided with origin, positive direction and unit length is called a real number line.

$$\underset{\substack{\bullet\\-3}}{\quad}\underset{\substack{\bullet\\-2}}{\quad}\underset{\substack{\bullet\\-1}}{\quad}\underset{\substack{\bullet\\O}}{\quad}\underset{\substack{\bullet\\1}}{\quad}\underset{\substack{\bullet\\2}}{\quad}\underset{\substack{\bullet\\3}}{\quad}\longrightarrow$$

The origin represents the number 0. By convention, the origin is denoted by O. A positive number a corresponds to a point on the right of the origin with a distance a to the origin; a negative number $-b$ ($b > 0$) corresponds to a point on the left of the origin with a distance b to the origin. There is a one-one correspondence between the real numbers and the points on the real number line — each real number corresponds to exactly one point on the real number line, and conversely, each point on the real number line represents exactly one real number.

If two points P and Q are symmetric about the origin, the numbers represented by them are opposites.

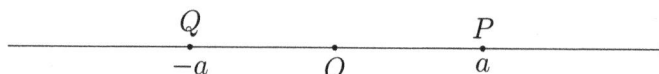

$$\overset{\textstyle Q}{\underset{\textstyle -a}{\bullet}}\qquad\qquad\overset{}{\underset{\textstyle O}{\bullet}}\qquad\qquad\overset{\textstyle P}{\underset{\textstyle a}{\bullet}}\longrightarrow$$

There is another operation on real numbers, called the absolute value. For any $a \in \mathbb{R}$, the absolute value of a, denoted by $|a|$, is defined to be the distance between the point representing a on the real number line and the origin. According to the definition we have the following

$$|a| = \begin{cases} a & \text{if } a \geq 0, \\ -a & \text{if } a < 0. \end{cases}$$

For example, $|0| = 0$, $|2| = 2$, $|-3| = 3$. As a distance, the absolute value is always nonnegative, i.e. $|a| \geq 0$. As a result $|a| = 0$ if and only if $a = 0$. Besides, $|a| = |-a|$ for all real number a. Conversely, if $|a| = |b|$, we can conclude $a = \pm b$ ($a = b$ or $a = -b$).

The multiplication and division of absolute values satisfy

$$|a||b| = |ab|;$$

$$\frac{|a|}{|b|} = |\frac{a}{b}| \text{ when } b \neq 0.$$

For example, $|-3| \cdot |2| = 3 \times 2 = 6$, $\dfrac{|-12|}{|-3|} = \dfrac{12}{3} = 4$.

At the moment, we have two terms that are never negative — squares and absolute values. If the sum of several such terms is zero, every term is zero. For example, if $a^2 + |b| = 0$, the only possibility is $a = b = 0$, since otherwise at least one of a^2 and $|b|$ is positive and the sum won't be zero.

Give any two numbers x and y, the distance between the points representing them on the real number line is $|x - y|$.

$$\overset{}{\underset{\textstyle y}{\bullet}}\qquad\qquad\overset{}{\underset{\textstyle O}{\bullet}}\qquad\qquad\overset{}{\underset{\textstyle x}{\bullet}}\longrightarrow$$

$$\text{dist}(A, B) = |x - y|$$

For example, the distance between the points denoting -2 and 4 is $|-2-4| = 6$.

The partial order on $\mathbb{R}$ is defined by $a > b$ if $a - b \in \mathbb{R}^+$. Using the real number line, $a > b$ if and only if the point representing a is on the right of the point representing b.

$$a > b$$

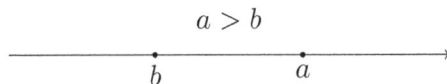

The other three order relations are defined accordingly. Positive numbers $> 0 >$ negative numbers. If $0 < a < b$, then $-b < -a < 0$, which should be clear from the following picture.

This means the order of negative numbers are opposite to the order of their absolute values. As a result, when comparing negative numbers, we can compare their absolute values first. For example, to compare -1.7 and $-\sqrt{3}$, since $\sqrt{3} \doteq 1.732 > 1.7$, immediately $-\sqrt{3} < -1.7$ by reversing $\sqrt{3} > 1.7$.

Exercises.

1. True or false.

(1) Positive rational numbers and negative rational numbers are all the rational numbers.

(2) Finite decimals are also common fractions.

(3) Some rational numbers are irrational.

(4) Rational numbers are real numbers.

(5) $\sqrt{2}$ is a rational number.

(6) The smallest number does not exist.

(7) The further a point is from the origin, the greater the number it represents.

(8) The points on the real number line only represent rational numbers.

(9) 0 is greater than all negative numbers, so 0 is the smallest positive number.

(10) There is no negative number between -1 and 0.

(11) If $a + |a| = 2a$, then a is positive.

(12) $-|a|$ is always negative.

(13) A number less than its absolute value must be negative.

(14) All irrational numbers have reciprocals.

(15) The sum of two numbers with different signs is less than the absolute value of their difference.

(16) The sum of two numbers never equals their difference.

(17) The difference of two numbers is less than the minuend.

(18) The difference of two numbers cannot be greater than their sum.

(19) If $a + b < a$, $a + b < b$, then $a, b < 0$.

(20) If $a < a + b < b$, then $ab < 0$.

(21) If $a + b > 0$, then $a, b > 0$.

(22) If $ab < 0$, $|a| \geq |b|$, then $|a + b| = |a| - |b|$.

(23) When subtracting a negative number, the result is greater than the minuend.

(24) Zero subtracting any number is negative.

(25) If $a, b < 0$, $|a + b| = |a| + |b|$.

(26) A negative number times a negative number is positive.

(27) A negative number plus a negative number equals the difference of their absolute values.

(28) The quotient of two numbers with different signs is negative.

(29) The product of 2015 negative numbers is positive.

(30) 1 to any power is 1; -1 to any power is -1.

2. Multiple choice.

(1) Which of the following is not correct?

    A. -3 is a negative rational number.   B. 0 is not an irrational number.

    C. -0.12 is a negative fraction.   D. π is a positive rational number.

(2) Which of the following is correct?

    A. Two opposite numbers are never equal.

    B. A number and its reciprocal are never equal.

    C. Two opposite numbers must have the same absolute value.

    D. A number and its reciprocal must have the same absolute value.

(3) If the reciprocals of two non-zero numbers have the same absolute value, then the two numbers

    A. are equal   B. have the same absolute value

    C. are opposites   D. have a negative product

(4) Adding '$-$' to the front of a number, we get

    A. a negative number   B. a nonpositive number

    C. a positive or negative number   D. the opposite of the number

(5) The origin and the points on the left of the origin on a real number line represent

    A. positive numbers   B. negative numbers

    C. nonpositive numbers   D. nonnegative numbers

(6) If the opposite of a number is nonpositive, this number must be

    A. positive   B. negative   C. nonpositive   D. nonnegative

(7) If $|x| = |y|$, then

    A. $x = y$ B. $x = -y$ C. $x = \pm y$ D. none of above

(8) If $|a| = a$, then

    A. $a \geq 0$ B. $a > 0$ C. $a < 0$ D. $a \leq 0$

(9) Which of the following is correct?

    A. Two opposite numbers differ by signs.

    B. The absolute value of a number is nonnegative.

    C. Two numbers with the same absolute values are equal.

    D. Smaller numbers have smaller absolute values.

(10) If two numbers are greater than their sum, then these two numbers must be

    A. both positive   B. both negative

    C. one positive and one negative   D. one of them is zero

(11) Which of the following is correct?

    A. For two negative numbers, the bigger one is further from the origin than the smaller one.

    B. The absolute value is always positive.

    C. Given any two numbers, the one with bigger absolute value is further from the origin.

    D. $-a$ is always nonpositive.

(12) A small number minus a big number, their difference must be

    A. positive   B. negative

    C. zero   D. zero or negative

(13) If two numbers are on the same side of the origin, their product must be

    A. positive   B. negative

    C. zero   D. zero or negative

(14) Which of the following is incorrect?

    A. Any number times zero is zero.

    B. Any number times one is itself.

    C. Any number times negative one is its opposite.

    D. Any number times its opposite is negative.

(15) If $a + b < 0$ and $ab < 0$, then

    A. $a > 0, b > 0$

    B. $a < 0, b < 0$

    C. a, b have different signs and the positive one has the bigger absolute value

    D. a, b have different signs and the negative one has the bigger absolute value

(16) Which of the following is correct?

    A. Given two numbers with different signs, their product has the same sign as the multiplicand with the bigger absolute value.

    B. Given two numbers with the same sign, the product has the same sign as two multiplicands.

    C. If the product is negative, the two multiplicands differ by signs.

    D. If the product is positive, the two multiplicands are both positive.

(17) According to the picture, which statement is correct?

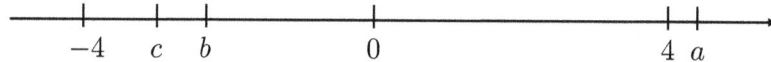

    A. $a + c < 0$ B. $-b - a > 0$

    C. $|-b + c| = b - c$ D. $b < -a$

(18) If $a + b > 0$, $ab < 0$, $a > b$, then

    A. $b < -a < a < -b$ B. $-a < b < -b < a$

    C. $a < -b < b < -a$ D. $-b < a < -a < b$

(19) If $\dfrac{a}{b} = 0$, then

    A. $a \neq 0$, $b = 0$ B. $a = b = 0$ C. $a = 0$, $b \neq 0$ D. $a \neq 0$, $b \neq 0$

(20) If $\dfrac{a}{b} = \dfrac{b}{a}$, then

    A. $a = b$ B. $a = -b$ C. $ab = 1$ D. $|a| = |b|$

(21) Which of the following is correct?

    A. Zero divided by any number is zero.

    B. Dividing by a number is the same as multiplying by its reciprocal.

    C. A nonzero number divided by its opposite is -1.

    D. The quotient is always smaller that the dividend.

(22) The quotient of two numbers on different sides of the origin must be

    A. negative   B. positive   C. zero   D. none of above

(23) Compare x and $2x$.

    A. $2x > x$ B. $2x = x$ C. $2x < x$ D. all are possible

(24) Which of the following is wrong?

    A. $a \cdot \dfrac{1}{a} = 1$ when $a \neq 0$.

    B. If $a \cdot b = 1$, then a and b are reciprocals of each other.

    C. $\dfrac{a}{-a} = -1$ for all number a.

    D. If $\dfrac{a}{b} = -1$, then a and b are opposites of each other.

(25) Which of the following is correct?

    A. Any number divided by itself is 1.

B. If $\dfrac{a}{b} = 0$, then $a = 0$ and $b = 1$.

C. $\dfrac{1}{m} < 1$ for all $m \neq 0$

D. The square of a number is nonnegative.

(26) Which of the following may not be negative?

    A. The product of a positive and a negative number

    B. The quotient of a positive and a negative number

    C. The sum of a positive and a negative number

    D. The product of odd number of negative numbers

(27) Which of the following is correct?

    A. If $a + b = 0$, then $\dfrac{a}{b} = -1$.   B. If $|a| = -a$, then $a < 0$.

    C. If $a > b > 0$, then $-a < -b < 0$.   D. If $a < b < 0$, then $\dfrac{1}{a} < \dfrac{1}{b} < 0$.

(28) The sign of the product is determined by

    A. the number of multiplicands

    B. the number of positive multiplicands

    C. the number of negative multiplicands

    D. the difference of the numbers of positive and negative multiplicands

(29) Which pair of numbers are equal?

    A. 3^2, 2^3 B. -2^3, $(-2)^3$ C. -3^2, $(-3)^2$ D. $(-1)^3$, $(-1)^4$

(30) Which of the following is wrong?

    A. $-6^2 = -36$ B. $(\pm\dfrac{1}{4})^2 = \dfrac{1}{16}$

    C. $(-4)^3 = -64$ D. $(-1)^{100} + (-1)^{200} = 0$

(31) If the sum of two numbers is 0, their quotient is

    A. 1   B. -1

    C. 0   D. can't be determined

(32) If a and b are opposite numbers, find the pair in which the two numbers are not opposite numbers

    A. $5a$ and $5b$ B. a^2 and b^2 C. $-a^2$ and $(-b)^2$ D. a^3 and b^3

3. Among the following numbers,

$$-1, \sqrt{2}, -\sqrt{3}, \dfrac{1}{\pi}, \dfrac{1}{2}, -\dfrac{1}{3}, 0, 2, 1\%, -1.24, \dfrac{9}{7}, 0.2\overline{43}, -0.6, -41, 127\%, -\pi, 4.13\overline{5}$$

using the set notation, find

    the set of negative numbers:

    the set of fractions:

the set of integers:

the set of nonnegative numbers:

the set of natural numbers:

the set of whole numbers:

the set of positive rational numbers:

the set of nonpositive irrational numbers:

4. Tell what is a real number line. Draw the points on the real number line representing the following numbers: $-3\frac{1}{4}$, 0, $1\frac{1}{2}$, -1, -2.5, $\frac{10}{3}$.

A real number line is _____

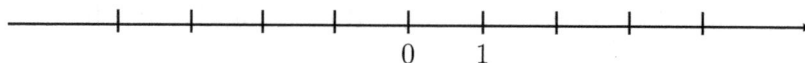

5. Fill in the blanks using $<$, $>$, $\leq$ or $\geq$.

(1) If $a < 0$, $b > 0$ and $a + b > 0$, then $|a|$____$|b|$.

(2) If $a > 0$, $b < 0$ and $|a| < |b|$, then $a + b$____0.

(3) If $a < 0$, $b \geq 0$, then $a - b$____0 and $b - a$____0.

(4) If a and b have different signs, then ab____0 and $|ab|$____0.

(5) If $ab < 0$ and $bc < 0$, then ac____0.

(6) If $\dfrac{ab}{c} < 0$ and $ac > 0$, then b____0.

(7) If $a < 0$, $b < 0$, then ab____0 and $\dfrac{ab}{a+b}$____0.

(8) If $a < 0$, $b < 0$ and $c > 0$, then $\dfrac{(a+b)(c-b)}{c}$____0.

(9) If $\dfrac{|a|}{a} = 1$, then a____0; if $\dfrac{|a|}{a} = -1$, then a____0.

(10) If $abc < 0$ and a, c have different signs, then b____0.

(11) If $-ab^2 \geq 0$, then a____0.

(12) For any $a \in \mathbb{R}$, a^2____0.

6. Assume $a, b \in \mathbb{R}$. Fill in the blanks using $<$, $>$, $\leq$, $\geq$ or $=$.

(1) When $|a + b|$____$|a - b|$, a and b have the same sign.

(2) $|a + b|$____$||a| - |b||$, if $ab < 0$.

(3) $|a - b|$____$|a| + |b|$, if $ab < 0$.

(4) If $a < 0$, $b > 0$, then $|a + b|$____$|a - b|$.

(5) When $|a + b|$____$|a - b|$, a and b have the same sign.

(6) When $|a + b|$____$|a - b|$, one of a and b is zero.

(7) $|a \pm b|$____$|a| + |b|$.

(8) $|a \pm b|$____$|a| - |b|$.

7. Evaluate.

(1) Find the value of $3a + 2b + c$ if $a = -\dfrac{3}{2}$, $b = \dfrac{2}{3}$ and $c = -\dfrac{1}{4}$.

(2) If $|x| = 4$, find x;

(3) If $|y| = 0$, find $|y + 3|$;

(4) If $|z| = -3$, find z.

(5) If a and 1 are opposite numbers, find $|a + 2|$;

(6) If $x = 2$, find $|x - 1| + |4 - x|$;

(7) If $a = -3$, find $|a - 5| - |a|$;

(8) If $x = 3$, find $x + |x - 2|$;

(9) If a and b are opposite numbers, find $|a - 3 + b|$.

(10) If the opposite of a is 0.3, $|b| = 2$. find $a + b$.

(11) If $|a| = 3$, $|b| = 4$, find all possible $a - b$.

(12) If $|a| = 2$, $|b| = 7$ and $ab < 0$, find all possible $a + b$.

(13) Assume $x + y = 0$ and $|x| = 4$, find $|x - y|$.

(14) If $|a| = 8$, $|b| = 5$ and $a + b > 0$, find all possible $a - b$.

(15) If m is the opposite of 6 and n is 2 less than the reciprocal of m, find $m - n$.

(16) If $|a| = 2$, $|b| = 1$ and $|c| = 0.5$, find all possible abc.

8. Simplify.

(1) $|-9| + |-6| - |-8| - |-7|$ (2) $|-7.2| - |-6.3| + |0.9|$

(3) $-1 + |4\frac{1}{2}| - |-5\frac{1}{8}| - |-3\frac{1}{4}|$ (4) $|7.2 - 6.4| - (|-0.8| - 0.4)$

9. Tell what properties are used in the following numerical identities.

(1) $(-3) + \pi = \pi - 3$ (2) $23 \cdot (-7) = -7 \cdot 23$

(3) $-\sqrt{2} + \sqrt{2} = 0$ (4) $\frac{7}{6} \cdot \frac{6}{7} = 1$

(5) $5 + 0 = 5$ (6) $1 \cdot (-\sqrt{7}) = -\sqrt{7}$

(7) $(-33 + 56) - 46 = -33 + (56 - 46)$ (8) $[(-23) \cdot 5] \cdot 2 = (-23) \cdot (5 \cdot 2)$

(9) $-5 \cdot 3 - (-5) \cdot 2 = -5 \cdot (3 - 2)$ (10) $(-1) \cdot (-5) = -(-5)$

(11) $0 \cdot \sqrt{0.2} = 0$ (12) $(-5) \cdot (-6) = 5 \cdot 6$

10. Distribute the negative sign and simplify.

(1) $-(a - b)$ (2) $-(-a + b)$

(3) $-(-a - b)$ (4) $-(-a + b - c - d)$

(5) $-(-a - b - c)$ (6) $-[a + b(-c)]$

(7) $-[(-a) - (-b + c))]$ (8) $-[-(-a) + (-b + c) - (-d)]$

(9) $-[+(-a + b - c)]$ (10) $-[-(-a - b) - c]$

11. Find the product.

(1) $a(-b + c)$ (2) $a(b - c)$

(3) $-a(-b - c)$ (4) $-a(b + c - d)$

12. Evaluate. Try to simplify the calculation as possible as you can.

(1) $1 - \frac{1}{6} - (+\frac{5}{6})$ (2) $-3 + 7\frac{1}{7} + (-8\frac{1}{7})$

(3) $43 - 19 - (+43) + 21$ (4) $-\frac{1}{3} + (-\frac{5}{7}) + \frac{7}{3} - \frac{22}{7} + \frac{62}{7}$

(5) $-2.3 + (-5.8) + (-3.2)$

(6) $1 - 2 + 3 - 4 + 5 - 6 + 7 - 8$

(7) $(-0.25) + (\frac{15}{2}) + (-\frac{7}{2}) + (-5.75)$

(8) $\frac{2}{5} - \frac{1}{2} + (-\frac{7}{4}) - (-\frac{13}{4})$

(9) $2.3 \times 91.1 \times 0 \times (-8.7)$

(10) $2 \cdot (-3.7) \cdot (-5)$

(11) $3.1 \times 5 + 4.7 \times 5 - 6.8 \times 5$

(12) $-24 \times (-\frac{1}{3} + \frac{3}{4} - \frac{5}{6})$

(13) $-99\frac{5}{7} \times 2.8$

(14) $(-24\frac{6}{7}) \div (-6)$

(15) $(-6) \div (-4) \cdot (-1\frac{1}{6})$

(16) $(-1.6) \cdot 25 \times (-0.4)$

(17) $[67 + (-123)] - 69$

(18) $99 \times (-47)$

13. Calculate.

(1) $(-48) \div 6 - (-25) \cdot (-4)$

(2) $2 + 42 \cdot (-8) \cdot 16 \div 32$

(3) $[2.45 + 3.5 \cdot (-0.3)] \div (-0.1) \cdot 12$

(4) $\frac{1}{18} \div (-\frac{2}{3} - \frac{1}{6} + \frac{1}{2})$

(5) $(\frac{5}{6} - \frac{3}{7} + \frac{1}{3} - \frac{9}{14}) \div (-\frac{1}{42})$

(6) $(-3)^2 \cdot [-\frac{2}{3} + (-\frac{5}{9})]$

(7) $-100 \div (-2)^3 - (-3) \div \frac{2}{3}$

(8) $-2\frac{1}{2} \cdot (-0.5)^3 \cdot 2^2 \cdot (-8)$

(9) $(-2)^2 - 2^2 - |-\frac{1}{4}| \cdot (10)^2$

(10) $0.25 \cdot (-2)^3 - [4 \div (-\frac{2}{3})^2 + 1]$

(11) $-1^4 \times (-\frac{1}{5})^2 \times (-\frac{5}{3}) + |0.8 - 1|$

(12) $\dfrac{|-3 + (-2)^2 \times (-3)| - \frac{3}{5} \times 3\frac{3}{4}}{-0.2^2 + (-4^2) \div (-8)}$

(13) $\dfrac{(2 - \frac{5}{2})^2 + 4 \times \frac{1}{2} + 3}{-\dfrac{4 + (-1)^5}{5} + \dfrac{21 - 3 \times 2^2}{7}}$

(14) $(-\frac{1}{3})^2 \div (-\frac{1}{4})^4 \times (-1)^9 - (1\frac{3}{8} + 2\frac{1}{3} - 3\frac{3}{4}) \times (-8)$

14. Order the numbers $-\sqrt{2}, -1.5, -1, -\sqrt{3}, -2, -\pi, -3$.

15. A and B are points on the real number line. If A represents -3 and the distance between A and B is 5, what numbers can the point B represent?

16. The points A, B on the real number line are as follows. Find the distance between A and B.

17. Find all integers whose absolute value is greater than 1 but no greater than 4.

18. If $x < 0$ and $y > 0$, find the smallest number among $x, y, x + y, x - y, y - x$.

19. If m and $3n$ $(n \neq 0)$ are opposite numbers, find $\dfrac{m}{n}$.

20. If $\left|\dfrac{a}{b}\right| = 1$, find the relation between a and b.

21. If $x < y < 0$, find $\dfrac{|x|}{x} + \dfrac{|xy|}{xy}$.

22. Find the condition for the number a so that $(1)a^2 > 0$, $(2)a^2 = 0$.

23. If $|a + 2| + (b - 3)^2 = 0$, find $a^3 - b^2$.

24. If $\dfrac{|a|}{a} + \dfrac{|b|}{b} + \dfrac{|c|}{c} = 1$, find $\left(\dfrac{abc}{|abc|}\right)^3 + \dfrac{ab}{|ca|}\dfrac{bc}{|ab|}\dfrac{ca}{|bc|}$.

25. If $-2 < x < 3$, simplify $|x + 2| + |x - 3|$.

Polynomials

The essential idea in elementary algebra is to represent numbers by variables. In this sense, the concept of variables lies in the core of elementary algebra. In fact, we have already used the idea of variables in the previous chapter. For example, $2+3 = 3+2$ specifies that the sum of 2 and 3 are commutative, but to show the general commutative law of addition, we write $a+b = b+a$. Here a, b are variables! Besides formulating the arithmetic rules, variables also allow us to formulate algebraic expressions, equations, inequalities and functions, which constitute the major topics of the rest of the book. Thus we can say that the whole subject of elementary algebra is about variables and their derivations.

From this chapter we come to the study of algebraic expressions — expressions built up from constants, variables, and a finite number of algebraic operations (addition, subtraction, multiplication, division and exponentiation by a rational exponent). Monomials and polynomials are the simplest algebraic expressions. The other two categories — rational expressions and radical expressions, will be introduced later.

3.1   Monomials and Polynomials

A monomial is a finite product of numbers and variables. For example, $2x$, $2x^2y^2$, 3, $-9a^3bc^5$ are all monomials. Particularly, a single number is a monomial. In a monomial, the number factor is called the coefficient, and the sum of the exponents of variables is called the degree of the monomial. For example, the coefficient of $-20x^2y$ is -20 and the degree is the exponent of x plus the exponent of y, which is $2 + 1 = 3$. The degree of a nonzero number, treated as a monomial, is defined to be 0. The number 0, treated as a monomial, is considered of degree $-\infty$(negative infinity). Here $-\infty$ is not a number, but it has the following rule of operation:

$$-\infty + n = n + (-\infty) = -\infty \text{ for any } n \in \mathbb{N},$$
$$-\infty + (-\infty) = -\infty.$$

If adding several monomials together, we get polynomials, i.e. polynomials are finite sum of monomials. For example, $2a-3b$, $4x^7-x^5+3$, x^2+y^2, $2x^2y+3xy-5y^3$ are all polynomials. A monomial, by default, is always a polynomial, so numbers are also polynomials. The monomials appearing in a polynomial are called terms. For example, the polynomial $2a - 3b$ has two terms: $2a$ and $-3b$. The degree of a polynomial is the highest degree of its terms. The constant terms are terms that are constant numbers(no variables involved). If there is no constant term, we treat it as zero. The number 0, treated as a polynomial, is called the zero polynomial. The zero polynomial can also be treated as a polynomial with arbitrary terms whose coefficients are all zero.

Example. Find the degrees and constant terms.

(1) $-2x^2y + 3xy^2 - 7x^2y^2 - x^3 - 9y^4 - 5$

(2) $6a^3 - 21a$

Solutions.

(1) The polynomial contains 5 terms: $-2x^2y$, $3xy^2$, $-7x^2y^2$, $-x^3$, $-9y^4$, -5. The degrees of these terms are 3, 3 ,4, 3, 4, 0. The greatest is 4, so the degree of the polynomial is 4. Among these terms, -5 is a constant number, so the constant term is -5.

(2) It is easy to see that the degree is 3. There is no constant term appeared, but we can treat it as 0 since $6a^3 - 21a$ is the same as $6a^3 - 21a + 0$, so the constant term is 0.

We can plug numbers into the variables of a polynomial. For example, plugging $x = 1, y = -1$ into $x^2 - y^2$, we get $1^2 - (-1)^2 = 1 - 1 = 0$. Such procedures are called evaluating.

Exercises.

1. True or false.

(1) The degree of $4x^2 - 2x^2y + 2xy^2 - 3y^2$ is 2.

(2) $-2.37a^2$, $-x^3y^7z^{11}$, $\frac{1}{3}$, a, $a + b$ are all monomials.

(3) $3x$, $-\frac{1}{x}$, $2x^2 + y^2$, $3x^2 + 2x - 1$ are all polynomials.

(4) The square of the difference of a and b is $a^2 - b^2$.

(5) $-|x|$ represents the absolute value of the opposite of x.

(6) $\frac{1}{\pi}$ is a polynomial.

(7) $\frac{2}{x} - \frac{3}{y}$ is a polynomial.

(8) When $x = -2$, $2 - |3 - x|$ is -3.

(9) $a^2 > a$.

(10) If $|a| < |b|$, then $a < b$.

(11) $3a > a$.

(12) $2n + 1$ represents an odd number if n is an integer.

2. Multiple choice.

(1) Which of the following is not a monomial?

    A. $-\frac{1}{2}$ B. $\frac{ab}{3}$ C. $-x^2y^3$ D. $x - 1$

(2) Which of the following is a monomial?

A. $\dfrac{a}{3} + 1$ B. 0   C. $c - d + e$ D. $\dfrac{3}{a}$

(3) Which of the following is correct?

A. 5 is not a polynomial.

B. $\dfrac{x + y}{2}$ is a monomial.

C. $x^2 y$ is a polynomial.

D. $x - \dfrac{3}{2}$ is a monomial.

(4) Which of the following is incorrect?

A. $-\dfrac{xy^2}{\pi}$ is not a monomial.

B. $\dfrac{x - y}{3}$ is a polynomial.

C. 0 is a polynomial.

D. $\dfrac{x + y}{x}$ is not a polynomial.

(5) Which of the following is incorrect?

A. 0 is a polynomial.

B. The coefficient and degree of a are both 1.

C. $\dfrac{1}{2} x^2 y^2$ is a monomial of degree 4.

D. The coefficient of $-\dfrac{2ab}{3}$ is $\dfrac{2}{3}$.

(6) Which of the following is correct?

A. The degree of b is 0.   B. b does not have a coefficient.

C. (-3) is a polynomial of degree 1.   D. -3 is a monomial.

(7) The degree of the polynomial $2^6 - 6x^3 y^2 + 3x^2 y^3 - x^4 - x$ is

A. 4   B. 5   C. 6   D. 10

(8) Which of the following is correct?

A. The coefficient of $\dfrac{1}{3} \pi x^2$ is $\dfrac{1}{3}$.   B. The coefficient of $\dfrac{1}{2} xy^2$ is $\dfrac{1}{2} x$.

C. The coefficient of $-5x^2$ is 5.   D. The coefficient of -1 is -1.

(9) Find the terms of $-x^2 - \dfrac{1}{2} x - 1$.

A. $x^2, \dfrac{1}{2} x, 1$ B. $x^2, -\dfrac{1}{2} x, -1$ C. $-x^2, -\dfrac{1}{2} x, 1$ D. $-x^2, -\dfrac{1}{2} x, -1$

(10) Find the coefficient of the term of degree 1 in $4x^2 - 4x^3 y - 2x + 12$.

A. 12   B. -2 C. -4 D. 4

(11) Which of the following is correct?

A. Numbers are monomials but not polynomials.

B. Monomials are not polynomials.

C. Polynomials include monomials but not numbers.

D. Both numbers and monomials are polynomials.

(12) Which of the following is not correct?

    A. A polynomial of degree 0 can have 0 as its only constant term.

    B. A polynomial of degree 1 can have 0 as its only constant term.

    C. A polynomial of degree 0 can have -1 as its constant term.

    D. A polynomial cannot have negative-integer or non-integer degrees.

3. Find expressions.

(1) the sum of five times of x and one seventh of y

(2) the number three fifths less than the sum of three times a and twice b

(3) the number two less than the opposite of t

(4) the difference of half of a and twice b

(5) the product of the reciprocal of -3 and the square of the opposite of a

(6) the reciprocal of the sum of the cubes of a and b

(7) the number three less than the absolute value of the difference of a and b

(8) the sum of twice the square of x and the square of b

(9) the reciprocal of the opposite of the absolute value of the sum of a and b

(10) the sum of two numbers whose product is 15 and one of which is m

(11) the product of two numbers whose sum is 18 and one of which is n

(12) one third of the difference of twice of a and the square of b

4. Tell which of the following are monomials and which of the following are non-polynomials.

$$a^2 - 1, \quad 0, \quad \frac{1}{3a}, \quad x + \frac{1}{y}, \quad -\frac{xy^2}{4}, \quad m, \quad , \frac{a+b}{2}, \quad \frac{x - 2y}{3}, \quad \pi, \quad 5a^3 b^6 c^6 d^2$$

$$\frac{3y^2}{x}, \quad \frac{b+1}{a}, \quad -a + \frac{b}{3}, \quad 4x^4 + 3x^2 + 1, \quad a + b, \quad -8, \quad 4 - \frac{2m}{3n}, \quad \sqrt{x+1}$$

Monomials:

Non-polynomials:

5. Find degrees and coefficients.

(1) $\dfrac{2m^3n^2}{3}$

(2) $15n^2$

(3) 0

(4) $x^7y^4z^5$

(5) $-\dfrac{3a^2b}{4}$

(6) -5

6. Find degrees and constant terms.

(1) $a^3 - a^2b + ab^2 - b^3$

(2) $3n^4 - 2n^2 + 1$

(3) $x - y$

(4) $a + 5$

(5) -4

(6) $5a - 3a^3b + ab^2 - 7$

(7) $3yx^2 - 1 - 6y^2x^5 - 4yx^3$

(8) $2x^3 + x^2 - 3xy - x^2y^2$

(9) 0

7. Find expressions accordingly.

(1) If the width and length of a rectangle are a and b separately, find the circumference and area of the rectangle.

(2) If x represents the width of a square, find the circumference and area of the square.

(3) If a disk has the length of diameter D, find its circumference and area.

(4) If the circumference of a square is s, find its area.

(5) If the sum of the diameters of two disks is 20 and the radius of one disk is r, find the

sum of their areas.

(6) In a trapezoid, the top base is a, the bottom base is 3 times the top base, and the height is 2 less than the bottom base. Find the area of the trapezoid.

8. Evaluate.

(1) $\dfrac{c}{a+b}$ when $a = 2, b = -3, c = 1$

(2) $2(a - 1)^2 - (b + 2)^2$ when $a = 1, b = -2$

(3) $(a - b)(b - c)(c - a)$ when $a = -3, b = 2, c = 1$

(4) $-3(a - 2b)^5 - 2(2a - b)^2$ when $a = -2, b = -1$

(5) $m_1 m_2 - m_1 - m_2 + 1$ when $m_1 = \dfrac{1}{4}, m_2 = \dfrac{2}{3}$

(6) $2x^3 + 3x^2 - x + 3$ when $x = 3\dfrac{1}{2}$

(7) $(|3x - 2y| - |2x - 3y|)^2$ when $x = -0.3, y = 0.2$

(8) $\dfrac{x^2 y^2 - xy}{1 + x + y}$ when $x = -2\dfrac{3}{4}, y = -\dfrac{1}{2}$

(9) $\dfrac{2k^2 - 4k - 1}{k^2 - k - 1}$ when $k = -\dfrac{3}{4}$

(10) $\dfrac{x^3 - y^3}{x - y}$ when $x = -4, y = 3$

(11) $(x + y)^2 - (x - y)^2$ when $x = -\dfrac{2}{3}, y = -\dfrac{3}{2}$

(12) $\dfrac{1}{\dfrac{1}{a} + \dfrac{1}{b}}$ when $a = -\dfrac{4}{3}, b = \dfrac{1}{4}$

(13) $\dfrac{x^2 + xy + y^2}{x^2 - xy + y^2}$ when $x = \dfrac{1}{3}, y = -\dfrac{1}{2}$

(14) $-3x^2 - 2xy + y^2$ when $x = -\dfrac{1}{2}, y = -\dfrac{1}{3}$

(15) $\dfrac{m + \dfrac{1}{m} - 1}{m^2 + \dfrac{1}{m}}$ when $m = -2$

9. Evaluate.

(1) $(a + b)^2 - 4ab$ if $a + b = -3, ab = -2$

(2) $5 - x + 3y$ if $x - 3y = -3$

(3) $(b + c) - (a - d)$ if $a - b = 3$ and $c + d = 2$

(4) $3a^2 + 6b + 2$ if $a^2 + 2b + 5 = 4$

(5) $\dfrac{3}{2}y^2 - y + 1$ if $3y^2 - 2y + 6 = 8$

(6) $\dfrac{a - b}{a + b} + \dfrac{3(a + b)}{a - b}$ if $\dfrac{a - b}{a + b} = 2$

3.2   Addition and Subtraction of Polynomials

We use capital letters to denote polynomials. For example, $P(x)$ can be used to denote a polynomial in one variable x; similarly, a polynomial in two variables a and b can be denoted by $Q(a, b)$. Two polynomials P and Q are equal, denoted by $P = Q$, if they have the same terms. For simplicity, the degree of some polynomial P is denoted by $\deg P$.

Two monomials or two terms are similar if they have the same powers of the same variables. For example, $-2x^2y^3z$ and $7x^2y^3z$ are similar, but xy^2 and $2xy$ are not, since the exponents of x are different. In other words, two similar terms can only differ by coefficients. 0 is similar to any monomial. For example, 0 is similar to $2ab$, since 0 equals $0ab$.

Similar terms can be added or subtracted by the distributive property. For example,

$$7x^2y^3z + (-2x^2y^3z) = 7(x^2y^3z) + (-2)(x^2y^3z) = [7 + (-2)](x^2y^3z) = 5x^2y^3z.$$

Similarly, $7x^2y^3z - (-2x^2y^3z) = [7-(-2)]x^2y^3z = 9x^2y^3z$. It is easy to see that when adding and subtracting similar terms, we only need to add or subtract the coefficients. Terms that are not similar cannot be combined.

Addition and subtraction of polynomials is defined in an obvious way. They can be simplified by combining similar terms. The only thing we need to pay attention to is to distribute the negative sign when doing subtractions.

Examples. Simplify.

(1) $(2xy - 3x^2y + 5) + (-2xy - 4x^2y + xy^2 - 2)$

(2) $(2a^2 + b^2) - (-a^2 + 2b^2)$

Solutions.

(1) $2xy$ and $-2xy$ are similar terms. So are $-3x^2y$ and $-4x^2y$, 7 and -2. xy^2 has no similar terms, so we just keep it. By the commutative and associative properties,

$$(2xy - 3x^2y + 5) + (-2xy - 4x^2y + xy^2 - 2)$$
$$=[2xy + (-2xy)] + [(-3x^2y) + (-4x^2y)] + xy^2 + [5 + (-2)]$$
$$=0 + (-7x^2y) + xy^2 + 3 = -7x^2y + xy^2 + 3.$$

(2) First, distribute the negative sign into the subtrahend. $(2a^2 + b^2) - (-a^2 + 2b^2) = (2a^2 + b^2) + (-1)(-a^2 + 2b^2) = (2a^2 + b^2) + [(-1)(-a^2) + (-1)(2b^2)] = (2a^2 + b^2) + (a^2 - 2b^2)$. Then it is easy to calculate

$$(2a^2 + b^2) + (a^2 - 2b^2) = 3a^2 - b^2.$$

Exercises.

1. Multiple choice.

(1) Which group of the following are similar terms?

    A. 6^2 and x^2 B. $4ab$ and $4abc$

    C. $2x^2y$ and $0.2xy^2$ D. mn and $-mn$

(2) Which of the following is correct?

    A. Any two numbers are similar terms.

    B. Any two monomials in x are similar.

    C. Polynomials in x and polynomials in y cannot be added.

    D. The difference of two monomials is again a monomial.

(3) Which group of the following are not similar terms?

    A. $3x^2y$ and $-\dfrac{1}{3}yx^2$ B. 1 and -2

    C. -10^3m^3n and $-0.01nm^3$ D. $\dfrac{1}{3}a^2b$ and $\dfrac{1}{3}b^2a$

(4) If the sum of $x^2 - 2x + 1$ and a polynomial is $3x - 2$, the polynomial being added is

    A. $x^2 - 5x + 3$ B. $-x^2 + x - 1$ C. $-x^2 + 5x - 3$ D. $x^2 - 5x - 13$

(5) If the polynomial $x^2 - 2axy - 3y^2 + \frac{1}{3}xy - 8$ does not contain the xy term, then a is

A. 0   B. $\frac{1}{3}$ C. $\frac{1}{6}$ D. $-\frac{1}{6}$

(6) If A and B are polynomials of degree 3, then the degree of $A - B$

A. must be 3   B. can be 4

C. may be smaller than 3   D. cannot be 0

(7) If A is a polynomial of degree 6 and B is of degree 4, then the degree of $2A + B$ is

A. 8   B. 6   C. 4   D. 2

(8) Which of the following is not correct?

A. For any two polynomials P, Q, $\deg P \pm Q \leq \deg P$

B. For any two polynomials P, Q, $\deg P \pm Q \leq \deg Q$

C. 0 is similar to any monomial.

D. For any two monomials M, N, since M and N are both similar to 0, M and N are similar terms.

2. Simplify.

(1) $2xy + 3xy$ (2) $(x^3 + y^3) - (-x^3 + 2y^3)$

(3) $a - [-3a - (a + b)]$ (4) $(5m^2 - 6n^2 + 3) - (2m^2 - 3n^2)$

(5) $6x - [3x^2 - (x - 1)]$ (6) $x - y - 2(x - y)$

(7) $-3a + [4b - (a - 3b)]$ (8) $(2x^2 + xy + 3y^2) - (x^2 - xy + 2y^2)$

(9) $(2x^2y + 3xy^2) - (6x^2y - 3xy^2)$ (10) $-3(a^2b + 2b^2) + (3a^2b - 13b^2)$

(11) $a^3b + (a^3b + 2c) - 2(a^2b + c)$ (12) $2(2a - 3b) + 3(2b - 3a)$

(13) $3(-ab + 2a) - (3a - b) + 3ab$ (14) $(x^4 - x^3 + 2x^2) + (3x^2 - 4x^4 + 2x - 2)$

(15) $\frac{1}{3}a - (\frac{1}{2}a - 4b - 6c) + 3(-2c + 2b)$ (16) $2(5a^2 - 2b^2 - 3c^2) - 3(-2a^2 + b^2 + 2c^2)$

(17) $2a^2 - [\frac{1}{2}(ab - a^2) + 8ab] - \frac{1}{2}ab$ (18) $\frac{2}{3}x^2y^2 - [5xy^2 - (4xy^2 - 3) + 2x^2y^2]$

(19) $3(a^2 + b^2) + 2a + 3b - 2(a + b) - 3a^2 - 2b^2 + 1$

(20) $6x^2y + 2xy - 8x^2y - 4x - 5xy + 2x^2y^2 - 6x^2y$

(21) $5(4x^2 - 4xy + y^2) - 4(x^2 + xy - 5y^2)$

(22) $(x^3 + 5x^2 + 4x - 3) - (-x^2 + 2x^3 - 3x - 1) + (4 - 7x - 6x^2 + x^3)$

(23) $(-3x^3 + 2x^2 - 5x + 1) - (5 - 6x - x^2 - x^3)$

(24) $(-x^2 + 4 + 3x^4 - x^3) - (x^2 + 2x - x^4 - 5)$

(25) $2(a^2 - ab - b^2) - 3(4a - 2b) + 2(7a^2 - 4ab + b^2)$

3. Simplify and evaluate.
(1) $3x^2 - 8x + x^3 - 5x^2 + 2x + 6x^2 + 7$ when $x = -4$

(2) $3a^2b - 3(a^2b - ab^2) - 3ab^2$ when $a = 105, b = 24$

(3) $3x^2 - [7x^2 - 2(x^2 - 3x) - 2x]$ when $|x| = 2$

(4) $2x^2 + 4x + 3x^3 - (x + 3x^3 - 2x^2)$ when $x = -3$

(5) $a^2b - 5ac - (3a^2c - a^2b) + (3ac - 4a^2c)$ when $a = -1, b = 2, c = -2$

(6) $ab - 2ab + 3b^2 + b^2 + 2ab$ when $a = -\dfrac{1}{2}, b = \dfrac{1}{2}$

(7) $2x^2 - 5xy + 2y^2 - x^2 - 4xy - 2y^2$ when $x = -1, y = 2$

(8) $5x^2 - (3y^2 + 5x^2) + (4y^2 + 3xy)$ when $x = -1, y = \dfrac{1}{3}$

(9) $-\dfrac{5}{3}xy^3 + 2x^3y - \dfrac{9}{2}x^2y - xy^3 - \dfrac{1}{2}x^2y - x^3y$ when $x = 1, y = 2$

(10) $5xyz - \{2x^2y - [3xyz - (4xy^2 - x^2y)]\}$ when $x = -2, y = -1, z = 3$

3.3 Rules of Powers

Given $a \in \mathbb{R}$ and $n \in \mathbb{Z}^+$, we have defined $a^n = \overbrace{a \cdots a}^{n}$. The definition of powers can be extended. When $a \neq 0$, define

$$a^0 = 1,$$
$$a^{-n} = \frac{1}{a^n}.$$

For example, $2^0 = 1$, $2^{-1} = \dfrac{1}{2} = \dfrac{1}{2}$, $(-2)^{-2} = \dfrac{1}{(-2)^2} = \dfrac{1}{4}$. In particular, when $a \neq 0$, a^{-1} is just the reciprocal $\dfrac{1}{a}$ of a. Similarly, $\left(\dfrac{a}{b}\right)^{-1} = \dfrac{b}{a}$ if $ab \neq 0$. At this moment, a power can have arbitrary integer exponents. In fact, real number exponents are also defined. We will extend the definition further after introducing radicals. Please pay attention that 0^0 is undefined (or D.N.E.).

Powers satisfy several rules. For any $a, b \in \mathbb{R}$ and $m, n \in \mathbb{Z}$, the following identities hold as long as all powers appeared below are well-defined (i.e. the base is nonzero when the exponent is nonpositive).

$$a^m \cdot a^n = a^{m+n}, \frac{a^m}{a^n} = a^{m-n},$$

$$(a^m)^n = a^{mn},$$

$$a^m \cdot b^m = (ab)^m, \frac{a^m}{b^m} = \left(\frac{a}{b}\right)^m.$$

These identities can be verified directly from the definition. For example, $2^3 \times 2^2 = (2 \times 2 \times 2) \times (2 \times 2) = 2^{3+2}$, $(2^2)^3 = 2^2 \times 2^2 \times 2^2 = 2^{2+2+2} = 2^2$, $(2 \times 3)^3 = (2 \times 3)(2 \times 3)(2 \times 3) = (2 \times 2 \times 2)(3 \times 3 \times 3) = 2^3 \times 3^3$. In general, the identities are proved by induction on exponents.

In the second formula $a^{m-n} = \dfrac{a^m}{a^n}$, if we let $m = n$, we get $a^0 = \dfrac{a^m}{a^m} = 1$; if we let $m = 0$, we get $a^{-n} = \dfrac{a^n}{a^n} = \dfrac{1}{a^n}$. This is the very hint how we should define the powers with negative and zero exponents. Since 0 can not be the denominator, the base a in a^0 and a^{-n} should not be 0.

The second formula can be easily derived from the first one by substituting n by $-n$. Similarly, if substituting b by $\dfrac{1}{b}$ in the fourth formula, we can get the last identity.

Examples. Assume the divisions below are legal.

(1) $8^3 = (2^3)^3 = 2^{3 \cdot 3} = 2^9$,

(2) $a^2 \cdot a^7 = a^{2+7} = a^9$,

(3) $(ab^3)^2 = a^2(b^3)^2 = a^2 b^{2 \cdot 3} = a^2 b^6$,

(4) $a^{2m+1} \div a^{m+2} = a^{(2m+1)-(m+2)} = a^{m-1}$,

(5) $\left(\dfrac{2a^2}{b^4 c^3}\right)^3 = \dfrac{(2a^2)^3}{(b^4 c^3)^3} = \dfrac{2^3 a^{2 \cdot 3}}{b^{4 \cdot 3} c^{3 \cdot 3}} = \dfrac{8a^6}{b^{12} c^9}$.

Exercises. Assume all exponents are integers.

1. True or false.

(1) $(3x + 2y)^5 (3x + 2y)^2 = (3x + 2y)^7$.

(2) If $a \neq 0$, $a^{5m} \div a^m = a^5$.

(3) $-p^2 \cdot (-p)^4 \cdot (-p)^3 = (-p)^7$.

(4) $t^m \cdot (-t)^{2n} = t^{m-2n}$.

(5) When $a \neq 0$, $(a^{n+1})^3 \div (a^n)^3 = a^3$.

(6) $(-2x^m y)^n = -2^n x^{mn} y^n$.

(7) Whenever $2x \neq y$, $(2x - y)^{2m+1} \div (y - 2x)^{2m} = 2x - y$.

(8) $(-b^4)^m = (-b^m)^4$.

(9) $a^1 = a$ for all $a \in \mathbb{R}$.

(10) $a^0 = 1$ for all $a \in \mathbb{R}$.

2. Multiple choice.

(1) x^{3m+1} can be written as

    A. $(x^3)^{m+1}$ B. $(x^m)^{3+1}$ C. $x \cdot x^{3m}$ D. $(x^m)^{2m+1}$

(2) Which of the following is correct?

    A. $a^m \cdot a^2 = a^{2m}$ B. $(a^3)^2 = a^5$

    C. $x^3 \cdot x^2 \cdot x = x^5$ D. $a^{3n-5} \div a^{5-n} = a^{4n-10}$

(3) When $x \neq 0$, $x^m \div (x^n \cdot x^p)$ equals

    A. x^{m+n+p} B. x^{m-n+p} C. x^{m-n-p} D. x^{m+n-p}

(4) $(4 \cdot 2^n)(4 \cdot 2^n)$ is

    A. $4 \cdot 2^{2n}$ B. $8 \cdot 4^n$ C. $4 \cdot 4^{2n}$ D. 2^{2n+4}

(5) If $a \neq 0$, then $|a|^{2n+1} \div a^{2n+1}$ is

    A. 1   B. -1 C. 0   D. ± 1

(6) If $x, y \neq 0$ are opposite numbers, n is a positive integer, then

    A. x^n, y^n are opposite numbers

    B. $(1/x)^n, (1/y)^n$ are opposite numbers

    C. $x^{2n}, -y^{2n}$ are opposite numbers

    D. $x^{2n-1}, -y^{2n-1}$ are opposite numbers

(7) If $a < 0$, $n \in \mathbb{Z}^+$, then $(a^n)^3$ is

    A. negative   B. nonnegative

    C. negative only when n is odd   D. negative only when n is even

(8) If $a < 0$ and $-(-a)^{2n}a^{2n+1} > 0$, then n is

    A. an odd number   B. an even number

    C. an integer   D. a natural number

(9) Which of the following is true for all $x \in \mathbb{R}$?

    A. $(x+3)^0 = 1$ B. $(x+3)^1 = x^1 + 3^1$

    C. $(x+3)^2 = x^2 + 3^2$ D. $(x+3)^9 = [(x+3)^3]^2$

(10) When $a \neq 0$, $(-a)^{-n}$ equals

    A. $-\dfrac{1}{a^n}$ B. $\dfrac{1}{a^n}$ C. $(-1)^n \dfrac{1}{a^n}$ D. $(-1)^{n+1} \dfrac{1}{a^n}$

(11) Suppose $a \neq 0$, $n \in \mathbb{N}$. Which of the following is correct?

    A. $(-a)^{-n} = a^{-n}$ B. $(-a)^{-2n} = -a^{-2n}$

    C. $(-a)^{-(2n+1)} = a^{-(2n+1)}$ D. $\left(\dfrac{1}{a^{-n}}\right)^{-1} = \dfrac{1}{a^n}$

3. Calculate.

(1) $(-2)^0$ (2) 1^{-2}

(3) $(-1)^{-3}$ (4) -2^{-5}

(5) 3^{-2}

(6) $(-2)^{-3}$

(7) $(-9)^{-2}$

(8) -4^{-3}

(9) 5^{-3}

(10) $(-5)^{-4}$

(11) $(-10)^{-2}$

(12) $(-10)^{-3}$

(13) $\dfrac{1}{7^{-2}}$

(14) $\dfrac{1}{-3^{-3}}$

(15) $\dfrac{-3}{(-2)^3}$

(16) $\dfrac{-3}{4^{-2}}$

4. Fill in the blanks.

(1) $(-a^{3n+1}) \div ($ $) = a^6$.

(2) $(x - y)^{n+5} = ($ $)(y - x)^2$.

(3) $(x^{(\ \ )})^3 = x^{9m}$.

(4) $16a^{8m}b^{12n}c^4 = ($ $)^4$.

(5) $16^{(\ \ )} = 2^{8m}$.

(6) If $a^m = -64$, $a^n = 16$, then $a^{m-n} = $ _____.

(7) If $x^{3n} = 3$, then $x^{6n} = $ _____.

(8) When $b \neq 0$, $\dfrac{-3a}{(-b)^{-5}} = $ _____.

(9) When $ab \neq 0$, $(a^{-2}b)^{-2} = $ _____, $(3a^{-1}b^2)^{-1} = $ _____, $\left(\dfrac{a}{b}\right)^{-1} = $ _____.

(10) If $t \neq 0$, $t^{-1} \div t^{-3} = $ _____, $\left(\dfrac{1}{t}\right)^{-p} = $ _____.

5. Assume all divisors are nonzero. Simplify.

(1) $a \cdot (-a^2) \cdot (-a^3)$

(2) $y^{n-2} \div y^{m-2}$

(3) $a^{2mn} \div a^{mn-1}$

(4) $(x - y)^6 \div (y - x)^3$

(5) $\left(-\dfrac{1}{2}xy^2\right)^4$

(6) $(x^m)^n \cdot (x^n)^m$

(7) $(x^m)^m \cdot (x^n)^n$

(8) $a^{24} \div [(-a^2)^3]^4$

(9) $[(-2)^2 \cdot (-2)^7] \div 8^3$

(10) $(-3)^{-5} \div 3^3$

(11) $(a^{-3}b^2)^3 (a \neq 0)$

(12) $(a \cdot a^5 \cdot a^4 \div a^7)^3$

(13) $x^{-2} \cdot x \div x^{-3} (x \neq 0)$

(14) $\left(\dfrac{b}{a}\right)^{-2} \cdot \left(\dfrac{a}{b}\right)^3 (ab \neq 0)$

(15) $(2^n \cdot 7^n \cdot 3^n \div 42^n)$

(16) $[(-a)^{2n}]^2 \cdot a^{2n} + (-a^{2n})^3$

(17) $(-a^{4n})^2 + (-a^{2n}) \cdot (-a^n)^6 + (a^n \cdot a^2)^n$ (18) $\{-y^3[-y^2(-y^4)^2]^3\}^2$

(19) $(n-m)^{3p} \cdot [(m-n)(m-n)^p]^5$ (20) $(x^2-x+1)^m(x^2-x+1)^{2n}(x^2-x+1)$

3.4 Multiplication of Polynomials

According to the rule $a^m \cdot a^n = a^{m+n}$ we can do multiplications of monomials. To multiply polynomials we apply the distributive property.

Examples.

(1) $(-2x^2y^3)(3xy^2z^2)$

(2) $(2x+y)(x+2y)$

(3) $2x(3x-2)(x^2-5x+10)$

Solutions.

(1) By the commutative and associative properties of multiplications, we can put powers of the same variables together as well as numbers.

$$\begin{aligned}
(-2x^2y^3)(3xy^2z^2) &= [(-2) \cdot 3](x^2x)(y^3y^2)(z^2) \\
&= -6x^{2+1}y^{3+2}z^2 \\
&= -6x^3y^5z^2.
\end{aligned}$$

(2) Treat the first part $(2x + y)$ as a single multiplicand A. By the distributive property, we get

$$
\begin{aligned}
(2x + y)(x + 2y) &= A(x + 2y) \\
&= Ax + A \cdot 2y \\
&= (2x + y)x + (2x + y)2y \\
&= (2x^2 + yx) + (4xy + 2y^2) \\
&= 2x^2 + 5xy + 2y^2.
\end{aligned}
$$

(3) The multiplications of numbers satisfy the associative property, so are the multiplications of polynomials. In this problem, we can multiply the last two polynomials first.

$$
\begin{aligned}
2x(3x - 2)(x^2 - 5x + 10) &= 2x[(3x - 2)(x^2 - 5x + 10)] \\
&= 2x[3x(x^2 - 5x + 10) - 2(x^2 - 5x + 10)] \\
&= 2x[(3x^3 - 15x^2 + 30x) + (-2x^2 + 10x - 20)] \\
&= 2x(3x^3 - 17x^2 + 40x - 20) \\
&= 6x^4 - 34x^3 + 80x^2 - 40x.
\end{aligned}
$$

Exercises.

1. Multiple choice.

(1) Which of the following is correct?

  A. $9a^3 \cdot 2a^2 = 18a^5$ B. $2x^5 \cdot 3x^4 = 5x^9$

  C. $3x^3 \cdot 4x^3 = 12x^3$ D. $3y^3 \cdot 5y^3 = 15y^9$

(2) Which of the following is incorrect?

  A. $(x + 1)(x + 4) = x^2 + 5x + 4$ B. $(m - 2)(m + 3) = m^2 + m - 6$

  C. $(y + 4)(y - 5) = y^2 + 9y - 20$ D. $(x - 3)(x - 6) = x^2 - 9x + 18$

(3) If $(x + q)(x + \dfrac{1}{5})$ does not contain the x term, then q is

  A. $\dfrac{1}{5}$ B. 5 C. -5 D. $-\dfrac{1}{5}$

(4) Which of the following is wrong?

  A. $[(a + b)^2]^3 = (a + b)^6$ B. $[(x + y)^{2n}]^5 = (x + y)^{2n+5}$

  C. $[(x + y)^m]^n = (x + y)^{mn}$ D. $[(x + y)^{m+1}]^n = (x + y)^{mn+n}$

(5) Which of the following is correct?

  A. $a^{m+1}a^2 = a^{2m+2}$ B. $c^{n+1}c^{n-1} = c^{2n-1}$

  C. $4x^{2n+2}(-\dfrac{3}{4}x^{n-2}) = -3x^{3n}$ D. $[-(-a)^2]^2 = -a^4$

(6) Which of the following is incorrect?

  A. $-(-3a^n b)^4 = -81a^{4n}b^4$

  B. $(a^{n+1}b^n)^4 = a^{4n+4}b^{4n}$

C. $(-2a^n)^2(3a^2)^3 = -54a^{2n+6}$

D. $(3x^{n+1} - 2x^n)5x = 15x^{n+2} - 10x^{n+1}$

(7) If A is a polynomial of degree 5, and B is of degree 3, then the degree of $A \cdot B$ is

    A. 2   B. 8   C. 15   D. 5

(8) Which of the following is correct?

    A. $(a + b)^2 = a^2 + b^2$ B. $a^m \cdot a^n = a^{mn}$

    C. $(-a^2)^3 = (-a^3)^2$ D. $(a - b)^3(b - a)^2 = (a - b)^5$

(9) If $(x^2 + px + 8)(x^2 - 3x + q)$ does not contain x^2 and x^3 terms, then

    A. $p = 0, q = 0$ B. $p = 3, q = -9$ C. $p = 3, q = 1$ D. $p = -3, q = 1$

(10) If $n \neq 0$ and $(x^n)^A = x^{2n}$, then A is

    A. n B. $n^2 - n$ C. $n^2 + n$ D. 2

2. Simplify.

(1) $(-4x^2y)(-x^2y)(\frac{1}{2}x^2y)$

(2) $(-3ab)(-a^2c)(6ab^2)$

(3) $(-4a)(2a^2 + 3a - 1)$

(4) $(-3xy)(5x^2y) + 6x^2(\frac{7}{2}xy^2 - 2y^2)$

(5) $(\frac{2}{3}ab^2 - 2ab + \frac{4}{3}b) \cdot \frac{1}{2}ab$

(6) $(3x^2)^3 - 7x^3[x^3 - x(4x^2 + 1)]$

(7) $(3m - n)(m - 2n)$

(8) $[-a^{2m}]^3 \cdot a^{3m} \cdot [(-a)^{5m}]^2$

(9) $(x + y)(x^2 - xy + y^2)$

(10) $5x(x^2 + 2x + 1) - (2x + 3)(x - 5)$

(11) $(x^2 + 3a^2)(\frac{1}{3}x^2 - a^2)$

(12) $(2a^2 - 1)(a - 4) - (a^2 + 3)(2a - 5)$

(13) $(-4xy^3)(-xy) + (-3xy^2)^2$

(14) $(3x^4 - 2x^2 + x - 3)(4x^3 - x^2 + 5)$

(15) $-2a^2(\frac{1}{2}ab + b^2) + 5ab(-a^2 + 1)$

(16) $5(x - 1)(x + 3) - 2(x - 5)(x - 2)$

(17) $(x + 3y + 4)(2x - y)$ (18) $y(y - 3(x - z)) + y[3z - (y - 3x)]$

(19) $2[(x + 2)(x + 1) - 3] - (x - 1)(x - 2) - 3x(x + 3)$

3. Evaluate after simplifying.

(1) $y^n(y^n + 9y - 12) - 3(3y^{n+1} - 4y^n)$ when $y = -3, n = 2$

(2) $(x - 2)(x - 3) + 2(x + 6)(x - 5) - 3(x^2 - 7x + 13)$ when $x = 3\dfrac{1}{2}$

(3) $x^2 \cdot x^{2n} \cdot (y^{n+1})^2$ when $x = -5, y = \dfrac{1}{5}$

(4) $-ab(a^2b^5 - ab^3 - b)$ when $ab^2 = -6$

4. Find the coefficient of the term x^3 in $(3x^4 - 2x^3 + x^2 - 8x + 7)(2x^3 + 5x^2 + 6x - 3)$.

5. Find the term of highest degree in $(4 + 2x - 3y^2)(5x + y^2 - 4xy)(xy - 3x^2 + 2y^4)$.

6. If $a^3 + b^3 = 1$ and $a(a^2 + 5b) + b(-3a + b^2) = 0.5$, find ab.

7. If $x^3 - 6x^2 + 11x - 6 = (x - 1)(x^2 + mx + n)$, find m and n.

8. Prove.

(1) If $x = b+c, y = c+a, z = a+b$, prove $(x-y)(y-z)(z-x)+(a-b)(b-c)(c-a) = 0$.

(2) Show $(2x+3)(3x+2) - 6x(x+3) + 5x + 16$ does not depend on x.

(3) Show $(m+1)(m-1)(m-2)(m-4) = (m^2 - 3m)^2 - 2(m^2 - 3m) - 8$.

9. If A, B are polynomials of degree m and n separately, find the degree of

(1) $A \pm B$ if $m > n$

(2) $A \pm B$ if $m = n$

(3) $A \cdot B$

3.5 Formulas for Polynomial Multiplications

The following formulas can be used in polynomial multiplications.

- $(a+b)(a-b) = a^2 - b^2$ (Difference of Squares)
- $(a \pm b)^2 = a^2 \pm 2ab + b^2$ (Perfect Squares)
- $(a \pm b)(a^2 \mp ab + b^2) = a^3 \pm b^3$ (Sum and Difference of Cubes)

These formulas can be proved directly by polynomial multiplications. For example,

$$(a+b)(a-b) = a(a-b) + b(a-b) = (a^2 - ab) + (ab - b^2) = a^2 - b^2,$$
$$(a+b)^2 = (a+b)(a+b) = a(a+b) + b(a+b),$$
$$= (a^2 + ab) + (ab + b^2) = a^2 + 2ab + b^2$$
$$(a-b)(a^2 + ab + b^2) = a(a^2 + ab + b^2) - b(a^2 + ab + b^2)$$
$$= (a^3 + a^2b + ab^2) - (a^2b + ab^2 + b^3) = a^3 - b^3.$$

The power of these formulas lies in substitutions. We can plug any algebraic expressions into a and b in these formulas.

Examples.

(1) $(2a - 3b)^2$

(2) $(a + b + c)(a + b - c)$

(3) $(x^2 - 2y^2)(x^4 + 2x^2y^2 + 4y^4)$

Solutions.

(1) Let $A = 2a, B = 3b$; by the perfect square formula

$$(2a - 3b)^2 = (A - B)^2$$
$$= A^2 - 2AB + B^2$$
$$= (2a)^2 - 2(2a)(3b) + (3b)^2$$
$$= 4a^2 - 12ab + 9b^2.$$

(2) Treat $a + b$ as a single term A; then $(a + b + c)(a + b - c) = (A + c)(A - c)$. Applying the difference of squares formula, $(A + c)(A - c) = A^2 - c^2 = (a + b)^2 - c^2$. After expanding $(a + b)^2$ by the perfect square formula, we get the answer is $a^2 + 2ab + b^2 - c^2$.

(3) Let $A = x^2, B = 2y^2$. Then $(x^2 - 2y^2)(x^4 + 2x^2y^2 + 4y^4)$ is of the form $(A - B)(A^2 + AB + B^2)$. By the formula of difference of cubes, we get $(x^2 - 2y^2)(x^4 + 2x^2y^2 + 4y^4) = A^3 - B^3 = (x^2)^3 - (2y^2)^3 = x^6 - 8y^6$.

The following are some more formulas for polynomial multiplications that may be needed in advanced math courses.

- $(a + b + c)^2 = a^2 + b^2 + c^2 + 2ab + 2bc + 2ca$.
- $(a \pm b)^3 = a^3 \pm 3a^b + 3ab^2 \pm b^3$.
- $(a + b)^n = \sum_{k=0}^{n} \binom{n}{k} a^{n-k} b^k$.
- $(a - b)^n = \sum_{k=0}^{n} (-1)^k \binom{n}{k} a^{n-k} b^k$.
- $(a - b)(a^{n-1} + a^{n-2}b + a^{n-3}b^2 + \cdots + ab^{n-2} + b^{n-1})$
  $= (a - b)(\sum_{k=0}^{n-1} a^{n-k-1} b^k) = a^n - b^n$ $(n \in \mathbb{Z}^+)$.
- $(a + b)(a^{2n} - a^{2n-1}b + a^{2n-2}b^2 - a^{2n-3}b^3 + \cdots - ab^{2n-1} + b^{2n})$
  $= (a + b)(\sum_{k=0}^{2n} (-1)^k a^{2n-k} b^k) = a^{2n+1} + b^{2n+1}$ $(n \in \mathbb{N})$.

Exercises.

1. True or false.

(1) $(x + 3y)(x - 3y)(x^2 - 9y^2) = x^4 - 81y^4$.

(2) $(2x^n - 3y^{m+1})^2 = 4x^{2n} - 6x^n y^{m+1} + 9y^{2m+2}$.

(3) $(a + \dfrac{1}{a})^2 = (a - \dfrac{1}{a})^2 + 4$.

(4) $(a - b)[(a + b)^2 - ab] = a^3 - b^3$.

(5) $(x + 1)(x^2 - 2x + 1) = x^3 + 1$.

(6) $(a - b)(a^2 - ab + b^2) = a^3 - b^3$.

(7) $(\dfrac{1}{2}a - \dfrac{1}{3}b)(\dfrac{1}{4}a^2 + \dfrac{1}{6}ab + \dfrac{1}{9}b^2) = \dfrac{a^3}{8} - \dfrac{b^3}{27}$.

(8) $(a - b)^2 = (b - a)^2$.

(9) $(a + b + c)(a - b + c) = a^2 - (b + c)^2$.

(10) If $(a - b)^2 = 1, a^2 + b^2 = 4$, then $ab = \dfrac{3}{2}$.

2. Multiple choice.

(1) Which of the following is correct?

    A. $(x - y)^2 = (-x - y)^2$ B. $(x - y)^2 = (y - x)^2$

    C. $(x - y)^2 = x^2 - y^2$ D. $(x - y)^2 = x^2 + y^2$

(2) For which does the formula of difference of squares not apply?

    A. $(m - n)(-m + n)$ B. $(x^3 - y^3)(x^3 + y^3)$

    C. $(-a - b)(a - b)$ D. $(c^2 - d^2)(d^2 + c^2)$

(3) Which of the following is correct?

    A. $(a - b)^2 = a^2 - ab + b^2$ B. $(a + 3b)^2 = a^2 + 9b^2$

    C. $(a + b)(a - b) = (b + a)(-b + a)$ D. $(x - 9)(x + 9) = x^2 - 9$

(4) For which does the formula of sum or difference of cubes apply?

    A. $(m + n)(m^3 + m^2 n + n^3)$ B. $(m - n)(m^2 + n^2)$

    C. $(x + 1)(x^2 - x + 1)$ D. $(x - 1)(x^2 - x + 1)$

(5) $(2a - 1)^2(4a^2 + 2a + 1)^2$ equals

    A. $64a^6 - 16a^3 + 1$ B. $64a^6 + 16a^3 + 1$

    C. $64a^6 - 1$ D. $64a^6 + 1$

(6) $(a - \frac{1}{3})(a + \frac{1}{3})(a^2 - \frac{1}{3}a + \frac{1}{9})(a^2 + \frac{1}{3}a + \frac{1}{9})$ is

    A. 0 B. $a^6 + \frac{1}{27^2}$ C. $a^6 - \frac{1}{27^2}$ D. $(s^3 - \frac{1}{27})^2$

(7) If $x + \frac{1}{x} = 3$, then $x^2 + \frac{1}{x^2}$ is

    A. 9 B. 11 C. 7 D. 1

(8) Which of the following is incorrect?

    A. $(3a - b)^2 = 9a^2 - 6ab + b^2$

    B. $(\frac{1}{2}x - y)^2 = \frac{1}{4}x^2 - xy + y^2$

    C. $(a + b - c)^2 = a^2 + b^2 + c^2 + 2ab - 2bc - 2ca$

    D. $(x + y)(x - y)(x^2 - y^2) = x^4 - y^4$

(9) $|5x - 2y| \cdot |2y - 5x|$ equals

    A. $(5x - 2y)^2$ B. $-(5x - 2y)^2$ C. $-(2y - 5x)^2$ D. $(5x)^2 - (2y)^2$

(10) If $x + y = 10, xy = 24$, then $x^2 + y^2$ is

    A. 52 B. 148 C. 58 D. 76

(11) $(-x - 1)(x^2 - x + 1)$ is

    A. $-x^3 + 1$ B. $-x^3 - 1$ C. $x^3 + 1$ D. $x^3 - 1$

(12) If $x - y = 8, xy = -15$, then $-2x^2 - 2y^2$ is

    A. -128 B. -68 C. 98 D. 128

3. Fill in the blanks.

(1) $a^2 + ($ $) + 4b^2 = (a - 2b)^2$

(2) $(a + b)^2 - ($ $) = (a - b)^2$

(3) $9a^4 + 12a^2b^2 + ($ $) = (3a^2 + 2b^2)^2$

(4) $(5x^3 + 2y^2)($ $) = 125x^9 + 8y^6$

(5) $[x^2 - ($ $)][($ $) + x^2y^2 + y^4] = x^6 - ($ $)$

(6) $4x^2 - 9x + ($ $) = ($ $)^2$

(7) $(x - 2a)(x + 2a)($ $) = x^6 - 64a^6$

(8) $[a - (b + c)^2][$ $] = a^2 - (b + c)^4$

4. Simplify.

(1) $(a - 2)(4 + 2a + a^2)$

(2) $(a + b)[(a - b)^2 + ab]$

(3) $(a + 2b)[(a + 2b)^2 - 6ab]$

(4) $(w + 2v)(w - 2v)$

(5) $(2x - 3y)(3y + 2x) - (4y - 3x)(3x + 4y)$

(6) $(x + y)(x - y)^2 - (x + y)(x^2 - xy + y^2)$

(7) $(a^2 - ab + b^2)(a^2 + ab + b^2)$

(8) $(2x - 3y)^2$

(9) $(3a + b)^2$

(10) $(a^2 + 9)^2 - (a + 3)(3 - a)(a^2 + 9)$

(11) $(2 - m + n)(2 + m - n) - (1 - m + n)(1 + m - n)$

(12) $(x - 1)(x + 1)(x^2 + 1)$

(13) $(x - y)^2(x + y)^2 - (x^2 + y^2)^2$

(14) $(x - \frac{1}{3})(x^2 + \frac{1}{3}x + \frac{1}{9})(x^2 - \frac{1}{3}x + \frac{1}{9})(x + \frac{1}{3})$

(15) $(y - x)^2 - 4(x - y)(y + x)$

(16) $(x + y + z)(x - y - z) - (x - y + z)(x + y - z)$

(17) $(9 - a^2)^2 - (3 - a)(3 + a)(9 + a^2)$

(18) $(x + y)^2 - 4(x + y)(x - y) + 4(x - y)^2$

(19) $(x + 5)^2(x - 5)^2 - (2x + 1)^2(2x - 1)^2$

(20) $(a^2 - a + 1)(a^2 + a + 1)(a^4 - a^2 + 1)$

(21) $(x^2 - y^2)(x + y)^2 - 2xy(x - y)(x + y)$

(22) $(2x + 1)(1 - 2x + 4x^2) - x(3x - 1)(3x + 1) + (x^2 + x + 1)(x - 1) - (x - 3)$

5. Evaluate

(1) $(a^2 + b^2)(a^2 - b^2) - (a + b)^2(a - b)^2$ when $a = 4, b = \dfrac{1}{4}$

(2) $(x^2 - 3x + 9)(3 + x) + (x - 2)(4 + 2x + x^2)$ when $x = -\dfrac{3}{2}$

(3) $[(a + \dfrac{b}{2})^2 - (a - \dfrac{b}{2})^2](2a - \dfrac{b}{2})(2a + \dfrac{b}{2})(4a^2 + \dfrac{b^2}{4})$ when $a = -1, b = 2$

(4) $[(a + b)(a - b)]^7 \div [(-a + b)(a + b)]^5 - 2(-ab)^2$ when $a = -2, b = 1$

(5) $\dfrac{a^2 + b^2}{2} - ab$ when $a - b = 2$

(6) $2(x - y)^2 - (y - x)^2 - (x + y)(y - x)$ when $x = 2a + 1, y = 1 - 2a$

(7) $a^2 + \dfrac{1}{a^2}$ when $a + \dfrac{1}{a} = 5$

(8) $a^3 + \dfrac{1}{a^3}$ when $a + \dfrac{1}{a} = 3$

6. Prove.

(1) $(a^2 + b^2)(x^2 + y^2) = (ax + by)^2 + (bx - ay)^2$.

(2) Dividing the square of an odd number by 4, the remainder is 1.

(3) The difference of the squares of two consecutive integers is odd.

(4) One plus the product of two consecutive odd numbers is a perfect square.

(5) $m^2 - n^2$ can be divided by 4 if m, n are both odd or even.

(6) The difference of the cubes of two consecutive odd numbers is a multiple of 8.

(7) $a^2 + (a+1)^2 + (a^2 + a)^2 = (a^2 + a + 1)^2$.

(8) If A, B, C are three consecutive odd numbers, then $B^2 - AC = 4$.

3.6 Division of Polynomials

For polynomials P, Q, R, when $R = P \cdot Q$, we have $R \div Q = P$. In this case, we say R can be divided by Q. In general, it is not easy to check if a polynomial can be divided by another one. For the special case that the divisor is a monomial, it is always ready to check if the division can be carried out as well as to find the quotient. If both the dividend and divisor are polynomials in one same variable, we can do the long division with remainders.

3.6.1 Dividing Monomials

According to the rule $a^m \div a^n = a^{m-n}$ for any real number a and natural numbers $m \geq n$, we can divide a monomial by a monomial when the exponents in the dividend are greater. To divide a polynomial by a monomial, we do the division term by term.

Examples.

(1) $(-5x^6 y^3) \div (2x^3 y^3)$

(2) $(2x^5 y^3 - 7x^2 y^4) \div (-xy^2)$

Solutions.

(1) $(-5x^6 y^3) \div (2x^3 y^3) = \dfrac{-5x^6 y^3}{2x^3 y^3} = \dfrac{-5}{2} \cdot \dfrac{x^6}{x^3} \cdot \dfrac{y^3}{y^3} = -\dfrac{5}{2} x^{6-3} y^{3-3} = -\dfrac{5}{2} x^3.$

(2) Directly

$$\begin{aligned}
(2x^5y^3 - 7x^2y^4) \div (-xy^2) &= \frac{2x^5y^3 - 7x^2y^4}{-xy^2} \\
&= \frac{2x^5y^3}{-xy^2} - \frac{7x^2y^4}{-xy^2} \\
&= -2x^4y + 7xy^2.
\end{aligned}$$

3.6.2 Long Division

We can also define divisions of polynomials in one same variable. To do this, we need to write such polynomials in descending form. A polynomial in one variable is

- ascending if the degrees of the terms are increasing. For example, $2 + 4x - 5x^2 - x^4$ is ascending.

- descending if the degrees of the terms are decreasing. For example, $-x^4 - 5x^2 + 4x + 2$ is descending.

The division of polynomials in one variable is called the long division with remainders. As the name indicates, we may get remainders besides the quotients. The method is demonstrated by the following example.

Example. $(2x + 1 - 4x^3 + 2x^4) \div (1 + 2x^2)$

Solutions. 1. Write both the dividend and divisor in descending order, i.e. in the form $(2x^4 - 4x^3 + 2x + 1) \div (2x^2 + 1)$.

2. If a degree in the middle is missed, make it up by placing a zero in its position. Here the term of degree 2 in the dividend $2x^4 - 4x^3 + 2x + 1$ is missed, so we rewrite it as $2x^4 - 4x^3 + 0x^2 + 2x + 1$. Similarly, we rewrite the divisor as $2x^2 + 0x + 1$.

3. Do the division as if you were doing division of integers.

(i) The term of highest degree in the dividend is $2x^4$, and in the divisor the term of highest degree is $2x^2$. Their quotient is $2x^4 \div 2x^2 = x^2$. The first term in the quotient is x^2. We put it in the following format.

$$
\begin{array}{r}
x^2 \\
2x^2 + 0x + 1 \overline{)\ 2x^4 - 4x^3 + 0x^2 + 2x + 1}
\end{array}
$$

(ii) Multiply x^2 by the divisor and subtract the product from the dividend.

$$
\begin{array}{r}
x^2 \\
2x^2 + 0x + 1 \overline{)\ 2x^4 - 4x^3 + 0x^2 + 2x + 1} \\
\underline{-2x^4 + 0x^3\ -\ x^2 } \\
-4x^3\ -\ x^2 + 2x
\end{array}
$$

(iii) Now we treat the remainder $-4x^3 - x^2 + 2x$ as a new dividend and repeat step (i) and (ii). $-4x^3 \div 2x^2 = -2x$, so the second term of the quotient is $-2x$.

$$\begin{array}{r}x^2 - 2x\\2x^2 + 0x + 1)\overline{2x^4 - 4x^3 + 0x^2 + 2x + 1}\\-2x^4 + 0x^3 - x^2\\\hline-4x^3 - x^2 + 2x\\4x^3 + 0x^2 + 2x\\\hline-x^2 + 4x + 1\end{array}$$

(iv) Repeat the procedure until the last term(constant term).

$$\begin{array}{r}x^2 - 2x - \frac{1}{2}\\2x^2 + 0x + 1)\overline{2x^4 - 4x^3 + 0x^2 + 2x + 1}\\-2x^4 + 0x^3 - x^2\\\hline-4x^3 - x^2 + 2x\\4x^3 + 0x^2 + 2x\\\hline-x^2 + 4x + 1\\x^2 + 0x + \frac{1}{2}\\\hline4x + \frac{3}{2}\end{array}$$

At this point, we get the quotient $x^2 - 2x - \dfrac{1}{2}$ with the remainder $4x + \dfrac{3}{2}$. By the formula

$$\text{Dividend} = \text{Divisor} \times \text{Quotient} + \text{Remainder}$$

we have $(2x + 1 - 4x^3 + 2x^4) = (1 + 2x^2) \times (x^2 - 2x - \dfrac{1}{2}) + (4x + \dfrac{3}{2})$.

From the procedure, we can see the degree of the remainder is always less than the degree of the divisor. If $\deg R(x) < \deg Q(x)$, immediately the quotient of $R(x) \div Q(x)$ is 0 with remainder $R(x)$, i.e. $R(x) = 0 \times Q(x) + R(x)$.

Given a polynomial $P(x)$ and a number a, suppose the quotient of $P(x)$ dividing $(x - a)$ is $Q(x)$, while the remainder is R. R must be a number since the degree of R is less than the degree of $x - a$, which must be 0 or $-\infty$(when $R = 0$). As $P(x) = Q(x)(x - a) + R$, we get $R = P(a)$ by plugging $x = a$ into the equation. For example, the remainder of $P(x) = x^3 - 2x + 1$ divided by $x + 2$ is $P(-2) = (-2)^3 - 2(-2) + 1 = -3$. Here we must treat $x + 2$ as $x - (-2)$.

A number a satisfying $P(a) = 0$ is called the solution to the equation $P(x) = 0$ or the root of the polynomial $P(x)$. By long division, a is a root of $P(x)$ if and only if $P(x)$ can be divided by $(x - a)$. For example, -1 is not a root of $Q(x) = x^3 - 3x + 2$, as $Q(-1) \neq 0$; 2 is a root of $x^2 - 3x + 2$, as when $x = 2$ it is easy to check that $x^2 - 3x + 2 = 0$. It is also easy to find out $(x^2 - 3x + 2) \div (x - 2) = x - 1$ by long division.

Exercise.

1. Find quotients.

(1) $5a^2b^2 \div 15ab^2$

(2) $4x^3y^2 \div 2xy$

(3) $9m^6n^3 \div 3n^3m^2$

(4) $4x^4y^2 \div (-2xy)^2$

(5) $-6x^4y^2z^4 \div (-\frac{1}{3}x^3yz^2)$

(6) $25x^3y^3 \div 5x^2y$

(7) $y^{m+2n+6} \div y^{m+n+2}$

(8) $(3ab^2)^3 \div 3a^2b^3$

(9) $(2a)^3b^4 \div 12a^2b^3$

(10) $(x^3)^3 \div [x(x^2)^2]$

(11) $(-2a^2)^3[-(-a)^4]^2 \div a^8$

(12) $(14a^3b^2 - 21ab^2) \div 7ab^2$

(13) $(x^ax^bx^cx^d)^m \div [(x^m)^a(x^m)^b]$

(14) $(16a^3 - 24a^2) \div (-8a^2)$

(15) $(x^4y + 6x^3y^2 - x^2y^3) \div 3x^2y$

(16) $(16x^2yz^4 - 18x^3y^2z^2 - 8xy^3z^3) \div (-3xyz^2)$

2. Write the polynomials in ascending form.

(1) $-5x^2 + 3x + 7x^4$

(2) $5y^2 - 5y + y^3 - 2y^4 - 10$

(3) $-2x - x^3 + 4 - 5x^2$

(4) $5a - 8a^7 - 12a^5 + 15a^3 + 21a^6 - 7 + 12a^2$

3. Write the polynomials in descending form.

(1) $-3 + 2x^2 + 11x$

(2) $-9 - 23z^4 - 65z^6 + 37z^2$

(3) $b^8 + 2 - 3b^2 - 4b^4 + 5b^9$

(4) $-100 + 22y^4 + 76y^3 - 35y^5 - 51y - 62y^2$

4. True or false.

(1) $(4a^2 + 4a + 3) \div (2a + 3) = 2a + 1$

(2) $(6x^2 + 9x + 15) \div (2x + 5) = 3x + 2$

(3) $(m^3 - 3m^2 - 9m + 22) \div (m - 2) = m^2 - m - 11$

(4) If the quotient of a polynomial divided by $a + 4$ is $a^3 + 2a - 4$, this polynomial is $a^5 - a$.

(5) If $4x^2 + x + a$ can be divided by $x - 1$, then $a = -5$.

(6) $(n + 3)(n - 3) - (n - 2)(n + 7)$ can be divided by 5 for any integer n.

(7) $(n + 3)(n - 2) - n(n - 6)$ can be divided by 6 for any integer n.

(8) The remainder of the square of an even number dividing 4 is 0.

(9) If the divisor is $6x^2 + 3x - 5$, the quotient is $4x - 5$ and the remainder is 8, then the dividend is $24x^3 - 18x^2 - 35x + 17$.

(10) $(a^2 - b^2)(a + b) \div (a + b)^2 = a - b$.

5. Multiple choice.

(1) If $(x^3 - 2x^2 + ax + 2) \div (x^2 - 4x + 1) = x + 2$, then $a =$

    A. -7 B. 7 C. $7x$ D. $-7x$

(2) If $(x^3 - 3x^2 - 9x + 23) = (x^2 - x - 11) \cdot N + 1$, then N is

    A. $x - 2$ B. $x + 2$ C. $-x - 2$ D. $-x + 2$

(3) If $x^3 - 3x^2 + ax + b$ can be divided by $x - 2$, which of the following is possible?

    A. $a = 9, b = 22$ B. $a = -9, b = 22$

    C. $a = 9, b = -22$ D. $a = -9, b = -22$

(4) If $9x^4 - 6x^2y^2 + y^4 = (3x^2 - y^2) \cdot N$, then $N =$

    A. $3x^2 + y^2$ B. $(3x)^2 - y^2$ C. $(3x)^2 + y^2$ D. $3x^2 - y^2$

(5) If $4x^3 + 9x^2 + mx + n$ can be divided by $x^2 + 2x - 3$, then

    A. $m = 10, n = 3$ B. $m = -10, n = 3$

    C. $m = 10, n = -3$ D. $m = -10, n = -3$

(6) If $x^2 + x + m$ can be divided by $x + 5$, it can also be divided by

    A. $x - 6$ B. $x + 6$ C. $x - 4$ D. $x + 4$

(7) $(a - b)(a + b)(a^4 + a^2b^2 + b^4) \div (b^6 - a^6)$

    A. 1 B. 0 C. -1 D. 2

(8) If $3x^4 - 2x^3 - 32x^2 + 66x + m$ can be divided by $x^2 + 2x - 7$, then $m =$

    A. 35 B. -32 C. -35 D. 32

6. Find the remainders. (Long division is not needed)

(1) $(4 + 2x - 5x^2) \div (x - 3)$ (2) $(x - 6x^3 + x^6 - 18) \div (x + 1)$

(3) $(2a^3 + 9a^2 - 3a + 3) \div (a + 3)$ (4) $(2x^3 + 5x^2 - 7x + 2) \div (x - \dfrac{1}{2})$

(5) $(x^5 - 2x^3 + 1) \div (3x + 6)$ (6) $(6x^4 - 3x^3 - 7x + 3) \div (2x - 1)$

7. Fill in the blanks.

(1) $($ $)(x^2 + xy + y^2) = x^3 - y^3$.

(2) If the quotient of $2x^3 + 6x^2 + 6x + 5$ divided by A is $x + 1$ with remainder $5x + 8$, then $A =$ _____.

(3) If $(2m^3 + bm^2 + 6m + 4) \div (m^2 + m + 1) = 2m + 4$, then $b =$ _____.

(4) If the remainder of $(3x^3 + nx + 20) \div (x^2 + 2x - 3)$ is $3x + 2$, then $n =$ _____.

(5) If $4x^3 + 9x^2 + mx + n$ can be divided by $x^2 + 2x - 3$, then $m =$ _____ , $n =$ _____ .

(6) The quotient and remainder of $(x^3 + 4x^2 + 5x + 2) \div$ _____ are both $x + 1$.

(7) If the remainder of $(3x^3 + nx + m) \div (x^2 + x - 2)$ is $3x + 2$, then $m =$ _____ , $n =$ _____ .

(8) If $x^2 - 3x - 2 = 0$, then $-x^3 + 11x + 8 =$ _____ .

(9) If $x^2 + kx + 14$ can be divided by $x - 2$, then $k =$ _____ .

(10) If $6x^2 + 13x + b$ can be divided by $3x - 1$, then $b =$ _____ .

8. Find quotients and remainders. Write the dividend as divisor times quotient plus remainder.

(1) $(4 + 2x - 5x^2) \div (x - 2)$

(2) $(x^5 - 4x^3 + 2x^2 - 1) \div (x + 2)$

(3) $(4x^3 + 2x^2 - 1) \div (2x - 4)$

(4) $(2x^4 - 5x^3 - 26x^2 - x + 28) \div (-3 + 2x^2 + x)$

(5) $(x^4 + 1) \div (x^3 + x^2 + x)$

(6) $(x - 6x^3 + x^6 - 18) \div (x^2 - 2)$

(7) $(6x^3 - 8x - 5) \div (3x + 2)$

(8) $(2x^3 + 3x - 3 + 9x^2) \div (4x + x^2 - 3)$

(9) $(6x^4 - 3x^3 - 7x - 3) \div (2x^2 - x - 2)$

(10) $(x^5 - 2x^4 - 4x^3 + 1) \div (x^3 + 2x + 1)$

(11) $(2x^4 + 7x^3 - 12x^2 - 27x) \div (2x^2 + 3x)$

(12) $(6x^6 - 4x^5 + 2x^4 - x - 5) \div (2x^4 - x - 3)$

9. P, D are polynomials. If the quotient of $P \div D$ is Q with remainder R, solve the following.

(1) If $D = 3x^2 + 2y$, $Q = 9x^4 - 6x^2y + 4y^2$, $R = x - 8y^3$, find P.

(2) If $D = 2x^3 - 3x^2 + 1$, $Q = x + 2$, $R = 6x^2 - 2$, find P.

(3) If $P = 4x^3 + 2x^2 - 1$, $D = 2x - 4$, $R = 39$, find Q.

(4) If $P = x^5 - 4x^3 + 2x^2 - 1$, $D = x + 2$, $Q = x^4 - 2x^3 + 2x - 4$, find R.

(5) If $P = 18x^4 + 82x^2 + 56 - 71x - 45x^3$, $Q = 6x^2 - 7x + 8$, $R = 16 - 4x$, find D.

Factoring

Factoring is the inverse operation of polynomial multiplication. The goal is to rewrite a polynomial in the form of a product of two or more polynomials. It is widely used in elementary algebra, especially in solving equations. Whether a polynomial can be factored or not strongly depends on the number system we are working with. For example, $x^2 + 1$ can be factored in the complex number system, but not in the real number system. In this chapter, we fix the number system for factoring to be the set of integers. Occasionally, we work with rational numbers, but we will state it explicitly.

Not every polynomial can be factored. Even for factorable polynomials, there is no universal method to do factorization. However, we do have some good methods to try for the ones with special structures. These methods are: 1) finding common factors; 2) using formulas; 3) cross products; 4) group factoring. Sometimes we can solve a problem by applying one single method, but most of the time we use combinations of them.

The rule of factoring is to factor completely, i.e. we should proceed until no further factors can be found.

4.1 Common Factor Method

The common factor method is always the first method to try for factoring. The idea comes from the distributive property of multiplication. For example, we have $a(b+c) = ab + ac$. If we reverse this equation, it becomes

$$ab + ac = a(b + c),$$

which makes $ab + ac$ factored. Here we treat a as the common factor. To comply with the rule of completion in factoring, we need to find all possible common factors.

Example. Factor $-3x^2y + 12x^2yz - 9x^3y^2$.

Solutions. The coefficient of the first term is negative, so we pull out the negative sign to get

$$-3x^2y + 12x^2yz - 9x^3y^2 = -(3x^2y - 12x^2yz + 9x^3y^2).$$

In $3x^2y - 12x^2yz + 9x^3y^2$, the greatest common factor of all three coefficients $3, -12, 9$ is 3; the lowest powers of x, y, z are $x^2, y, 1$. As a result, we get a common factor $3x^2y$. We can pull it out directly, or pull out 3 and x^2y in separate steps.

$$-3x^2y + 12x^2yz - 9x^3y^2 = -(3x^2y - 12x^2yz + 9x^3y^2)$$
$$= -3(x^2y - 4x^2yz + 3x^3y^2)$$
$$= -3\left(x^2y \cdot \frac{x^2y - 4x^2yz + 3x^3y^2}{x^2y}\right)$$
$$= -3x^2y(1 - 4z + 3xy).$$

Exercises.

1. True or false.

(1) The factorization $x^2(a - b)^2 - xy(b - a)^2 = (a - b)^2(x^2 + xy)$ is complete.

(2) $8(a - b)^3 + 4(b - a)^2 = 4(a - b)^2(2a - 2b + 1)$.

(3) $(a - b)(m - 2) - (b - a)(n + 3) = (a - b)(m + n + 1)$.

(4) $9a^2b + 12ab^2 - 21a^2b^2 = 3ab(3a + 4b - 7ab)$.

(5) $(m + n)(x - y) - (m - n)(x + y)$ can be factored by finding a common factor.

(6) $x(8 - x) + y(x - 8)$ can be factored by finding a common factor.

(7) $2(a - b)^3a - 5(b - a)^3b = (a - b)^3(2a - 5b)$.

(8) $p(a - b)(m - n) - q(b - a)(m - n) = (p + q)(a - b)(m - n)$.

(9) $m(a - b)^2 - n(b - a)^2 = (m + n)(a - b)^2$.

(10) $x(x + y - z) - y(z - x - y) = (x + y - z)(x + y)$.

2. Multiple choice.

(1) $2y(x - y)^2 - (y - x)^3$ can be factored as

    A. $(x + y)(x - y)^2$ B. $(3y - x)(x - y)^2$

    C. $(x - 3y)(y - x)^2$ D. $(y - x)^3$

(2) The common factor in $-4a^3 + 16a^2 + 12a$ is

    A. a B. $-4a^2$ C. -4 D. $-4a$

(3) $(m - n)^3 + m(m - n)^2 + n(n - m)^2$ can be factored into

    A. $2(m - n)^3$ B. $2m(m - n)^2$ C. $-2n(n - m)^2$ D. $2(n - m)^3$

(4) The common factor in $a^2x + ay - a^3xy$ is

    A. a B. a^2 C. ax D. ay

(5) $(x - y)(a - b) - (y - x)^2 + (y - x)$ equals

    A. $(x - y)(x - y + a - b - 1)$ B. $(x - y)(a - b - x - y - 1)$

    C. $(x - y)(y - x + a - b - 1)$ D. $(x - y)(y - x + a - b + 1)$

(6) The common factor in $6a(a - b)^2 - 8(a - b)^3$ is

    A. a B. $(a - b)^2$ C. $8a(a - b)$ D. $2(a - b)^2$

3. Factor.

(1) $3x - 6$ (2) $5a^2 + 15a$

(3) $2\pi R - 2\pi r$

(4) $6x^2y + 9x^3y^2$

(5) $-m^3 - 5m^2 + 15m$

(6) $-12x^2y^2 + 6x^3y^3 - 4xy$

(7) $6a^2b - 3a^3b - 12a^4b^2$

(8) $(-x)^3 + (-x)^4 + 2x^2$

(9) $a^{m+1}b + a^mc^2$

(10) $-8m^4 - 24m^3 + 48m^2$

(11) $-3xy - 15xyz + 27abxy$

(12) $-3x^2yz^2 + 12x^3y^2z^2 + 9x^2yz^3$

(13) $18a^3bc - 45a^2b^2c^2 + 27a^2b^3c$

(14) $7abx + 21ab^2y - 49a^2bxy$

4. Factor.

(1) $0.5x(a - b) - 0.25(b - a)$

(2) $2x(a^2 + b^2) - 3y(a^2 + b^2)$

(3) $a(x - y) + 2b(y - x)$

(4) $3(a - b)^2 - (b - a)^3$

(5) $a(x - y) - b(x - y) - c(x - y)$

(6) $20x(a - b)^2 - 4y(a - b)^3$

(7) $a(a - 3) + b(a - 3) + c(3 - a)$

(8) $x^2y(a - 2b) - 2xy(2b - a)$

(9) $a(ab + bc + ca) - abc$

(10) $m(m + 2n)(m - 2n) - m(m + 2n)^2$

(11) $(2x + 1)y^2 + (2x + 1)^2y$

(12) $x(m - n) + y(n - m) - z(m - n)$

(13) $m(a - m)(a - n) - n(m - a)(a - n)$

(14) $(x + 2y)(3x^2 - 4y^2) - (x + 2y)^2(x - 2y)$

(15) $-a(a - x)(x - b) + ab(a - x)(b - x)$

(16) $(2m + 3n)(2m - n) - 4n(2m - n)$

(17) $(x + y)(a + b) - (y - z)(a + b) + (z + x)(a + b)$

(18) $(a - 2b)(3a + 4b) + (2a - 4b)(2a - 3b)$

4.2　Formula Method

Sometimes we can factor polynomials by applying formulas directly. These formulas are the same formulas for polynomial multiplications, but we now swap the two sides.

- Difference of Squares: $a^2 - b^2 = (a + b)(a - b)$
- Perfect Squares: $a^2 \pm 2ab + b^2 = (a \pm b)^2$
- Sum and Difference of Cubes: $a^3 \pm b^3 = (a \pm b)(a^2 \mp ab + b^2)$

Before applying the formulas, we should pull out all possible common factors to simplify the problem. In order to apply the right formula, we also need to recognize the special structures involved.

Examples. Factor.

(1) $x^7 y - xy^7$

(2) $(x + y)^2 - 12(x + y)z + 36z^2$

Solutions.

(1) First notice xy is a common factor, so we pull it out to get $x^7y - xy^7 = xy(x^6 - y^6)$. The last term $x^6 - y^6 = (x^3)^2 - (y^3)^2$ is a difference of squares, so we factor it as $(x^3 + y^3)(x^3 - y^3)$, but both $x^3 + y^3$ and $x^3 - y^3$ can be factored by the formulas of sum or difference of cubes. As a result

$$
\begin{aligned}
x^7y - xy^7 &= xy(x^6 - y^6) \\
&= xy(x^3 + y^3)(x^3 - y^3) \\
&= xy(x + y)(x^2 - xy + y^2)(x - y)(x^2 + xy + y^2).
\end{aligned}
$$

(2) Treat $x + y$ as a single term A and $36z^2$ as the square of $B = 6z$; then $12(x + y)z = 2(x + y)(6z) = 2AB$, hence $(x + y)^2 - 12(x + y)z + 36z^2 = A^2 - 2AB + B^2$ has the structure of a perfect square; immediately

$$
\begin{aligned}
(x + y)^2 + 12(x + y)z + 36z^2 &= (A - B)^2 \\
&= [(x + y) - 6z]^2 \\
&= (x + y - 6z)^2.
\end{aligned}
$$

Other formulas for factoring that may be used in advanced math courses are listed below.

- $a^2 + b^2 + c^2 + 2ab + 2bc + 2ca = (a + b + c)^2$.
- $a^3 \pm 3a^b + 3ab^2 \pm b^3 = (a \pm b)^3$.
- $\sum_{k=0}^{n} \binom{n}{k} a^{n-k} b^k = (a + b)^n$.
- $a^n - b^n = (a - b)(a^{n-1} + a^{n-2}b + a^{n-3}b^2 + \cdots + ab^{n-2} + b^{n-1})(n \in \mathbb{Z}^+)$.
- $a^{2n+1} + b^{2n+1} = (a + b)(a^{2n} - a^{2n-1}b + a^{2n-2}b^2 - a^{2n-3}b^3 + \cdots - ab^{2n-1} + b^{2n})(n \in \mathbb{N})$.
- $a^3 + b^3 + c^3 - 3abc = (a + b + c)(a^2 + b^2 + c^2 - ab - bc - ca)$.

Exercises.

1. Factor by the difference of squares formula.

(1) $m^2 - 1$

(2) $x^2 - 4y^2$

(3) $a^2 - 16$

(4) $25 - 4b^2$

(5) $36p^2 - 49q^2$

(6) $x^2 - 9(x + y)^2$

(7) $(x + 2y)^2 - (x - y)^2$

(8) $8a^5 - 72a^3$

(9) $m^4 - m^2n^2$

(10) $25a^4x^{10} - 9b^6y^8$

(11) $4(a + b + c)^2 - 81(a - b + 2c)^2$

(12) $144a^2 - 256b^2$

(13) $16x^{16} - y^4z^4$

(14) $25a^2b^4c^{16} - 1$

(15) $x^4 - y^4$

(16) $(a - b)^4 - 81b^4$

(17) $81a^4 - 16b^4$

(18) $b^2 - (a - b + c)^2$

(19) $a^{m-1} - a^{m+1}$

(20) $(m + n)^2 - 9(m - n)^2$

2. Factor by the perfect squares formulas.

(1) $a^2 - 2a + 1$

(2) $t^2 + 4t + 4$

(3) $x^2 - 4xy + 4y^2$

(4) $9m^2 - 6m + 1$

(5) $16x^2 - 40xy + 25y^2$

(6) $1 - 6ab^3 + 9a^2b^6$

(7) $x^4 + 2x^2 + 1$

(8) $4a^2 - 4a + 1$

(9) $2x^3 - 4x^2y + 2xy^2$

(10) $x^2y^2 - 14xy + 49$

(11) $xm^3n^2 + 2xm^2n + xm$

(12) $(x + y)^2 + 4(x + y) + 4$

(13) $2ab - a^2 - b^2$

(14) $-(x - y)^2 + 10(x - y) - 25$

(15) $4x^2 - 20xy + 25y^2$

(16) $x^4 - 2x^2y^2 + y^4$

(17) $(2x - 3y)^2 - 2(3x - 2y) + 1$

(18) $(x^2 + x)^2 + 4(x^2 + x) + 4$

(19) $16(a - b)^2 - 24(a^2 - b^2) + 9(a + b)^2$

(20) $a^2(b + c)^2 - 2ab(a - c)(b + c) + b^2(a - c)^2$

3. Factor by the sum and difference of cubes formulas.

(1) $x^3 + 8$

(2) $1 - y^3$

(3) $64m^3 - 1$

(4) $8x^3 + 27$

(5) $a^3 - b^6$

(6) $27x^3 - 8y^3$

(7) $(2x + y)^3 + (x + 2y)^3$

(8) $1 - 8(a - b)^3$

(9) $(2a + 1)^3 - a^3$

(10) $(x + 1)^3 - (x - 1)^3$

(11) $(m - n)^3 + (2n - m)^3$

(12) $\dfrac{1}{8}x^3 - \dfrac{1}{27}y^3$

(13) $8a^3 - 125b^6x^3$

(14) $(x + y)^3 + 8$

(15) $(z - x)^3 - (y + x)^3$

(16) $(x + y)^3 + 125$

(17) $8(x + y)^3 - 1$

(18) $a^9 - b^9$

4. Factor.

(1) $x^2(a - 2b) + y^2(2b - a)$

(2) $(a^2 + b^2)^2 - 4a^2b^2$

(3) $(a^2 + b^2)^2 - 6ab(a^2 + b^2) + 9a^2b^2$

(4) $27a^3(8x^3 - y^3) - 64b^3(y^3 - 8x^3)$

(5) $(x^2 + y^2 - 1)^2 - 4x^2y^2$

(6) $3x^4 - 6x^2 + 3$

(7) $a^4x^2 + 4a^2x^2y + 4x^2y^2$

(8) $(x^2 - 2x)^2 + 2x(x^2 - 2x) + 1$

(9) $(x - y)^2 + 12(y - x)z + 36z^2$

(10) $(ax + by)^2 + (ay - bx)^2 + 2(ax + by)(ay - bx)$

(11) $3x^4 - 48y^4$

(12) $(a^2 + \dfrac{1}{a^2})^2 - 4$

(13) $x^9 + y^9$

(14) $x^{12} - y^{12}$

(15) $x^6(x^2 - y^2) + y^6(y^2 - x^2)$

(16) $\dfrac{1}{9}m^3 - m$

5. Factor using the formula $a^3 \pm 3a^2b + 3ab^2 \pm b^3 = (a \pm b)^3$.

(1) $x^3 + 3x^2 + 3x + 1$

(2) $8 - 12b + 6b^2 - b^3$

(3) $y^5 - 3y^4 + 3y^3 - y^2$

(4) $a^6 - 3a^4b^2 + 3a^2b^4 - b^6$

(5) $6x^5y^4 + 18x^4y^3 + 18x^3y^2 + 6x^2y$

(6) $(a + b)^6 + 3(a + b)^4(a - b)^2 + 3(a + b)^2(a - b)^4 + (a - b)^6$

4.3 Cross Product Method

Polynomials of the form $ax^2 + bx + c(a \neq 0)$ are called trinomials, i.e. a polynomial in one variable of degree 2. We can try the cross product method to factor them. If a can be factored as $a_1 \cdot a_2$ and c can be factored as $c_1 \cdot c_2$ such that $b = a_1c_2 + a_2c_1$, then $ax^2 + bx + c = (a_1x + c_1)(a_2x + c_2)$. We can draw a cross sign for assistance.

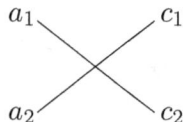

$$a_1 \diagdown \quad \diagup c_1$$
$$\diagup\diagdown$$
$$a_2 \diagup \quad \diagdown c_2$$

$$a = a_1a_2 \quad a_1c_2 + a_2c_1 = b \quad c = c_1c_2$$

The idea here is to factor a and c to make the sum of the cross products b. We may have to try several times to get a combination that works. The method also applies to trinomial-like polynomials.

(1) $2x^2 + x - 6$

(2) $x^2 - 5xy + 6y^2$

Solutions.

(1) 2 can only be factored as 1×2, but there are different choices to factor -6. $-6 = -1 \times 6 = -6 \times 1 = -2 \times 3 = -3 \times 2$. After trials, we figure out that the following picture works,

so $2x^2 + x - 6 = (2x - 3)(x + 2)$.

(2) Here we treat x as the only variable, $-5y$ as the coefficient of x and $6y^2$ as the constant term. y^2 must be factored as $y \cdot y$, so $6y^2$ can be factored as $6y \times y$, $2y \times 3y$, $(-6y) \times (-y)$ or $(-2y) \times (-3y)$. After trials we see the following diagram

has the sum of the cross products $-5y$, so $x^2 - 5xy + 6y^2 = (x - 2y)(x - 3y)$.

Exercises.

1. Factor.

(1) $x^2 + 5x - 6$

(2) $x^2 - 5x - 6$

(3) $x^2 + 5x + 6$

(4) $x^2 - x - 30$

(5) $x^2 + 18x - 144$

(6) $x^2 + 30x + 144$

(7) $2x^2 - 7x - 15$

(8) $18x^2 - 19x + 5$

(9) $6x^2 + 5x - 6$

(10) $6x^2 - 5x - 6$

(11) $6x^2 - 13x + 6$

(12) $6 + 11a - 35a^2$

(13) $x^2 - 4x - 77$ (14) $6x^2 - 7x - 3$

(15) $6 + 5x^2 - 17x$ (16) $6x^2 - x + 15$

(17) $12 + 5x - 2x^2$ (18) $x^2 - 22x - 75$

(19) $6x^2 + x - 35$ (20) $6x^2 - 37x + 35$

2. Factor.

(1) $x^2 - 3xy - 28y^2$ (2) $x^2 - 4xy + 3y^2$

(3) $20x^2 - 43xy + 14y^2$ (4) $-3x^2 + 10xy - 3y^2$

(5) $x^2 + 13xy - 30y^2$ (6) $6a^2 - ab - 35b^2$

(7) $6a^2 - 11ab - 35b^2$ (8) $6a^2 - 29ab - 35b^2$

(9) $6a^2 - 31ab + 35b^2$ (10) $-35m^2n^2 + 37mn + 6$

(11) $x^2y^2 + 6xy - 16$ (12) $18m^4 - 21m^2n^2 + 5n^4$

3. Factor.

(1) $x^4 + 2x^2 - 8$ (2) $3x^4 + 6x^2 - 9$

(3) $x^5 - 2x^3 - 8x$ (4) $x^4 - 13x^2 + 36$

(5) $x^2(x^2 + y^2) - 2y^4$

(6) $(x^2 + 2)^2 - 9x^2$

(7) $(m - n)^2 - 3(m - n) + 2$

(8) $(x^2 - 7x)^2 + 10(x^2 - 7x) - 24$

(9) $(x^2 + x)^2 - 14(x^2 + x) + 24$

(10) $(m^2 + 3m)^4 - 8(m^2 + 3m)^2 + 16$

(11) $x^2 - (a + b)x + ab$

(12) $abx^2 - (a^2 + b^2)x + ab$

(13) $(a - b)x^2 + 2ax + a + b$

(14) $x^2 + (m - \dfrac{1}{m})x - 1$

(15) $(a + b)x^2 + (b^2 - a^2 - 1)x + a - b$ (16) $x^4 - 2x^2(y^2 + z^2) + (y^2 - z^2)^2$

(17) $x^2 - y^2 + 3x - 7y - 10$

4.4 Group Factoring Method

If none of the above three methods works, we should try group factoring. We need to arrange the terms into groups and make sure some group can be factored. Then we can try to see what will happen after factoring each group.

Examples. Factor.

(1) $2x^2 + 2xy - 3x - 3y$

(2) $x^2 - y^2 - z^2 - 2yz - 2x + 1$

Solutions.

(1) There is no common factor for all terms; no formulas apply either. We have to try grouping. The first two terms have a common factor $2x$, while the last two terms have a common factor -3, so we try to group them according to this observation.

$$2x^2 + 2xy - 3x - 3y = (2x^2 + 2xy) + (-3x - 3y)$$
$$= 2x(x + y) - 3(x + y).$$

Now it is easy to see $(x + y)$ is a common factor, so immediately we get

$$2x^2 + 2xy - 3x - 3y = 2x(x + y) - 3(x + y)$$
$$= (x + y)(2x - 3).$$

(2) Notice $x^2 - 2x + 1$ is a perfect square and $-y^2 - z^2 - 2yz$ is a perfect square after pulling out the negative sign. Then it becomes a difference of two squares.

$$x^2 - y^2 - z^2 - 2yz - 2x + 1 = (x^2 - 2x + 1) - (y^2 + 2yz + z^2)$$
$$= (x - 1)^2 - (y + z)^2$$
$$= [(x - 1) - (y + z)][(x - 1) + (y + z)]$$
$$= (x - y - z - 1)(x + y + z - 1).$$

Sometimes the grouping method does not apply directly. We may need to split a term or add canceling terms. The following are two examples.

Examples. Factor.

(1) $x^4 + x^3 - 2$

(2) $x^4 + x^2y^2 + y^4$

Solutions.

(1) Here we split 2 into two 1's and make the following grouping.

$$x^4 + x^3 - 2 = x^4 + x^3 - 1 - 1$$
$$= (x^4 - 1) + (x^3 - 1)$$

Since both $x^4 - 1$ and $x^3 - 1$ has a factor $(x - 1)$, we may pull out this common factor.

$$x^4 + x^3 - 2 = (x^4 - 1) + (x^3 - 1)$$
$$= (x - 1)(x + 1)(x^2 + 1) + (x - 1)(x^2 + x + 1)$$
$$= (x - 1)[(x^3 + x^2 + x + 1) + (x^2 + x + 1)]$$
$$= (x - 1)(x^3 + 2x^2 + 2x + 2).$$

(2) This expression is very close to a perfect square. We may add x^2y^2 to complete the square, but we also need to subtract it to keep the expression invariant. Then the expression naturally splits into two parts.

$$x^4 + x^2y^2 + y^4 = x^4 + 2x^2y^2 + y^4 - x^2y^2$$
$$= (x^4 + 2x^2y^2 + y^4) - x^2y^2$$
$$= (x^2 + y^2)^2 - (xy)^2.$$

It is clear that the last expression is the difference of two squares. Immediately, we have

$$x^4 + x^2y^2 + y^4 = (x^2 + y^2)^2 - (xy)^2$$
$$= (x^2 + xy + y^2)(x^2 - xy + y^2).$$

Exercises.

   1. Factor by grouping.

     (1) $2xm + 2xn - 3ym - 3yn$ (2) $2am + 4an - bm - 2bn$

     (3) $x^2 - 2x + x - 2$ (4) $pq - 7p + 3q - 21$

     (5) $2hf - 2h + f^2 - f$ (6) $3xy + 2y - 3x - 2$

     (7) $mn + m^2 + 4n + 4m$ (8) $2px - py - 2qx + qy$

     (9) $nx - ny - 2x + 2y$ (10) $3ab + bc - 3ac - c^2$

   2. Factor in the set of rational numbers.

     (1) $x^2 - \dfrac{2}{3}xy + \dfrac{1}{9}y^2$ (2) $\dfrac{1}{8} + x^3$

     (3) $\dfrac{a^2}{4} - ab + b^2$ (4) $a^4 - 3a^3 + \dfrac{9}{4}a^2$

     (5) $4x^2 - \dfrac{1}{9}$ (6) $x^2 - xy - 1 + \dfrac{y^2}{4}$

   3. Factor.

     (1) $3mx^3y^2 + 6mx^2y + 3mx$ (2) $-(x - y)^2 + 10(x - y) - 25$

     (3) $16a^3 + 2b^3 - 2a - b$ (4) $x^3 - y^3 - x(x^2 - y^2) + y(x - y)^2$

     (5) $x^2 - 4 + y(y - 2x)$ (6) $a^2 - 2ab + b^2 - 2a + 2b + 1$

     (7) $x^2y^2 - x^2 - y^2 + 1$ (8) $a^2b + b^2c + c^2a - ab^2 - bc^2 - ca^2$

(9) $x^2 - 4xy + 4y^2 + x - 2y - 6$ (10) $(ay + bx)^3 + (ax + by)^3 - (a^3 + b^3)(x^3 + y^3)$

4. Factor.

(1) $x^2 + y^2 - x^2y^2 - 4xy - 1$ (2) $(x^2 + x + 1)(x^2 + x + 2) - 6$

(3) $x^4 + x^3 - 4x - 16$ (4) $x^3 - 9x + 8$

(5) $x^2 - 2x + 2y - y^2$ (6) $x^3 + 3x^2 - 4$

(7) $a^2 + (a + 1)^2 + (a^2 + a)^2$ (8) $x^9 + x^6 + x^3 - 3$

(9) $x^3 + 6x^2 + 11x + 6$ (10) $a^3 - a^5 - 2a^6 - a^7$

(11) $a^4b^2 + a^2b^2 + b^2$ (12) $4x^4 + 1$

(13) $x^5 - x^4 + x^3 - x^2 + x - 1$ (14) $x^4 + 324$

(15) $x^3 + 3x^2 + 4x + 2$ (16) $9a^2 - 2b - b^2 - 6a$

(17) $a^2 + b^2 + c^2 + 2ab - 2bc - 2ca$ (18) $x^4 + y^4 + z^4 + 2x^2y^2 + 2y^2z^2 + 2z^2x^2$

5. Prove the formula $a^3 + b^3 + c^3 - 3abc = (a + b + c)(a^2 + b^2 + c^2 - ab - bc - ca)$. Use the formula to factor $a^3 + b^3 - 3ab + 1$. (Hint: $(a + b)^3 = a^3 + 3a^2b + 3ab^2 + b^3$; replace $a^3 + b^3$ by $(a + b)^3 - 3ab(a + b)$; then factor by grouping).

Linear Equations and Inequalities

People use variables to build algebraic expressions. We have learned one category of them—polynomials in the previous two chapters. There is a question: why do ancient algebraists play with these expressions? To answer this question, we need to understand the main task of the elementary algebra. The major problems that ancient algebraists worked on are to solve equations and inequalities. To solve inequalities, one has to solve a corresponding equation. As a result, we may say the ultimate task of elementary algebra is to solve equations.

Corresponding to different types of algebraic expressions, there are different forms of equations: polynomial equations, rational equations and radical equations. Polynomial equations are fundamental. We always solve other forms of equations by solving an equivalent system of polynomial equations and inequalities. However, only polynomial equations of lower degrees (e.g. 1 and 2) are easy to solve. In this chapter, we shall learn how to solve the simplest polynomial equations and inequalities: linear equations and inequalities in one variable.

5.1   Equations and their Propositions

Equations are statements that assert the equality of two expressions. If two expressions are equal, we use the equal sign '$=$' to connect them and make it an equation. For example, $5 - (-3) = 8$, $3x^2 + x - 1 = 0$ are both equations. In the second example, x is an unknown. For equations with unknowns, the values of the unknowns which make the equality holds are called solutions to the equation. For example, $x = -2$ is a solution to $x^2 = 4$. The set of all solutions of an equation is its solution set. For example, the solution set to $x^2 = 1$ is $\{-1, 1\}$. There are also equations true for all possible values of unknowns, which are called identities. For example $5 - (-3) = 8$, $x^2 - 1 = (x+1)(x-1)$ are both identities. Some equations may not have solutions, e.g. $x + 1 = x + 2$. The solution sets of these equations are all $\emptyset$. Two equations are equivalent if they have the same solution set.

Equations have the following basic propositions. For any real numbers a, b, c,

- $a \pm c = b \pm c$ if and only if $a = b$;

- when $c \neq 0$, $ac = bc$ if and only if $a = b$.

These propositions are the basic tools to solve equations.

Exercises.

1. True or false.

(1) $2a + 5b + 3c = 5$ is an equation.

(2) $2a - 3b + 4a = 6a - 3b$ is an equation but not an identity.

(3) 3 is a solution to the equation $x^2 - 2x = 3$.

(4) 0 and -3 are both solutions to $x^3 - 3x^2 = 0$.

(5) $x^2 + 1 = 0$ has no solutions.

(6) The unknowns in $a + b = c$ are a, b, c.

(7) $x + y = 5$ has infinitely many solutions.

(8) $y^2 - 1 = 0$ has two solutions.

(9) $|a| = -1$ has two solutions.

(10) $-3 + 4 = -1$ is an identity.

(11) For any fixed number a, $x = y$ and $(a^2 + 1)x = (a^2 + 1)y$ are equivalent equations.

(12) If $|x| - |y| = 0$, then $x = y$.

(13) If $a + b = a + c$, then $b = c$.

(14) If $ab = ac$, then $b = c$.

(15) $|x| + |y| = 0$ and $x^2 + y^2 = 0$ are equivalent.

(16) Dividing both sides of an equation by the same nonzero number, the equality still holds.

2. Multiple choice.

(1) Which of the following is an equation?

    A. $ab + 3a$ B. $2 \neq 3$ C. $8 > 5$ D. $xyz = 4$

(2) Which of the following is a solution to $6x^2 - 13x + 6 = 0$?

    A. $\dfrac{1}{2}$ B. 1   C. $\dfrac{3}{2}$ D. 2

(3) Which of the following is not a solution to $x^3 - x = 0$?

    A. -2 B. -1 C. 0   D. 1

(4) Which is not a solution to $2x + 5y = 11$?

    A. $x = -2, y = 3$ B. $x = -1, y = \dfrac{9}{5}$ C. $x = 2, y = \dfrac{7}{5}$ D. $x = \dfrac{13}{6}, y = \dfrac{4}{3}$

(5) How many positive integer solutions to $3x + 4y = 25$ are there?

    A. 0   B. 1

    C. 2   D. Infinitely many

(6) Which is an identity?

    A. $a + b = c$ B. $|x| = x$

    C. $a(b + c) = ab + ac$ D. $x + 7 = 14$

(7) Which of the following is correct?

    A. If $xy = xz$, then $x = z$.   B. If $|x| = |y|$, then $x = y$.

    C. If $x^2 = y^2$, then $x = y$.   D. If $\dfrac{x}{y} = \dfrac{z}{y}$, then $x = z$.

3. Find equations accordingly. Do not solve them.

(1) The difference between 7 times x and 6 is -9.

(2) The difference between 5 times y and 7 equals the sum of 3 times y and 8.

(3) One third of x plus 4 equals the difference of twice x and 9.

(4) The difference of the absolute value of 4 times x and -3 is 2.

(5) The difference between the opposite of x and 7 times 7 is 8.

(6) The absolute value of the difference of 3 times a and -4 is 9.

(7) The difference of 7 times x and 1 is $\dfrac{2}{3}$ less than 3 times the sum of twice x and 1.

4. Tell the reason for the transformations.

(1) $\dfrac{x}{3} = 2 \Longrightarrow x = 6$

(2) $5x - 4 = 2x \Longrightarrow 3x - 4 = 0$

(3) $3x - 5 = 12 - 7x \Longrightarrow 3x + 7x = 12 + 5$

(4) $4x - 3 = x + 2 \Longrightarrow 3x = 5$

(5) $\dfrac{3x}{7} = \dfrac{x}{7} + 1 \Longrightarrow 3x = x + 7$

(6) $\dfrac{x+4}{0.3} - \dfrac{2x-1}{0.7} = 10 \Longrightarrow \dfrac{x+4}{3} - \dfrac{2x-1}{7} = 1$

(7) $\dfrac{3x+1}{4} = 2 - \dfrac{5x+3}{4} \implies 3x + 1 = 8 - 5x - 3$

5. Tell if the transformations are correct. If wrong, write a correct one.

(1) $-\dfrac{1}{3}x = 6 \implies x = -2.$

(2) $(x-2) = 2(x-2) \implies 1 = 2$

(3) $\dfrac{3x}{2} - 1 = \dfrac{2x}{3} \implies 9x - 1 = 4x$

(4) $\dfrac{x}{0.2} - \dfrac{y}{0.3} = 0.4 \implies \dfrac{x}{2} - \dfrac{y}{3} = 4$

(5) $\dfrac{2a}{3} = 3 - \dfrac{a-1}{4} \implies 8a = 36 - 3a - 3$

(6) $\dfrac{x+1}{yz} = \dfrac{13-5x}{yz} \implies x + 1 = 13 - 5x$

5.2 Linear Equations in One Variable

If the expressions on the two sides of an equation are both polynomials, such equations are called polynomial equations. For example, $2x + 3 = 5 - 9x$, $x^2 + y^2 = 1$, $x^3 + 4x^2 - 1 = 0$ are all polynomial equations. If the highest degree of the polynomials involved is exactly 1, such polynomial equations are called linear equations, e.g. $2x - 3 = 1$, $2x - y = 5$. In this section we are only interested in linear equations in one variable, i.e. with only one unknown.

A linear equation in one variable can always be simplified into its standard form

$$ax = b$$

where a, b are constants. There are three possibilities:

- If $a \neq 0$, there is only one solution $x = \dfrac{b}{a}$, i.e. the solution set is $\{\dfrac{b}{a}\}$;

- If $a = 0, b \neq 0$, there is no solution, so the solution set is the empty set $\emptyset$;

- If $a = b = 0$, any real number is a solution, and the solution set is $\mathbb{R}$.

According to the basic propositions of equations, we get equivalent equations if

- adding or subtracting the same expression on both sides of an equation;

- multiplying or dividing the same nonzero expression on both sides of an equation.

From the first proposition, $a = b+c$ and $a-c = b$ are equivalent. The rule can be summarized as: in order to get equivalent equations, when moving a term from one side to the other, change the sign of the term. For example, in the equation $2x - 3 = x$, we can move -3 from the left side to the right side to get $2x = x + 3$ by changing -3 into 3. The following examples show how to solve linear equations in one variable.

Examples. Solve the equations.

(1) $3(y + 2) - 2 = 2y - (y + 6)$

(2) $\dfrac{3x - 2}{4} = 1 + \dfrac{4 - 2x}{3}$

Solutions.

(1) i) Simplify both sides of the equation to get $(3y+6)-2 = 2y-y-6$, i.e. $3y+4 = y-6$. ii) Move all variables to one side and all constants to the other side: $3y - y = -6 - 4$. We need to change the sign if moving a term from one side to the other. iii) Simplify to get the standard form: $2y = -10$. iv) If the coefficient of the unknown is nonzero, divide this coefficient on both sides: $\dfrac{2y}{2} = \dfrac{-10}{2}$. As a result, the solution is $y = -5$.

(2) If there are fractions involved, we first clear these fractions. The least common denominator of all fractions is 12, so we multiply both sides by 12, i.e.

$$12(\frac{3x - 2}{4}) = (1 + \frac{4 - 2x}{3})12.$$

The equation simplifies to $3(3x - 2) = 12 + 4(4 - 2x)$. Then following the steps in Example (1), we get

$$9x - 6 = 12 + 16 - 8x$$
$$9x + 8x = 12 + 16 + 6$$
$$17x = 34$$
$$x = 2.$$

We should form the habit of checking the correctness of solutions. Here take the second problem as an example. When $x = 2$, the left side is $\dfrac{3 \times 2 - 2}{4} = 1$; the right side is $1 + \dfrac{4 - 2 \times 2}{3} = 1$. Two sides are equal, which verifies that $x = 2$ is the solution.

Exercises.

1. True or false.

(1) A linear equation in one variable always has a solution.

(2) $x + y = z$ is not a linear equation.

(3) $x + y = 2$ has infinitely many solutions.

(4) $x^2 = x$ is a linear equation.

(5) $x^5 - x^3 + x - 1 = 0$ is a polynomial equation but not a linear equation.

(6) $\dfrac{x}{2^3} - \dfrac{1 - 2x}{2^2} = 1$ is not a linear equation.

(7) $x = 3$ is not a solution to the equation $5x - 1 = \dfrac{x - 1}{2} + 13$.

(8) The equations $\dfrac{3x - 1}{5} = \dfrac{x + 7}{2}$ and $3x - 1 = x + 7$ share the same solutions.

(9) $x = 3$ is not a solution to $5x - 11 = \dfrac{x - 51}{12} + 13$.

(10) $x = 0$ is a solution to $\dfrac{5(x - 3)}{6} = \dfrac{x - 6}{2} - \dfrac{x}{3}$.

(11) The equation $1 - \dfrac{5 - x}{3} = \dfrac{x}{2}$ and $6 - 10 - 2x = 3x$ share the same solutions.

(12) There are infinitely many solutions to the equation $2x + 3 = 2(x + 1)$.

(13) The solution to the equation $4y - 3(a - y) = 6y - 7a - y$ is $y = 2a$.

(14) There is no solution to $x = -x$.

(15) $\dfrac{4 - 3x}{12} - \dfrac{1 - 5x}{4} = \dfrac{1 + 3x}{6} + \dfrac{6x - 1}{12}$ has infinitely many solutions.

2. Multiple choice.

(1) $ax = b$ has

    A. one solution   B. infinitely many solutions

    C. no solutions   D. all are possible

(2) If the solution of $\dfrac{2x + a}{2} = 4(x - 1)$ is $x = 3$, then a is

    A. 2   B. 22   C. 10   D. -2

(3) Find the solution to $\dfrac{x}{3} - 1 = 5 - \dfrac{x - 1}{6}$.

    A. $\dfrac{7}{3}$ B. $\dfrac{5}{3}$ C. $\dfrac{35}{3}$ D. $\dfrac{37}{3}$

(4) If the value of the algebraic expression $\dfrac{3m + 5}{7}$ is 3, then the value of m is

    A. $\dfrac{16}{3}$ B. $-\dfrac{16}{3}$

    C. $\dfrac{26}{3}$ D. cannot be determined

(5) If $5m + \dfrac{1}{4}$ and $5\left(m - \dfrac{1}{4}\right)$ are opposite numbers, then m is

    A. 0   B. $\dfrac{3}{20}$ C. $\dfrac{1}{20}$ D. $\dfrac{1}{10}$

(6) The solution to $\dfrac{x}{4} + \dfrac{x + 1}{3} = \dfrac{7x + 4}{12}$ is

    A. 0   B. infinitely many solutions

    C. no solution   D. none of the above

(7) If $3a^3b^{2x}$ and $\frac{1}{3}a^3b^{4(x-\frac{1}{2})}$ are similar terms, then x is

    A. $x = -1$ B. $x = -\frac{1}{3}$ C. $x = \frac{1}{3}$ D. $x = 1$

(8) If $y = 1$ is a solution to $2 - \frac{1}{3}(m-y) = 2y$, then the solution to $m(x-3) - 2 = m(2x-5)$ is

    A. $x = -10$ B. $x = 0$

    C. $x = \frac{4}{3}$ D. none of the above

(9) If $\frac{y-2}{6}$ is greater than $\frac{y}{3}$ by 1, then y is

    A. -3 B. 4 C. -8 D. -1

(10) If one half of a number is 5 greater than two thirds of it, this number is

    A. -30 B. 30 C. -60 D. 60

3. Solve the following equations and find their solution sets.

(1) $5x - 11 = 24$ (2) $-6x - 7 = -2x - 33$

(3) $2x = -(5 - x)$ (4) $3(9 - y) - 7(7 - y) + 10 = 12 - 5(y - 1)$

(5) $3(y + 2) - 2 = 2y - (y + 5)$ (6) $3(4x - 2) = 2(4x - 2) - 6$

(7) $2x - 3 = 2(x - 3) + 3$ (8) $4(y + 1) = -2(4 - y)$

(9) $3(y - 4) - 6 = 0$ (10) $5(x - 1) = 3(x + 1)$

(11) $5 - [5x - 3(2x + 1)] - (1 - x) = 0$ (12) $43[2(x + 1) - 8] - 20 - 7 = 1$

(13) $8x + 4 = 3(4x - 3) - 2(-3x - 4)$ (14) $3[x - 2(x - 1)] = 2(1 - x)$

(15) $3x - 3 - [4x + (x - 1)] - 2x = 8$ (16) $x - 2[x - 3(x + 3) - 5] = 3[2x - 8(4 - x)] - 2$

(17) $10x + 3 = 3(2x + 1) - (3 - x)$ (18) $3(x - 2) = 3x + 1$

(19) $3x - 4(2x - 1) = 3(x + 2) - 2(3x - 2)$ (20) $2x - (x + 7) = 13 - 2(3x - 5)$

4. Solve the equations.

(1) $3x = \dfrac{1}{4}$

(2) $\dfrac{1}{6}y = 12$

(3) $-\dfrac{5}{7}x = -\dfrac{10}{21}$

(4) $\dfrac{0.5x}{0.6} + \dfrac{0.07x - 0.01}{0.03} = 1$

(5) $-1.25x = \dfrac{1}{8}$

(6) $x - \dfrac{x - 1}{3} = 1 - \dfrac{x + 3}{5}$

(7) $\dfrac{x - 2}{3} = \dfrac{x - 3}{2}$

(8) $y - \dfrac{y - 1}{4} = 2 - \dfrac{y + 2}{3}$

(9) $65\%(y - 1) = 37\%(y + 1) - 0.82$

(10) $\dfrac{x + 17}{5} - \dfrac{3x - 7}{4} = -3$

(11) $\dfrac{1}{3}[\dfrac{1}{2}x - \dfrac{1}{4}(\dfrac{4}{3}x - \dfrac{2}{3})] = 1 - \dfrac{x}{3}$

(12) $\dfrac{x - 2}{6} - \dfrac{x + 2}{3} = \dfrac{x - 1}{2} + 1$

(13) $3x - \dfrac{2(x+3)}{3} = 16 - \dfrac{x+2}{2} - \dfrac{x-2}{6}$ (14) $2(0.3x - 4) - 5(0.2x + 3) = 9$

(15) $\dfrac{0.2x + 0.5}{0.5} - \dfrac{0.03 + 0.02x}{0.03} = \dfrac{x-5}{2}$ (16) $\dfrac{5y - 1}{6} - 0.9 = y + \dfrac{1 - 3y}{5}$

(17) $3 - \dfrac{5 - 2y}{7} = 4 - \dfrac{4 - 5y}{4} + \dfrac{y + 2}{2}$ (18) $0.2(y - 4) = \dfrac{2}{7}(y - 6) - \dfrac{y}{4}$

(19) $\dfrac{1}{2}[x - \dfrac{1}{2}(x - 1)] = \dfrac{2}{3}(x - 1)$ (20) $\dfrac{1}{2}\{x - \dfrac{1}{3}[x - \dfrac{1}{4}(x - \dfrac{2}{3}) - \dfrac{3}{2}]\} = x + \dfrac{3}{4}$

(21) $x - \dfrac{1 - \dfrac{3}{2}x}{4} - \dfrac{2 - \dfrac{x}{4}}{3} = 2$ (22) $\dfrac{3x - \dfrac{x+1}{10}}{0.2} + 1.5 = \dfrac{2x - \dfrac{1-x}{10}}{0.3} - 1$

5. Solve equations for the indicated variable.

(1) Given $a = \dfrac{b + c}{2}$, find b.

(2) Given $C = 2\pi(R + r)$, find r.

(3) Given $S = 2rh\pi + 2r^2\pi$, find h.

(4) Given $S = \dfrac{1}{2}(a + b)h$, find b.

(5) Given $S = \dfrac{1}{2}ab$, find a.

(6) Given $V = \dfrac{1}{3}\pi r^2 h$, find h.

(7) Given $S = 2\pi r h$, find r.

(8) Given $\dfrac{1}{R} = \dfrac{1}{R_1} + \dfrac{1}{R_2}$, find R.

(9) Given $S = vt + \dfrac{1}{2}at^2$, find a.

(10) Given $F = \dfrac{mv^2}{r}$, find r.

(11) Given $C = \dfrac{5}{9}(F - 32)$, find F.

(12) Given $y = ax^2 + 2bx + c$, find b.

(13) Given $a = \dfrac{1}{b} + \dfrac{1}{c}$, find c.

6. Solve the following equations for x.

(1) $(b - 1)x = b(b + x)$

(2) $m(x - m) = n(x + n)$ $(m \neq n)$

(3) $\dfrac{b + x}{a} + 1 = \dfrac{x - a}{b}$ $(a \neq b)$

(4) $a - cx = bx - x$ $(b + c \neq 1)$

(5) $\dfrac{x-1}{m} - \dfrac{x-2m}{m^2} = 1 \ (m \neq 0, 1)$ (6) $\dfrac{x}{2m} - \dfrac{x}{2n} = 1 \ (m \neq n)$

7. The sum of three consecutive odd numbers is 39. Find the three odd numbers.

8. It takes Tina 3 hours to finish a school-assigned project, while it takes Tommy 5 hours to finish the project. If Tina and Tommy work on the project together, how long will it take?

9. Taylor and Ryan practice running together. Taylor runs 7m/s; Ryan runs 6.5m/s. If Taylor lets Ryan run 2 seconds ahead of her, how long will it take Taylor to catch up with Ryan?

10. Mars University provides students dorms for housing. If each room holds 4 students, 1 room will be left. If each room holds 3 students, 5 students will be without rooms. How many students are there? How many dorm rooms in total?

11. Ge has two Christmas candles, a thick one and a thin one, and they have the same height. It takes 5 hours to burn out the thick candle, and 4 hours to burn out the thin one. Ge lit the two candles together. After a certain time, the thick candle is 4 times longer than the thin candle. How long have they burned so far?

12. Tina and Tommy are typing an essay for school. It takes Tina 20 days to finish the typing by herself ,and it takes 12 days if Tina and Tommy type together. Now Tina and Tommy plan to type the first 7 days together; then Tommy will type the rest of the essay all by himself. How long will it take to finish the typing?

13. Tommy needs to do a part time job to pay for his new bike. He made the contract to deliver 800 vases for a flower shop. The delivery fee for each vase is 0.5 dollars. If Tommy breaks one vase, he will not be paid for that one. Instead he will be charged 2.5 dollars for compensation. After the job was done, Tommy made 289 dollars. How many vases did Tommy break?

14. A slow train travels from A to B at the speed of 60km/h. After 2 hours, a fast train travels from B to A at the speed of 90km/h. Given that the distance between A and B is 250km, find the distance from A to the place where two trains meet.

15. Lulu and Nunu are cousins. Two years from now, the age ratio between Lulu and Nunu will be $5 : 4$. However, 4 years ago, Lulu's age was twice as Nunu's. How old are they right now?

16. Tina loves math. One day she used a string to make a square; then she used the same string to make a circle. Given that the radius of the circle is shorter than the side length of the square by $3(\pi - 2)$ meters, find the length of the string.

17. A tourist ship in Hawaii travels back and forth between Oahu Island and Maui Island. If it follows the sea current, it will take 3 hours. If it goes against the sea current, it will take 30 minutes more. Given that the ship travels 26 km/h in still water, find the speed of the current flow.

5.3 Equations involving Absolute Values

If absolute values appear in an equation, the strategy is to simplify the absolute values and make it an ordinary equation. By the definition of absolute values, if $|x| = a$, we have three possibilities:

- $x = \pm a$ if $a > 0$;

- $x = 0$ if $a = 0$;

- x has no solution if $a < 0$ as $|x|$ is always nonnegative.

From this, we can solve the following three types of equations.

Examples. Solve the equations.

(1) $|2x + 5| = 7$

(2) $|2x - 3| = |3x - 2|$

(3) $|x - 7| = 2x + 4$

Solutions.

(1) Treat $2x + 5$ as a single term whose absolute value is 7. This term must be 7 or -7, so $2x + 5 = 7$ or $2x + 5 = -7$. Solving these two equations we get $x = 1$ or -6.

(2) If $|a| = |b|$ we must have $a = b$ or $a = -b$, so from $|2x - 3| = |3x - 2|$ we get $2x - 3 = 3x - 2$ or $2x - 3 = -(3x - 2)$. Solving these two equations we get $x = -1$ or $x = 1$, i.e. $x = \pm 1$.

(3) First, $x - 7 = \pm(2x + 4)$. Solving these equations we get $x = -11$ or 1. Besides, as an absolute value $2x + 4$ must be nonnegative, so $x = 1$ is the only solution.

As $|a| = \pm a$ depends on the sign of a, to simplify $|a|$ we should distinguish the cases $a > 0$, $a = 0$ and $a < 0$. If more than one absolute value appear in an equation other than the above case, to simplify the absolute values we need to discuss all different possible situations. Take the following as an example.

Example. Solve $|2x - 3| - |3x + 7| = 1$.

Solutions. First, let us make two absolute values zero. If $2x - 3 = 0$ we get $x = \dfrac{3}{2}$. If $3x + 7 = 0$ we get $x = -\dfrac{7}{3}$. These two numbers divide the real number line into three parts.

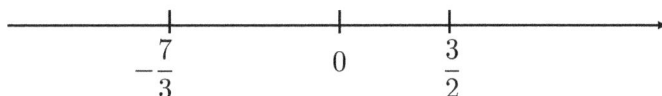

Then we can discuss as follows:

- If $x < -\dfrac{7}{3}$, both $2x - 3$ and $3x + 7$ are negative, so $|2x - 3| = -(2x - 3)$ and $|3x + 7| = -(3x + 7)$. As a result, the equation simplifies to $-(2x - 3) - [-(3x + 7)] = 1$. Solving this equation we get $x = -9$. Notice that $-9 < -\dfrac{7}{3}$, so it is in the assumed interval. Thus, $x = -9$ is a solution.

- If $-\dfrac{7}{3} \le x \le \dfrac{3}{2}$, $2x - 3 \le 0$ but $3x + 7 \ge 0$, then $|2x - 3| = -(2x - 3)$ and $|3x + 7| = 3x + 7$. As a result, the equation becomes $-(2x - 3) - (3x + 7) = 1$. Solving this equation we get $x = -1$. Notice that $-\dfrac{7}{3} \le -1 \le \dfrac{3}{2}$, so -1 is in the assumed interval and $x = -1$ is a solution.

- If $x > \dfrac{3}{2}$, both $2x - 3$ and $3x + 7$ are positive, so $|2x - 3| = 2x - 3$ and $|3x + 7| = 3x + 7$. As a result, the equation simplifies to $(2x - 3) - (3x - 7) = 1$. Solving this equation we get $x = 3$. Notice $3 > \dfrac{3}{2}$ and it is in the assumed interval, so $x = 3$ is a solution.

To summarize, the solution set to this equation is $\{-9, -1, 3\}$. Here we omit the procedure of plugging the solutions back into the original equation, and leave it to the readers.

Exercises.

1. True or false.

(1) If $|x - 5| = 5$, then $x = 0$.

(2) If $(a + 3)^2$ and $|b - 1|$ are opposite numbers, then $\dfrac{-a^2}{a + b} = \dfrac{9}{2}$.

(3) Given $|2y - 1| = 0$, then $y^2 - |y| = 2$.

(4) $|x| = 6$ and $x = 6$ are equivalent equations.

(5) $|x - 3| = -\dfrac{1}{2}$ has no solutions.

(6) $|2x - 1| + 3 = 2$ has 2 solutions.

(7) $|-x - 1| = 0$ and $|x + 1| = 0$ share the same solutions.

(8) If $\dfrac{|a| - |b|}{|a + b|} = 0$, then $a = b$ or $a = -b$.

(9) If $\dfrac{|m| - |n|}{2mn} = 0$, then $m = \pm n$.

(10) $|x| = -|x|$ has no solutions.

(11) $|x| = |-x|$ has infinitely many solutions.

2. Multiple choice.

(1) $\dfrac{|x - 2|}{3} - 5 = -1$ has

    A. no solutions   B. 1 solution

    C. 2 solutions   D. infinitely many solutions

(2) $|x + 1| + 1 = 1$ has

    A. no solutions   B. 1 solution

    C. 2 solutions   D. infinitely many solutions

(3) $|1 - x| = |x - 1|$ has

    A. no solutions   B. 1 solution

    C. 2 solutions   D. infinitely many solutions

(4) The solution of $|2 - 4x| = -x - 1$ is

    A. 1 or $\dfrac{1}{5}$ B. 1   C. $\dfrac{1}{5}$ D. no solutions

(5) $\dfrac{1}{2}|3x - 1| + 1 = 0$ has

A. no solutions   B. 1 solution

C. 2 solutions   D. infinitely many solutions

(6) $|2x - 1| + |3x - 2| = 0$ has

A. no solutions   B. 1 solution

C. 2 solutions   D. infinitely many solutions

3. Solve the equations.

(1) $|x| = 9$

(2) $|2x - 1| = 2$

(3) $7|x| - |-x| = 18$

(4) $2 + 3|2x - 4| = 0$

(5) $|x + 3| - 2 = 4$

(6) $\dfrac{2|x| - 1}{3} = 5$

(7) $|\dfrac{1}{2}x + 7| - 6 = 7$

(8) $|x - 2| - |x - 8| = 0$

(9) $|5x - 4| = |3x + 8|$

(10) $|2x - 3| = -|3x - 2|$

(11) $|1 - 2x| = -4|2x - 1|$

(12) $|7x - 1| = 2|-x + 4|$

(13) $2|3x + 4| = 5x - 14$

(14) $|x + 3| - 2 = 2x - 1$

(15) $2|x| - |x - 2| = 4$

(16) $|x - 2| - |x - 5| = 2$

(17) $3|2x + 3| + 2|3x - 2| = 17$

(18) $|4x - 3| - |2x + 3| = 6$

4. Find values.

(1) If 2 is a solution to $3|m| - x = 6x + 4$, find m.

(2) Given $|m - n + 4| + (n - 3)^2 = 0$, find m, n, and $m^2 - n^2$.

(3) Given $|x - 1| + 2|y + 3| = 0$, find $x + y$ and $|x| + |y|$.

(4) Given $|a - 3| + (b + 2)^2 = 0$, find $a + b$ and ab.

(5) If $x = 0$ is the solution of $|x - k| = \dfrac{1}{3}$, find k.

(6) Given $|x + \dfrac{1}{2}| + |y - 2| = 0$, find $x + y$.

(7) If $(a - \dfrac{1}{2})^2 + |b - \dfrac{1}{2}| = 0$, find $4ab$.

(8) If $|a + 1| + |b + 2| + |c - 3| = 0$, find $b(a + c)$.

5.4 Inequalities and their Propositions

We have met inequality signs when ordering numbers. If connecting two expressions by an inequality sign, we get an inequality. In fact, inequalities are statements that assert the inequality of two expressions. There are five types of inequalities:

$$a > b, \quad a \geq b, \quad a < b, \quad a \leq b, \quad a \neq b.$$

For example, $3 + 5 > 6$ and $x + 2 \leq -1$ are both inequalities. For inequalities with unknowns, the values of the unknowns which make the inequality hold are called the solutions to the inequality. The set of all solutions is its solution set. Two inequalities are equivalent if they have the same solution set. Inequalities have the following basic propositions: for any real numbers a, b, c,

- if $a > b$, $b > c$, then $a > c$;
- if $a > b$, then $a \pm c > b \pm c$;
- if $a > b$ and $c > 0$, then $ac > bc$.
- if $a > b$ and $c < 0$, then $ac < bc$.

Similar propositions hold if replacing '>' by '$\geq$' and '<' by '$\leq$', or reversing each inequality sign. We may freely add or subtract expressions on both sides of an inequality. The new inequality is always equivalent to the original one. For example, $3x + 2 \leq 2y + 1$ is equivalent to $3x + 1 \leq 2y$ by subtracting 1 on both sides. However, when multiplying negative numbers to both sides of an inequality, in order to get an equivalent inequality, we need to reverse the inequality sign. For example, from $a > b$, we get $3a > 3b$ by multiplying both sides by 3, but $-2a < -2b$ by multiplying both sides by -2, in the same time, switching the sign to '>'.

It is convenient to use interval notations to denote the solution sets of inequalities. For any real numbers a, b such that $a < b$, we define

- $(a, b) = \{x | a < x < b\}$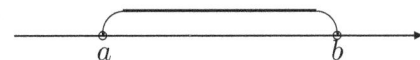
- $[a, b] = \{x | a \leq x \leq b\}$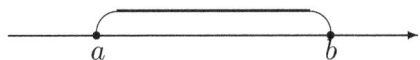
- $[a, b) = \{x | a \leq x < b\}$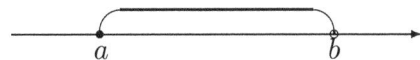
- $(a, b] = \{x | a < x \leq b\}$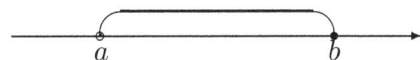
- $(a, +\infty) = \{x | x > a\}$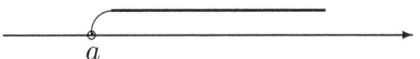
- $[a, +\infty) = \{x | x \geq a\}$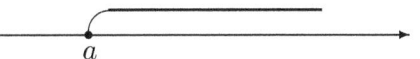

- $(-\infty, a) = \{x | x < a\}$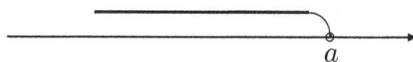

- $(-\infty, a] = \{x | x \leq a\}$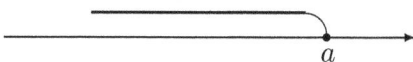

As subsets of $\mathbb{R}$, we can apply the union and intersection operations. From the following example, it should be clear that the intersection operation models the AND operation in logic, while the union operation models the OR operation.

Examples. Find the set of numbers satisfying the given condition.

(1) $x > 3$ and $x \leq 6$

(2) $x \geq 3$ or $x \leq 2$

(3) $-3 \leq x < 5$ or $2 < x < 7$

(4) $-3 < x < 5$ and $1 < x \leq 7$

Solutions.

(1) The condition $x > 3$ determines the interval $(3, +\infty)$; the condition $x \leq 6$ determines the interval $(-\infty, 6]$. x satisfies both of the conditions if and only if it lies in both of the intervals, which means x must be in the intersection of them. Drawing a picture below, it is easy to see the subset required is corresponding to the shaded area, i.e. the required subset is $(3, +\infty) \cap (-\infty, 6] = (3, 6]$.

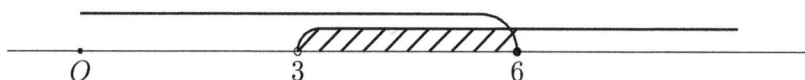

(2) The condition $x \geq 3$ or $x \leq 2$ requires x lies either in $[3, +\infty)$ or $(-\infty, 2]$, i.e. x is in the union of $[3, +\infty)$ and $(-\infty, 2]$. The required set is $\{x | x \geq 3$ or $x \leq 2\} = [3, +\infty) \cup (-\infty, 2]$. Drawing the picture below, we can see the result can not be simplified.

(3) The set is $[-3, 5) \cup (2, 7) = [-3, 7)$.

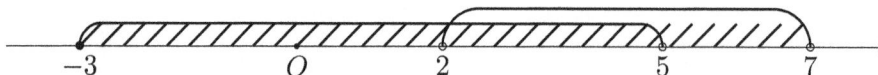

(4) The set is $(-3, 5) \cap (1, 7] = (1, 5)$.

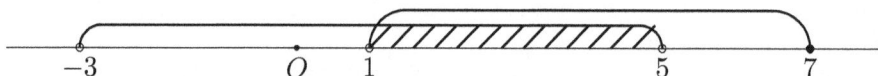

Exercises.

1. True or false.

(1) If $a > b$, then $a \geq b$.

(2) If $a \leq b$, then $-(-3a) \geq -(-3b)$.

(3) If $2x \geq y + 3$, then $6x - 9 \geq 3y$.

(4) If $x \leq y$, then $x < y$.

(5) If $a < 0$, $ax < -b$, then $x > b/a$.

(6) If two integers m, n satisfy $m > n$, then $m \geq n + 1$.

(7) If $a^2 \geq b^2$, then $a + 1 \geq b + 1$.

(8) $1 + 2a \geq 1 + a$.

(9) If $a < b$, then $3 - 2a < 3 - 2b$.

(10) If $ac^2 > bc^2$, then $a > b$.

(11) If $ac^2 \geq bc^2$, then $a \geq b$.

(12) If $a > b$, then $ac^2 > bc^2$.

(13) If $a \geq b$, then $ac^2 \geq bc^2$.

(14) If $a - b > a + b$, then $b < 0$.

(15) $2x - 3 < 4$ and $2x < 4 - 3$ are equivalent.

(16) If $a < 0$, then $2a + b < b$.

(17) If $y > 0$, then $x - y < x + y$.

(18) If $3 - 2x > -y$, then $4x + 2y \leq -6$.

2. Multiple choice.

(1) The condition to get $ax \geq ay$ from $x > y$ is

    A. $a > 0$ B. $a < 0$ C. $a \geq 0$ D. $a \leq 0$

(2) The condition to get $ax \geq ay$ from $x \leq y$ is

    A. $a > 0$ B. $a < 0$ C. $a \geq 0$ D. $a \leq 0$

(3) If a is an integer, the relation between a and $3a$ is

    A. $a > 3a$ B. $a < 3a$

    C. $a = 3a$ D. cannot be determined

(4) If $a < 0$ and $b = -a$, which of the following is correct?

    A. $b - \dfrac{1}{b} > 0$ B. $a^2 - b \geq 0$ C. $a + |b| = 0$ D. $\dfrac{1}{a} - \dfrac{1}{b} = 0$

(5) Which of the following is correct?

    A. $3 - a > 2 - a$ B. $a > -a$ C. $a > \dfrac{1}{a}$ D. $4a < 5a$

(6) If $a > b$, which of the following is correct?

    A. $\dfrac{1}{a} - 1 > \dfrac{1}{b} - 1$ B. $\dfrac{1}{a} - 2 < \dfrac{1}{b} - 2$ C. $2 - a < 2 - b$ D. $-a > -b$

(7) If $x < y < 0$, which of the following is correct?

    A. $\dfrac{1}{x} > \dfrac{1}{y}$ B. $\dfrac{1}{x} < \dfrac{1}{y}$ C. $|x| < |y|$ D. $x^2 < y^2$

(8) Which of the following asserts a is non-negative?

    A. $a > 0$ B. $|a| \geq 0$ C. $a < 0$ D. $a \geq 0$

(9) If $a < b$, which of the following is correct?

    A. $a + 5 > b + 5$ B. $a - b > b - b$ C. $a - a < b - a$ D. $a + 7 < b - 7$

(10) Given $a > b$ and $ac > bc$, then

    A. $c > 0$ B. $c < 0$ C. $c = 0$ D. $c \geq 0$

(11) If $a > 0$ and $ab \geq 0$, then

    A. $b > 0$ B. $b = 0$ C. $b < 0$ D. $b \geq 0$

(12) If $\dfrac{a}{c^2} < \dfrac{b}{c^2}$, then

    A. $a < b$ B. $a > b$ C. $a \leq b$ D. $a \geq b$

(13) The condition to make $6m < -5m$ true is

    A. $m > 0$ B. $m < 0$ C. $m \neq 0$ D. $m \geq 0$

(14) The condition to get $x > \dfrac{b}{a}$ from $ax < b$ is

    A. $a \leq 0$ B. $a \geq 0$ C. $a > 0$ D. $a < 0$

(15) Which of the following pairs are not equivalent?

    A. $5x \geq 15$ and $x \geq 3$ B. $3x - 8 < x + 2$ and $x < 5$

    C. $-4x > 28$ and $x < -7$ D. $x - 7 < 2x + 8$ and $x > 15$

(16) If $a \leq b$, for the two statements (1)$\dfrac{a}{5} \leq \dfrac{b}{5}$, (2) $3c - a \geq 3c - b$

    A. only (1) is correct   B. only (2) is correct

    C. both (1) and (2) are correct   D. none of them are correct

(17) The only condition to ensure $a + b < 0$ and $ab < 0$ is

    A. $a > 0$, $b > 0$

    B. $a < 0$, $b < 0$

    C. a and b have opposite signs

    D. only one of a and b is negative and its absolute value is greater

(18) If a and b are both negative, then

    A. $\dfrac{a}{b} > 1$ B. $\dfrac{a}{b} < 1$ C. $\dfrac{1}{a} < \dfrac{1}{b}$ D. $ab > -1$

3. Fill in the blanks.

(1) $|a|$ is nonnegative, which can be denoted by _____ .

(2) If $-\dfrac{x}{3} > -3$, then $-\dfrac{x}{3} \times (-3)$ _____ $-3 \times (-3)$.

(3) If $a < b$, then $a + 5$ _____ $b + 5$.

(4) If $a > b$, then $3a$ _____ $3b$.

(5) If $a \leq b$, then $-4a$ _____ $-4b$.

(6) If $a \geq b$, then $\dfrac{a}{3}$ _____ $\dfrac{b}{3}$.

(7) If $a \leq b$, then $-\dfrac{a}{7}$ _____ $-\dfrac{b}{7}$.

(8) If $a > b$, then $b - a$ _____ 0.

(9) If $a < b$, then $b - a$ _____ 0.

(10) If $a \leq b$, then $-(2 - 3a)$ _____ $-(2 - 3b)$.

(11) If $a < b$, $a + b < 0$, then $|a|$ _____ $|b|$.

(12) If $ax > b$ and $ab^2 < 0$, then x _____ $\dfrac{b}{a}$.

(13) If $-\pi x \geq 0$, then x _____ 0.

(14) If $-\dfrac{x}{4} > -\dfrac{2x}{3}$, then $3x$ _____ $8x$.

4. Find interval notations and sketch the graphs.

(1) $-2 < x < 0$ \qquad\qquad (2) $-6 \leq y < 4$

(3) $-1 \leq a \leq 3$ \qquad\qquad (4) $-4 < b \leq -2$

(5) $a > 1$ \qquad\qquad (6) $a \leq -1$

(7) $a \geq 2$ \qquad\qquad (8) $a < 4$

5. Find the results.

(1) $(-5, -1) \cap (-3, 0]$ \qquad\qquad (2) $[2, 5] \cap (1, 5)$

(3) $(-1, 3] \cap (3, 5)$ \qquad\qquad (4) $[-4, 0] \cap [0, 4]$

(5) $(-6, +\infty) \cap (-8, -3]$ \qquad\qquad (6) $(-\infty, -1] \cap [1, +\infty)$

(7) $[-2, 5) \cup (-2, 7]$ \qquad\qquad (8) $(3, +\infty) \cup [-2, 3]$

(9) $(-7, 7] \cup (-3, 3)$ \qquad\qquad (10) $[2, 4) \cup (3, 6)$

6. Tell the reasons for the transformations.

(1) $x > y \Longrightarrow 4x - 3 > 4y - 3$

(2) $a < b \Longrightarrow a - 3 < b + 3$

(3) $x \leq y \Longrightarrow a^2 x \leq a^2 y$ where $a \neq 0$

(4) $7x \geq 7y \Longrightarrow 3 - 2x \leq 3 - 2y$

(5) $2a > b - 4 \Longrightarrow -a - 2 < -b/2$

5.5 Linear Inequalities in One Variable

The concepts of polynomial inequalities and linear inequalities are analogous to polynomial equations and linear equations. If we change the equal sign in a polynomial equation (linear equation) by an inequality sign, we get a polynomial inequality (linear inequality). Using the propositions of inequalities, a linear inequality in one variable can always be simplified into its standard form $ax \neq b$, $ax > b$, $ax \geq b$, $ax < b$ or $ax \leq b$ where a, b are constants.

First, let us solve $ax \neq b$. The solution set of such an inequality is the complement of the solution set of the equation $ax = b$, i.e. $\mathbb{R} - \{x | ax = b\}$.

Secondly, consider
$$ax > b.$$

- If $a > 0$, we get $x > \dfrac{b}{a}$ and the solution set is $(\dfrac{b}{a}, +\infty)$;
- If $a < 0$, we get $x < \dfrac{b}{a}$ and the solution set is $(-\infty, \dfrac{b}{a})$;
- If $a = 0, b \leq 0$, there is no solution, so the solution set is the empty set $\emptyset$;
- If $a = 0, b < 0$, any real number is a solution, so the solution set is $\mathbb{R}$.

For a third example, consider
$$ax \leq b.$$

- If $a > 0$, we get $x \leq \dfrac{b}{a}$ and the solution set is $(-\infty, \dfrac{b}{a}]$;
- If $a < 0$, we get $x \geq \dfrac{b}{a}$ and the solution set is $[\dfrac{b}{a}, +\infty)$;
- If $a = 0, b < 0$, there is no solution, so the solution set is the empty set $\emptyset$;
- If $a = 0, b \geq 0$, any real number is a solution, so the solution set is $\mathbb{R}$.

The other two cases can be discussed similarly.

According to the basic propositions of inequalities, we get equivalent inequalities if

- adding or subtracting the same expression to both sides of an inequality;

- multiplying or dividing both sides of an inequality by the same positive expression;

- multiplying or dividing both sides of an inequality by the same negative expression as well as reversing the inequality sign.

From the first rule, in order to get equivalent inequalities, when moving a term from one side to the other side of an inequality, change the sign of the term. For example, we can move x in $3x - 7 < x - 5$ from the right side to the left side to get $(3x - 7) - x < -5$. The following example shows how to solve linear inequalities in one variable.

Example. Solve $\dfrac{1-x}{6} + \dfrac{4-2x}{3} \leq \dfrac{5}{2}$.

Solutions. To clear the denominator, multiply both sides of the inequality by the least common denominator 6.

$$6(\frac{1-x}{6} + \frac{4-2x}{3}) \leq (\frac{5}{2})6,$$
$$(1-x) + 2(4-2x) \leq 15.$$

Simplify the left side to get $-5x + 9 \leq 15$. Move the constant term to the right side: $-5x \leq 15 - 9$, i.e. $-5x \leq 6$. Notice when moving 9 from the left side to the right side, we change it from positive to negative. Dividing the coefficient -6 of x, we get $x \geq -\dfrac{6}{5}$. Here we must change '$\leq$' into '$\geq$' as we are dividing a negative number. As a result, the solution set is $[-\dfrac{6}{5}, +\infty)$.

Two or more inequalities together form a system of inequalities. A number is a solution to the system if it satisfies all inequalities in the system. Consequently, the solution set is the intersection of the solution sets to each inequality in the system.

Example. Solve the system of inequalities.

$$\begin{cases} 2x + 1 > 3 \\ 4 - 2x \leq 10 \\ 3x - 5 \leq 13 \end{cases}$$

Solutions. The solution sets to each inequality in the system are $(1, +\infty)$, $[-3, +\infty)$ and $(-\infty, 6]$ separately. The solution set to the system is their intersection $(1, +\infty) \cap [-3, +\infty) \cap (-\infty, 6] = (1, 6]$.

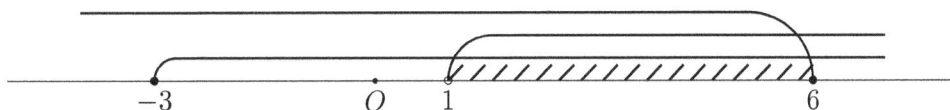

An inequality of the form $P \leq Q \leq R$ is equivalent to a system of inequalities $\begin{cases} P \leq Q \\ Q \leq R \end{cases}$. Hence we can solve it accordingly.

Exercises.

1. True or false.

(1) $x + y > 5$ is a linear inequality.

(2) $\dfrac{1}{x} < 0$ is a linear inequality.

(3) $\dfrac{2(y+1)}{3} \geq \dfrac{3(y-2)}{4}$ is a linear inequality in one variable.

(4) $x(3x+1) > 3x(2x-1)$ is a linear inequality.

(5) There are 5 positive integer solutions to $23 - 5x > -7$.

(6) $x - y \leq 1$ has infinitely many solutions.

(7) 0 is a solution to $3 + 2x < 5 - 6x$.

(8) There are finitely many positive solutions to $x - 3 < 0$.

(9) When $a = 0, b < 0$, there is no solution to $ax > b$.

(10) When $a = 0, b < 0$, there are infinitely many solutions to $ax < b$.

(11) If the solution to $ax > b$ is $x < \dfrac{b}{a}$, then $a < 0$.

(12) Any linear inequality in one variable has either no solutions or infinitely many solutions.

2. Multiple choice.

(1) Which of the following is a linear inequality?

A. $x^2 > 0$

B. $\dfrac{x}{y} < 0$

C. $3a + 2b \leq 6 - c$

D. $-\dfrac{2}{x} > -\dfrac{3}{y}$

(2) If $|m - 5| = 5 - m$, then

A. $m \geq 5$ B. $m > 5$ C. $m \leq 5$ D. $m < 5$

(3) If the solution to $2x + 3k = 5$ is positive, then

A. $k > \dfrac{5}{3}$ B. $k < \dfrac{5}{3}$ C. $k > \dfrac{5}{2}$ D. $k < \dfrac{5}{2}$

(4) If $\dfrac{y}{3} - 3 > \dfrac{y}{2} - 3$, then

A. $y > 0$ B. $y < 0$ C. $y \leq 0$ D. $y \geq 0$

(5) If $\dfrac{|x+1|}{x+1} = -1$, then

A. $x > -1$ B. $x \geq -1$ C. $x < -1$ D. $x \leq -1$

(6) The nonnegative integer solutions to $\dfrac{x+5}{2} - 1 > \dfrac{3x+2}{3}$ are

A. 1

B. 0 and 1

C. 0, 1 and 2

D. non of the above

(7) If $a < b < 0$, the solution to $ax < b$ is

A. $x < \dfrac{b}{a}$ B. $x > \dfrac{b}{a}$ C. $x < -\dfrac{b}{a}$ D. $x > -\dfrac{b}{a}$

(8) If $a < b < 0$, then

A. $\dfrac{1}{a} < \dfrac{1}{b}$ $\qquad$ B. $\dfrac{a}{b} < 1$ $\qquad$ C. $\dfrac{a}{b} > 1$ $\qquad$ D. $ab < 1$

3. Solve the inequalities. Write the solution set in interval notation.

(1) $2x - 19 \le 7x + 31$ $\qquad\qquad\qquad$ (2) $2(3x - 1) - 3(4x + 5) \le x - 4(x - 7)$

(3) $2(x - 1) - x > 3(x - 1) + 3x - 5$ $\qquad$ (4) $3[y - 2(y - 7)] \ge 4y$

(5) $15 - (7 + 5x) \le 2x + (5 - 3x)$ $\qquad$ (6) $\dfrac{x + 1}{2} > \dfrac{2x - 1}{3}$

(7) $\dfrac{x}{3} - \dfrac{x - 1}{2} < 1$ $\qquad\qquad\qquad$ (8) $2(x + 1) - \dfrac{x - 2}{3} > \dfrac{7x - 2}{2}$

(9) $\dfrac{x - 1}{3} - \dfrac{2x + 5}{4} > -2$ $\qquad\qquad$ (10) $\dfrac{2x - 1}{3} - \dfrac{2x + 5}{12} \le \dfrac{6x - 7}{4} - 1$

(11) $\dfrac{3}{2}x - 7 < \dfrac{1}{6}(8x - 1)$ $\qquad\qquad$ (12) $3 - \dfrac{3x}{2} < \dfrac{5}{8} - \dfrac{4x - 3}{6}$

(13) $x - \dfrac{1}{2}[x - \dfrac{1}{2}(x - 1)] > \dfrac{2}{3}(x - 1)$ $\qquad$ (14) $m - \dfrac{m - 1}{2} \ge 2 - \dfrac{m + 2}{5}$

(15) $x - \dfrac{x - 1}{2} + \dfrac{5x - 8}{6} < -\dfrac{x + 1}{3} + 2$ $\quad$ (16) $\dfrac{2x + 1}{4} > \dfrac{15x - 2}{6} - \dfrac{6x + 4}{3}$

(17) $\dfrac{1.8 - 8x}{1.2} - \dfrac{1.3 - 3x}{2} > \dfrac{5x - 0.4}{0.3}$ $\quad$ (18) $\dfrac{1}{5}(0.5 - \dfrac{x}{2}) - \dfrac{1}{4} \le \dfrac{x}{2} + 2.1$

4. Solve the systems of inequalities. Write the solution set in interval notation.

(1) $\begin{cases} 5x + 1 \geq 3(x + 1) \\ \dfrac{x - 1}{2} \leq \dfrac{2x - 1}{5} \end{cases}$
   (2) $\begin{cases} 2x - 4 < x + 2 \\ 3x - 2 \geq x + 8 \end{cases}$

(3) $\begin{cases} 3x - 6 < 2(2x - 1) \\ 3(2 + x) \geq 2(2x - 1) \\ 5(1 + x) \leq 10 - 2(2 - 3x) \end{cases}$
   (4) $5 \leq \dfrac{2x - 1}{5} < 7$

(5) $-x + 3 \leq 5 - 2x \leq x - 4$
   (6) $-3x + 7 < -2x - 1 \leq \dfrac{5 - 3x}{2}$

5.6   Inequalities involving Absolute Values

To solve inequalities involving absolute values, the strategy is still to simplify the absolute values. First let us investigate the four basic types of inequalities.

- $|x| > a$
  - If $a \geq 0$, x should satisfy $x > a$ or $x < -a$, so the solution set is $(-\infty, -a) \cup (a, +\infty)$;
  - If $a < 0$, any number x is a solution, so the solution set is $\mathbb{R} = (-\infty, +\infty)$.
- $|x| \geq a$
  - If $a > 0$, x should satisfy $x \geq a$ or $x \leq -a$, so the solution set is $(-\infty, -a] \cup [a, +\infty)$;
  - If $a \leq 0$, any number x is a solution, so the solution set is $\mathbb{R} = (-\infty, +\infty)$.
- $|x| < a$
  - If $a > 0$, x should satisfy $-a < x < a$, so the solution set is $(-a, a)$;
  - If $a \leq 0$, there is no solution, so the solution set is the empty set $\emptyset$.
- $|x| \leq a$
  - If $a \geq 0$, x should satisfy $-a \leq x \leq a$, so the solution set is $[-a, a]$;
  - If $a < 0$, there is no solution, so the solution set is the empty set $\emptyset$.

To solve inequalities of the form $|x| < |y|$ and three other analogies, we need to discuss the sign of y and simplify it to the above cases. For other inequalities having more than one absolute values, we should discuss the signs of terms inside the absolute values.

Examples. Solve the inequalities.

(1) $|2x - 1| > 3$

(2) $|2x + 1| \leq x + 5$

(3) $|2x + 4| > |x - 5|$

(4) $|x - 2| + |x + 3| < 9$

Solutions.

(1) Applying the results for basic types, the inequality is equivalent to $2x - 1 > 3$ or $2x - 1 < -3$. Solving these inequalities we get $x > 2$ or $x < -1$, so the solution set is their union $(-\infty, -1) \cup (2, +\infty)$.

(2) It is still of the basic type and we can apply the results directly.

• If $x + 5 < 0$, $|2x + 1| \leq x + 5$ has no solutions.

• If $x + 5 \geq 0$, the original inequality is equivalent to $-(x + 5) \leq 2x + 1 \leq x + 5$. As a result, we get a system of inequalities:

$$\begin{cases} 2x + 1 \leq x + 5 \\ -(x + 5) \leq 2x + 1 \\ x + 5 \geq 0 \end{cases} .$$

Notice that x also needs to satisfy the assumption $x + 5 \geq 0$, so we should include it into the system. Solving each inequality of the system, we get the solution sets are $(-\infty, 4]$, $[-2, +\infty)$ and $[-5, +\infty)$ separately. The solution set of the system is their intersection $[-2, 4]$.

Combining the two cases together, the final solution set is $[-2, 4]$.

(3) We should discuss the sign of $2x + 4$ or $x - 5$. Take $x - 5$ for example. Letting $x - 5 = 0$ we get $x = 5$. There are two possibilities:

• When $x < 5$, $|x - 5| = -(x - 5) = 5 - x$. The original inequality becomes $|2x + 4| > 5 - x$, which is equivalent to $2x + 4 > 5 - x$ or $2x + 4 < -(5 - x)$. Solving these inequalities we get $x > \dfrac{1}{3}$ or $x < -9$. With the assumption $x < 5$ the solution set should be $\{x | x < -9$ or $x > \dfrac{1}{3}\} \cap \{x | x < 5\} = (-\infty, -9) \cup (\dfrac{1}{3}, 5)$.

• When $x \geq 5$, $|x - 5| = x - 5$. The original inequality becomes $|2x + 4| > x - 5$, which is equivalent to $2x + 4 > x - 5$ or $2x + 4 < -(x + 5)$. Solving these inequalities we get $x > -9$ or $x < -3$, which means x can be any real number (since $(-9, +\infty) \cup (-\infty, -3) = (-\infty, +\infty)$). With the assumption $x \geq 5$ in advance, the solution set is $[5, +\infty)$.

The two cases have an 'OR' relation, so the solution set is the union of the two $(-\infty, -9) \cup (\dfrac{1}{3}, 5) \cup [5, +\infty) = (-\infty, -9) \cup (\dfrac{1}{3}, +\infty)$. We can also solve the inequality by making discussions on the sign of $2x + 4$, and the solution set would be the same. Interested readers can try by yourselves.

(4) We should discuss the signs of both $x - 2$ and $x - 3$. If $x - 2 = 0$, $x = 2$; if $x + 3 = 0$, $x = -3$. We make the following discussions.

- When $x \leq -3$, both $x + 2$ and $x + 3$ are nonpositive, so $|x - 2| = -(x - 2)$, $|x + 3| = -(x + 3)$, and the inequality simplifies to $-(x - 2) - (x + 3) < 9$, whose solution is $x > -5$. The solution set in this case is $(-5, +\infty) \cap (-\infty, -3] = (-5, -3]$.

- When $-3 < x < 2$, $x - 2$ is negative but $x + 3$ is positive, so $|x + 2| = -(x + 2)$, $|x + 3| = (x + 3)$, and the inequality simplifies to $-(x - 2) + (x + 3) < 9$, which is true for all x, so the solution set is $(-3, 2)$.

- When $x \geq 2$, both $x + 2$ and $x + 3$ are nonnegative, so $|x + 2| = (x + 2)$, $|x + 3| = (x + 3)$, and the inequality simplifies to $(x - 2) + (x + 3) < 9$, whose solution is $x < 4$. The solution set in this case is $(-\infty, 4) \cap [2, +\infty) = [2, 4)$.

Finally, the solution set is the union of the three $(-5, -3] \cup (-3, 2) \cup [2, 4) = (-5, 4)$.

Exercises.

1. True or false.

(1) If $|a| > |b|$ and $ab < 0$, then $a + b > 0$.

(2) $|x - 1| \leq 0$ has only one solution.

(3) $|2x - \pi| > 0$ only has positive solutions.

(4) If $|a| < |b|$, then $a < b$.

(5) If $|a| \leq |b|$, then $a^2 \leq b^2$.

(6) $|x|$ is always greater than $|x - 1|$.

(7) $|3x - 2| < -1$ has solutions.

(8) $|x + 1| + |x - 1|$ is always greater than or equal to 2.

(9) $-2 \leq |x + 1| - |x - 1| \leq 2$ is true for all $x \in \mathbb{R}$.

(10) $|x^2 + 1| < 1$ has infinitely many solutions.

2. Multiple choice.

(1) Which of the following is true?

    A. If $a > b$, then $a^2 > b^2$.   B. If $|a| > b$, then $a^2 > b^2$.

    C. If $a > |b|$, then $a^2 > b^2$.   D. If $a \neq |b|$, then $a^2 \neq b^2$.

(2) The solution set of $|x - 5| > 0$ is

    A. $\{5\}$ B. $\mathbb{R}$ C. $\emptyset$ D. $\{x | x \neq 5\}$

(3) If the solutions set of $|ax + b| \leq c$ is not empty, then

    A. $c > 0$ B. $c \geq 0$ C. $c < 0$ D. $c \leq 0$

(4) If $|x - 1| < 3$, then $|x - 4|$ equals

    A. $x - 4$ B. $4 - x$

    C. can't be determined   D. non of above

(5) The solution set to $|x + 2| + 1 < 0$ is

    A. $\{x | 1 < x < 3\}$ B. $\{x | x < 1 \text{ or } x > 3\}$

    C. $\mathbb{R}$ D. $\emptyset$

(6) Which of the following is correct?

    A. $|x| \leq a$ always has solutions.   B. $|x| < a$ always has solutions.

    C. $|x| > a$ is equivalent to $x^2 > a^2$.   D. $|x| < a$ is equivalent to $x^2 < a^2$.

3. Solve the inequalities.

(1) $|2x + 3| \geq 2$

(2) $|4x - 7| \leq 0$

(3) $|5 - 3x| > -\pi$

(4) $|\dfrac{4 - 3x}{5}| < 1$

(5) $|16 - 7x| < -\dfrac{1}{10^5}$

(6) $|-2x + 7| \leq 13$

(7) $|3x - 4| > |4x - 3|$

(8) $|2x - 3| \leq |x|$

(9) $|2x + 1| \leq |2x - 4|$

(10) $|6 - 3x| < |2x - 4|$

(11) $|3x + 1| > 2x - 3$

(12) $|4 - x| \leq 3x$

(13) $|7 - 2x| \geq 4x - 5$

(14) $|x + 1| - |x - 1| \geq 2$

(15) $|x + 3| - |x - 5| > 9$

(16) $|x - 2| - |x + 1| < 3$

(17) $|-2x + 3| - |x + 1| < 5$

(18) $|3x + 1| + |5x - 2| \leq 7$

(19) $|2 - x| - |2x + 5| > 2x$

(20) $|3x - 2| - |2x + 3| < x + 1$

4. Solve the system of inequalities.

$$\begin{cases} |x + 1| \geq 1 \\ |2x - 1| < 3 \end{cases}$$

Rational Expressions and Rational Equations

In this chapter, we continue to introduce the second category of algebraic expressions—rational expressions. In Chapter 2, we learned that the quotient of two integers may not be an integer any more, which leads to the discovery of rational numbers. Applying the same idea to polynomials, it is natural to have fractions of polynomials. Such expressions are called rational expressions. As we are dealing with fractions, we should keep in mind that the denominators can never be zero.

Corresponding to rational expressions, there are also rational equations and inequalities. To solve rational equations, we transform them into equivalent systems of combined polynomial equations and inequalities. The method of solving rational inequalities is also ready to develop, but we prefer to delay the introduction till the readers have more knowledge on polynomial equations.

6.1 Rational Expressions

A rational expression is the quotient of two polynomials in which the denominator is not the zero polynomial. For example,

$$\frac{x}{2}, \qquad \frac{x^2 + 2x - 3}{x^3 - 1}, \qquad \frac{xy^3 - x^3y + 1}{x - y}$$

are all rational expressions. In general a rational expression is of the form $\frac{P}{Q}$ where P, Q are polynomials and $Q \neq 0$. All polynomials are rational expressions as any polynomial P can be written as a quotient $\frac{P}{1}$. Particularly, numbers can be considered as rational expressions.

A rational expression is well-defined only for the values of variables which make the denominator nonzero. For example, $\frac{2x}{x - 5}$ makes sense only when $x - 5 \neq 0$, i.e. when $x \neq 5$.

The fundamental principle of fractions asserts

$$\frac{a}{b} = \frac{a \cdot c}{b \cdot c}$$

as long as $c \neq 0$. According to this principle, rational expressions can be simplified by canceling common factors.

Examples. Simplify.

(1) $\dfrac{9x^2yz}{15xy^3z^2}$

(2) $\dfrac{x^2 - 1}{x^2 - 4x + 3}$

Solutions.

(1) $3xy^2z$ is a common factor of the denominator and numerator. Canceling this factor we get

$$\frac{9x^2yz}{15xy^3z^2} = \frac{3xy^2z \cdot 3x}{3xy^2z \cdot 5yz} = \frac{3x}{5yz}.$$

(2) Both the denominator and numerator can be factored. $x^2 - 1 = (x + 1)(x - 1)$, $x^2 - 4x + 3 = (x - 1)(x - 3)$. $x - 1$ is a common factor. Doing a cancellation we get

$$\frac{x^2 - 1}{x^2 - 4x + 3} = \frac{(x - 1)(x + 1)}{(x - 1)(x - 3)} = \frac{x + 1}{x - 3}.$$

A rational expression is in lowest terms or reduced form if its denominator and numerator have no common factors. When simplifying rational expressions, we should always simplify them to their lowest terms.

Exercises.

1. True or false.

(1) $\dfrac{x^2}{x}$ is not a rational expression.

(2) If A and B are polynomials, then $\dfrac{A}{B}$ is a rational expression.

(3) $\dfrac{(x - 1)(x - 2)}{x - 2} = 0$ when $x = 1$ or 2.

(4) $\dfrac{x}{y} = \dfrac{x^2}{xy}$.

(5) $\dfrac{2a}{b} = \dfrac{2ac}{bc}$ if $c \neq 0$.

(6) $-\dfrac{2 - a + b}{c} = \dfrac{a + b - 2}{c}$.

(7) $\dfrac{(x - 1)(x^2 + 1)}{x - 1} \neq 0$ can never assume the value 0.

(8) $\dfrac{1}{x} = \dfrac{5 + y + x^2}{(5 + y + x^2)x}$ for all numbers $x \neq 0, y \in \mathbb{R}$.

(9) When $a > 5$, $\dfrac{1}{y} = \dfrac{y^2 + a - 5}{(y^2 + a - 5)y}$ holds for any $y \neq 0$.

(10) $\dfrac{|x| - 1}{x - 1} = 0$ when $x = \pm 1$.

2. Multiple choice.

(1) Which of the following is not a rational expression?

A. $\dfrac{x}{2}$ 　　　B. $-x$ 　　　C. $\dfrac{x+y-1}{y}$ 　　　D. $\dfrac{\sqrt{x}}{x^2+y^2}$

(2) Which of the following is not a rational expression?

A. $\sqrt{3}$ 　　　B. $\dfrac{1}{\sqrt{x}}$ 　　　C. $\dfrac{x}{x}$ 　　　D. $\dfrac{\sqrt{2}}{x^2+y^2}$

(3) $\dfrac{x^2-9}{x^2+1}$ is well-defined for

A. $x \neq \pm 1$ 　　　　　　　　　B. $x \neq \pm 3$

C. any real number x 　　　　　D. $x \neq 0$

(4) $\dfrac{m+1}{m-n}$ is well-defined when

A. $m = n$ 　　　B. $m \neq -1$ 　　　C. $m \neq n$ 　　　D. $m, n \in \mathbb{R}$

(5) $\dfrac{x-y}{x^2+1}$ is well-defined for

A. $x \neq \pm 1$ 　　　　　　　　　B. $x = y$

C. $x \neq y$ 　　　　　　　　　　　D. any real numbers x, y

(6) If $\dfrac{1}{a^2+b^2}$ is well-defined, then

A. a, b are any real numbers 　　　B. $a \neq 0$

C. $b \neq 0$ 　　　　　　　　　　　　D. None of the above

(7) $\dfrac{2}{(m-3)(m-2n)}$ makes sense when

A. $m \neq 3$ 　　　　　　　　　　　B. $m \neq 3$ and $n \neq -\dfrac{3}{2}$

C. $m \neq 3$ and $m \neq 2n$ 　　　D. $m, n \in \mathbb{R}$

(8) $\dfrac{|x|-2}{-x+2}$ equals 0 when

A. $x = \pm 2$ 　　　B. $x = 2$ 　　　C. $x = -2$ 　　　D. $x = 0$

(9) $\dfrac{a^2-a-2}{a+1}$ is 0 when

A. $a = -1$ 　　　B. $a = 2$ or -1 　　　C. $a = 2$ 　　　D. $a = 1$

(10) $\dfrac{(x-3)(x^2-16)}{x-4} = 0$ when

A. $x = -4$ 　　　B. $x = 3$ or 4 　　　C. $x = 3$ or ± 4 　　　D. $x = 3$ or -4

3. Find when the rational expression is well-defined.

(1) $\dfrac{x}{x}$ 　　　　　　　　　　　　(2) $\dfrac{p^2-9}{p-3}$

(3) $\dfrac{y^2 - y}{y^2}$

(4) $\dfrac{2a + 1}{2a - 1}$

(5) $\dfrac{x^6 - x^4 + x - 12}{20x - 36}$

(6) $\dfrac{2b}{|b| + 2}$

(7) $\dfrac{(r + 1)(r - 1) - 15}{(r + 4)(r - 3)}$

(8) $\dfrac{-24}{r^2 - 1}$

4. Simplify.

(1) $\dfrac{15x^2 y^3}{25x^3 y}$

(2) $\dfrac{-12a^2 b^3 c}{28a^5 bc^3}$

(3) $\dfrac{46xy^3}{69x^2 y^3 z}$

(4) $\dfrac{3x + 6y}{9x - 3y}$

(5) $\dfrac{5c - 9}{27 - 15c}$

(6) $\dfrac{a^2 - ab}{a^2 - b^2}$

(7) $\dfrac{a^2 - 4}{a^2 + 3a - 10}$

(8) $\dfrac{9 - a^2}{a^2 + 4a - 21}$

(9) $\dfrac{16 - y^2}{y^2 + y - 20}$

(10) $\dfrac{3y + y^2 + 2}{y^2 + 4y + 3}$

(11) $\dfrac{25 - b^2}{b^2 + 4b - 45}$

(12) $\dfrac{x^2 + 2x - 3}{2x^2 - 2x}$

(13) $\dfrac{x^2 - 5x + 6}{x^2 + 2x - 8}$

(14) $\dfrac{6x^2 - 5x - 6}{8x^2 - 6x - 9}$

(15) $\dfrac{2p^2 - pq - 15q^2}{10p^2 + 23pq - 5q^2}$

(16) $\dfrac{xy + 3y - 2x - 6}{2xz + 6z + x + 3}$

(17) $\dfrac{4x^2 - 4xy + y^2}{2x - y}$

(18) $\dfrac{3x^3 + 3y^3}{4x^2 - 4xy + 4y^2}$

(19) $\dfrac{x^3 - 1}{x^4 - 1}$

(20) $\dfrac{2x^2 + 5x + 3}{2x^2 + x - 3}$

(21) $\dfrac{a^3 - 8b^3}{2a^2 - ab - 6b^2}$

(22) $\dfrac{(x + y)^2 - z^2}{x + y + z}$

5. Find when the rational expression is zero.

(1) $\dfrac{x^2}{x}$

(2) $\dfrac{2x - 3}{|x| - 7}$

(3) $\dfrac{|s| - 4}{s - 4}$

(4) $\dfrac{|t - 3|}{t}$

(5) $\dfrac{x - y}{x - 1}$

(6) $\dfrac{x^2 - 4}{x - 2}$

6.2   Multiplication and Division of Rational Expressions

The rules of multiplying and dividing rational expressions are the same as the rules of multiplying and dividing rational numbers. Given any two rational expressions $\dfrac{P_1}{Q_1}$ and $\dfrac{P_2}{Q_2}$, where P_1, P_2, Q_1, Q_2 are polynomials and $Q_1, Q_2 \neq 0$, their product and quotient are defined as follows.

- **Multiplication.** $\dfrac{P_1}{Q_1} \cdot \dfrac{P_2}{Q_2} = \dfrac{P_1 P_2}{Q_1 Q_2}$, i.e. multiplying numerators by numerators and denominators by denominators;

- **Division.** If $P_2 \neq 0$, $\dfrac{P_1}{Q_1} \div \dfrac{P_2}{Q_2} = \dfrac{P_1}{Q_1} \cdot \dfrac{Q_2}{P_2}$, i.e. multiplying the reciprocal of the divisor.

Usually we simplify rational expressions before the actual multiplication and division to reduce the complexity.

Examples. Simplify.

(1) $\dfrac{4x - 4y}{10xy} \cdot \dfrac{15x^2 y^2}{x^2 - y^2}$

(2) $\dfrac{3a^2 + 12}{a - 3} \div \dfrac{a^4 - 16}{a^2 - a - 6}$

Solutions.

(1) As $4x - 4y = 4(x - y)$, $x^2 - y^2 = (x + y)(x - y)$, the original product becomes $\dfrac{4(x - y)}{10xy} \cdot \dfrac{15x^2 y^2}{(x - y)(x + y)}$. The factors $2, (x - y)$ and $5xy$ can be canceled as follows:

$$\frac{4(x - y)}{10xy} \cdot \frac{15x^2 y^2}{(x - y)(x + y)}$$
$$= \frac{4 \times 15x^2 y^2}{10xy} \cdot \frac{x - y}{(x - y)(x + y)}$$
$$= \frac{6xy}{1} \cdot \frac{1}{x + y}$$
$$= \frac{6xy}{x + y}.$$

So $\dfrac{4(x - y)}{10xy} \cdot \dfrac{15x^2 y^2}{(x - y)(x + y)} = \dfrac{6xy}{x + y}$.

(2) By changing the division into multiplying the reciprocal, we get

$$\frac{3a^2 + 12}{a - 3} \div \frac{a^4 - 16}{a^2 - a - 6}$$
$$= \frac{3a^2 + 12}{a - 3} \cdot \frac{a^2 - a - 6}{a^4 - 16}.$$

Notice that $3a^2 + 12 = 3(a^2 + 4)$, $a^4 - 16 = (a^2 + 4)(a + 2)(a - 2)$, $a^2 - a - 6 = (a - 3)(a + 2)$, so the original expression equals

$$\frac{3(a^2 + 4)}{(a - 3)} \cdot \frac{(a - 3)(a + 2)}{(a^2 + 4)(a + 2)(a - 2)}.$$

After canceling common factors $a - 3$, $a + 2$ and $(a^2 + 4)$, the answer is $\dfrac{3}{a - 2}$.

Exercises.

1. True or false.

(1) $\dfrac{(a-b)^2(p-q)^3}{(b-a)^3(q-p)^5} = \dfrac{1}{(b-a)(p-q)^2}$.

(2) $\dfrac{x+1}{x-1} \div \dfrac{x+4}{x-5}$ is well-defined when $x \neq 1$ and $x \neq 5$.

(3) $\dfrac{(2m-3n)^3 \cdot (a-2b)}{(3n-2m)^2 \cdot (2b-a)^5}$ simplifies to $\dfrac{3n-2m}{(2b-a)^4}$.

(4) $\dfrac{(x-1)(x-2)(3-x)}{(1-x)(x-3)(2-x)}$ simplifies to -1.

(5) For all rational numbers x, $\dfrac{1+x}{1+|x|} \cdot (1-x) = x-1$.

(6) $[-\dfrac{1}{(x-y)^2}]^n = \dfrac{1}{(x-y)^{2n}}$.

2. Multiple choice.

(1) Among the rational expressions $\dfrac{b}{2a}$, $\dfrac{a+b}{a^2-b^2}$, $\dfrac{x-y}{x^2-y^2}$, $\dfrac{x+y}{x-y}$, how many of them cannot be simplified further?

    A. 1   B. 2   C. 3   D. 4

(2) Which of the following is in lowest terms?

    A. $\dfrac{2(m-n)^3}{n-m}$ B. $\dfrac{17a^2b}{19xy^2}$ C. $\dfrac{x^{14}}{x^5}$ D. $\dfrac{2x-4y}{4y-2x}$

(3) Which of the following is correct?

    A. $\dfrac{(a^2)^3}{a^3} = a^2$ B. $\dfrac{a^2-b^2}{a-b} = a-b$

    C. $\dfrac{(m-n)^3}{(n-m)^3} = -1$ D. $\dfrac{(m-n)^2}{(n-m)^2} = -1$

(4) If $x = m+n$, $y = m-n$, then $-\dfrac{(y-x)^2}{xy}$ is

    A. $\dfrac{2n}{m^2-n^2}$ B. $-\dfrac{2n}{m^2-n^2}$ C. $-\dfrac{4n^2}{m^2-n^2}$ D. $\dfrac{4n^2}{m^2-n^2}$

(5) If $2a = 3b$, then $\dfrac{2a^2}{3b^2}$ equals

    A. 1   B. $\dfrac{2}{3}$ C. $\dfrac{3}{2}$ D. $\dfrac{9}{6}$

(6) Which of the following is correct?

    A. The lease common denominator of $\dfrac{a^n}{(a+1)^2}$ and $\dfrac{a+1}{a^2(a^2-1)}$ is $[a(a+1)(a-1)]^2$.

    B. $\dfrac{4x^2-4xy+4y^2}{x-y} = 2x-2y$.

    C. $\dfrac{(m^2-n^2)^2}{(m+n)^2}$ simplifies to $(m-n)^2$.

D. $\dfrac{3a}{4b} = \dfrac{3a+3}{4b+4}$.

(7) Which of the following is correct?

A. For any $x \in \mathbb{Q}$, $\dfrac{3a}{4b} = \dfrac{3ax}{4bx}$.

B. $\dfrac{(y-x)^2}{(x-y)^2}$ simplifies to -1.

C. The quotient of $x^2 - x + \dfrac{1}{4}$ and $x - \dfrac{1}{2}$ is $x - \dfrac{1}{2}$.

D. $\dfrac{x^2}{y} \div \dfrac{x}{y^2} = \dfrac{x}{y}$.

(8) Which of the following is correct?

A. $\dfrac{(a^2 + b^2)^2}{a^2 - b^2} = a^2 + b^2$

B. $\dfrac{-(a^2 - b^2)^2}{(a-b)^2} = (a+b)^2$

C. $\dfrac{-a-b}{-b-a} = 1$

D. $\dfrac{(a-b)^2}{b-a} = -(b-a)$

(9) Which of the following is correct?

A. A rational expression is zero when the numerator is zero.

B. A rational expression does not change if multiplying the numerator and denominator by the same algebraic expression.

C. If the signs of the numerator and the denominator of a rational expression change at the same time, the rational expression does not change.

D. When $x < 1$, the rational expression $\dfrac{|2-x|+x}{2}$ is not well-defined.

(10) $\dfrac{x^n(y^{2n}-1)}{x^{n+1}(y^n+1)} (n \in \mathbb{N})$ simplifies to

A. $\dfrac{y^{2n}-1}{x(y^n+1)}$

B. $\dfrac{x^n y^n - 1}{x^{n+1}}$

C. $\dfrac{y^n+1}{x}$

D. $\dfrac{y^n-1}{x}$

3. Answer the following questions.

(1) If $\dfrac{|x-1|}{(x-1)^2} = -\dfrac{1}{x-1}$, find the condition for x.

(2) If $\dfrac{a^2-1}{|a-1|} = 0$, find a.

(3) If n is a positive integer, simplify $\dfrac{(1-a)^n}{(a-1)^{n+1}}$.

(4) If x is a negative number, simplify $\dfrac{1+x}{1+|x|} \cdot (1-x)$.

(5) If $\dfrac{a+b}{a-b} = 1$, find a and b.

4. Simplify.

(1) $\dfrac{12x^3yz}{3y^4} \cdot \dfrac{6xy^3z}{28x^2yz^3}$

(2) $\dfrac{5ab^3}{9xy^2z} \div \dfrac{10a^2b}{12x^3yz^2}$

(3) $20x^4y^2 \div \dfrac{5x^3}{2y}$

(4) $\dfrac{2x^4}{y^2} \cdot \dfrac{y^6}{x^3} \div (3xy^3)$

(5) $(x^2 - 2x + 1) \cdot \dfrac{x+1}{x^2 - 1}$

(6) $\dfrac{x^2 - 9y^2}{3x^2y^3} \cdot \dfrac{6xy}{x + 3y}$

(7) $(2xy - x^2) \div \dfrac{x - 2y}{xy}$

(8) $\dfrac{x^3 - 1}{x + 1} \div (x - 1)$

(9) $\dfrac{x^2 + 7x + 10}{x^2 - x + 1} \cdot \dfrac{x^3 + 1}{x^2 + 4x + 4}$

(10) $\dfrac{2x - 2y}{x^2 + xy} \cdot \dfrac{x^4 - x^2y^2}{3x^2 - 3xy}$

(11) $\dfrac{x^2 - 16y^2}{x^2 + 3xy - 4y^2} \cdot \dfrac{x - y}{x + y}$

(12) $\dfrac{4x^2 - 4xy + y^2}{2x - y} \div (4x^2 - y^2)$

(13) $\dfrac{x^2 - 6x + 8}{x^2 - 5x + 6} \div \dfrac{x^2 - 3x - 4}{x^2 - 2x - 3}$

(14) $-\dfrac{6x^2 + 11x + 3}{6x^2 - 13x + 5} \cdot \dfrac{8x^2 + 6x - 5}{8x^2 + 22x + 15}$

(15) $\dfrac{4a^2 + 4ab + b^2}{15a^3b^2} \cdot \dfrac{20a^2b + 25ab^2}{8a^2 + 14ab + 5b^2}$ (16) $\dfrac{3x^2 - 20xy + 25y^2}{2x^2 - 7xy - 15y^2} \div \dfrac{9x^2 - 3xy - 20y^2}{12x^2 + 28xy + 15y^2}$

(17) $\dfrac{9n^2 - 12n + 4}{3n^2 + 12n - 36} \cdot \dfrac{8n^2 + 48n}{6n^2 - 4n}$ (18) $(x^2 - 9) \cdot \dfrac{x^2 + 6x + 9}{(x + 3)^4} \div \dfrac{x^3 - 27}{x^3 + 27}$

5. Simplify.

(1) $\dfrac{1 + a - b - ab}{1 - a^2 - b^2 + a^2b^2} \cdot \dfrac{1 + a^2b^2 - a - b^2}{1 + b}$

(2) $\dfrac{x^2 - 3x - 10}{x^3 - 1} \cdot \dfrac{2x^2 + 2x + 2}{25 - 10x + x^2} \cdot \dfrac{x^2 - 6x + 5}{x^2 - 4}$

(3) $\dfrac{x^6 - 64y^6}{x^2 + 3xy - 10y^2} \cdot \dfrac{x^3 - 8y^3}{x^4 + 4x^2y^2 + 16y^4} \div \dfrac{3x^2 + 5xy - 2y^2}{2x^2 + 9xy - 5y^2} \cdot \dfrac{1}{x^2 + 2xy + 4y^2}$

(4) $\dfrac{(a^2 - b^2)^2}{a^3 + b^3} \div \dfrac{(a + b)^2}{a^2 - ab + b^2} \cdot \dfrac{1}{(b - a)^2}$

(5) $\dfrac{x^3 + 3x^2 - 4}{x^3 + 5x^2 + 2x - 8} \cdot \dfrac{x + 4}{x^3 + x^2 - 4x - 4}$

(6) $\dfrac{x^2 - 4x + 4}{x^2 - x - 6} \cdot \dfrac{x^2 - 3x - 10}{x^2 - 9} \div \dfrac{x - 2}{x^2 - 6x + 9}$

(7) $\dfrac{x^3 - y^3}{x^3 + y^3} \cdot \dfrac{x + y}{x - y} \div \dfrac{x^2 + xy + y^2}{x^2 - xy + y^2}$

(8) $\dfrac{6x^2 - 15x}{x^2 - x - 2} \div \dfrac{6x^2 - 7x - 20}{x^2 - 4} \cdot \dfrac{3x^2 + 7x + 4}{x^2 + 2x}$

(9) $\dfrac{x^2 - 2ax + a^2}{x^2 - a^2} \cdot \dfrac{2x^2 + 3ax + a^2}{x} \div \dfrac{2x + a}{ax}$

(10) $\dfrac{x^2 + y^2 - 2xy - z^2}{a^2 - 9 + 4b^2 + 4ab} \div \dfrac{x^2 - xy - xz}{2a + 4b + 6} \div \dfrac{x - y + z}{a + 2b - 3}$

6. Evaluate.

(1) $\dfrac{a^3 + 4b^3 + c^3}{b(a^2 + c^2)}$ when $a + c = 2b$

(2) $\dfrac{3(a - 1)(b - 1)(c - 1)}{(a - 1)^3 + (b - 1)^3 + (c - 1)^3}$ when $a + b + c = 3$

(3) $\dfrac{(x+1)(y+1)(z+1)}{(1-x)(1-y)(1-z)}$ when $x = \dfrac{a-b}{a+b}$, $y = \dfrac{b-c}{b+c}$, $z = \dfrac{c-a}{c+a}$, and $abc \neq 0$.

6.3 Addition and Subtraction of Rational Expressions

Adding or subtracting rational expressions is similar to adding or subtracting rational numbers. Given any two rational expressions $\dfrac{P_1}{Q_1}$ and $\dfrac{P_2}{Q_2}$, where P_1, P_2, Q_1, Q_2 are polynomials and $Q_1, Q_2 \neq 0$, generally their sum and difference are defined by

$$\frac{P_1}{Q_1} \pm \frac{P_2}{Q_2} = \frac{P_1 Q_2 \pm P_2 Q_1}{Q_1 Q_2}.$$

To reduce the complexity, we always try to find the least common denominator (LCD) and rewrite the rational expressions to have their LCD as denominators. The method to find the LCD is to factor the denominators and collect all common and different factors. For example, to find the LCD of $\dfrac{2x}{x^2-4}$ and $\dfrac{2}{x^2-5x+6}$, we factor the denominators: $x^2-4 = (x+2)(x-2)$, $x^2-5x+6 = (x-2)(x-3)$. The common factor is $(x-2)$ and the different factors are $x+2$ and $x-3$. Then the least common denominator of $\dfrac{2x}{x^2-4}$ and $\dfrac{1}{x^2-5x+6}$ should be $(x+2)(x-2)(x-3)$. If the rational expressions are not in reduced form, we need to simplify them first. The following examples demonstrates how to add and subtract rational expressions.

Examples. Simplify.

(1) $\dfrac{3z}{2x^3 y} - \dfrac{5}{4xy^2}$

(2) $\dfrac{2x}{x^2-4} + \dfrac{2}{x^2-5x+6}$

Solutions.

(1) The least common denominator of $2x^3 y$ and $4xy^2$ is $4x^3 y^2$. We get this by finding the higher (highest, if there are more than 2 terms) power of x and y as well as the least common multiples of the coefficients. $\dfrac{3z}{2x^3 y} = \dfrac{3z(2y)}{4x^3 y^2}$, $\dfrac{5}{4xy^2} = \dfrac{5x^2}{4x^3 y^2}$, so

$$\frac{3z}{2x^3 y} - \frac{5}{4xy^2} = \frac{6yz}{4x^3 y^2} - \frac{5x^2}{4x^3 + y^2}$$
$$= \frac{6yz - 5x^2}{4x^3 y^2}.$$

(2) By the analysis before the example, the LCD of $\dfrac{2x}{x^2-4}$ and $\dfrac{1}{x^2-5x+6}$ is $(x+2)(x-$

$2)(x-3)$. By the fundamental principle of fractions,

$$\frac{2x}{x^2-4} = \frac{2x}{(x+2)(x-2)}$$
$$= \frac{2x(x-3)}{(x+2)(x-2)(x-3)}$$
$$= \frac{2x^2-6x}{(x+2)(x-2)(x-3)}$$
$$\frac{1}{x^2-5x+6} = \frac{1}{(x-2)(x-3)}$$
$$= \frac{x+2}{(x+2)(x-2)(x-3)}.$$

As a result,

$$\frac{2x}{x^2-4} + \frac{1}{x^2-x-6} = \frac{2x^2-6x}{(x+2)(x-2)(x-3)} + \frac{x+2}{(x+2)(x-2)(x-3)}$$
$$= \frac{(2x^2-6x)+(x+2)}{(x+2)(x-2)(x-3)}$$
$$= \frac{2x^2-5x+2}{(x+2)(x-2)(x-3)}$$
$$= \frac{(2x-1)(x-2)}{(x+2)(x-2)(x-3)}$$
$$= \frac{2x-1}{(x+2)(x-3)}.$$

Here we take two more steps to simplify the answer. It is a convention that all rational expressions in solutions should be simplified to their reduced forms.

Exercises.

1. True or false.

(1) $\dfrac{2m+3n}{m^2 n} - \dfrac{m-3n}{m^2 n} = \dfrac{1}{mn}$.

(2) $\dfrac{x^2}{x+y} - \dfrac{y^2}{x+y} = x-y$.

(3) $\dfrac{b^2-2ab}{(a-b)^2} - \dfrac{a^2}{(b-a)^2} = 1$.

(4) $\dfrac{x}{x-y} + \dfrac{y}{x+y} - \dfrac{2xy}{y^2-x^2} = 1$.

(5) $\dfrac{(x+y)^2}{xy} - \dfrac{(x-y)^2}{xy} = 2$.

(6) $\dfrac{m^3}{m-1} - (m^2+m+1) = \dfrac{1}{m-1}$.

(7) $\dfrac{x}{x+a} - \dfrac{x}{x-a} = \dfrac{2ax}{x^2-a^2}$.

(8) $m - n - \dfrac{m^3 + n^3}{m^2 + mn + n^2} = 0.$

(9) $x + 8 - \dfrac{x^2 + 4}{x - 2} = 6.$

(10) $\dfrac{a^3}{(a - b)^3} + \dfrac{b^3}{(b - a)^3} = \dfrac{a^2 + ab + b^2}{(a - b)^2}.$

2. Multiple choice.

(1) The least common denominator of $\dfrac{12}{m^2 - 9}$ and $\dfrac{2}{3 - m}$ is

    A. $(m - 3)(m + 3)$ B. $(m + 3)(m - 3)^2$

    C. $(m^2 - 9)(3 - m)$ D. $m - 3$

(2) $\dfrac{x^3 + a^3}{x + y} - \dfrac{a^3 - y^3}{x + y}$ simplifies to

    A. $\dfrac{x^3 - y^3}{x + y}$ B. $x - y$ C. $x^2 - xy + y^2$ D. $x^2 + xy + y^2$

(3) The least common denominator of $\dfrac{2}{x - x^2}$ and $\dfrac{x^2 + 3}{x^2 + x - 6}$ is

    A. $(x - 2)(x + 3)$ B. $(1 - x)(x - 2)(x + 3)$

    C. $x(x - 1)(x - 2)(x + 3)$ D. $x(1 - x)(x - 3)(x + 2)$

(4) The least common denominator of $\dfrac{3}{2x - 6}, \dfrac{1 - x}{x^2 - 2x - 3}$ and $\dfrac{2x}{x^2 - 5x + 6}$ is

    A. $2(x - 2)(x - 3)(x + 1)$ B. $2(x + 2)(x - 3)(x + 1)$

    C. $(x - 2)(x - 3)(x + 1)$ D. $2(x + 3)(x + 2)(x - 1)$

(5) Which of the following is correct?

    A. $\dfrac{1}{2a} + \dfrac{1}{2b} = \dfrac{1}{2(a + b)}$ B. $\dfrac{b}{a} + \dfrac{b}{c} = \dfrac{2b}{ac}$

    C. $\dfrac{c}{a} - \dfrac{c + 1}{a} = \dfrac{1}{a}$ D. $\dfrac{1}{a - b} + \dfrac{1}{b - a} = 0$

(6) $-\dfrac{1 - x}{a} + \dfrac{x}{a}$ equals

    A. $-\dfrac{1}{a}$ B. $\dfrac{1 - 2x}{a}$ C. $\dfrac{1}{a}$ D. $\dfrac{2x - 1}{a}$

(7) $\dfrac{x^3}{x - 1} - x^2 - x - 1$ equals

    A. 1 B. $-\dfrac{1}{x - 1}$ C. $\dfrac{1}{x - 1}$ D. $\dfrac{x^2 + x + 1}{x - 1}$

(8) If $2x - 3y - z = 0$, $x + 3y - 14z = 0$ and $z \neq 0$, then $\dfrac{x^2 + 3xy}{y^2 + z^2}$ is

    A. 7 B. 2 C. 0 D. -2

(9) If $a + c = 2b$, $\dfrac{a^3 + 4b^3 + c^3}{b(a^2 + c^2)}$ equals

A. 6   B. 3   C. 1   D. 2

(10) $\dfrac{x+2y}{x^2-y^2} + \dfrac{y}{y^2-x^2} - \dfrac{2x}{x^2-y^2}$ equals

    A. $-\dfrac{1}{x+y}$ B. $-\dfrac{1}{x-y}$ C. $\dfrac{3y-x}{x^2-y^2}$ D. $\dfrac{1}{x-y}$

(11) If a is an integer, $b \neq 0$, $|a| \neq |b|$, then $[a-b+\dfrac{a(b-a)}{a+b}][\dfrac{a+1}{a-b} \cdot (1+\dfrac{a}{b})]$ is an

    A. odd number   B. even number   C. integer   D. 1

3. Simplify the following rational expressions.

(1) $\dfrac{2a-3}{8} - \dfrac{4a+5}{12}$ (2) $\dfrac{3y}{2x} - \dfrac{2x}{3y}$

(3) $\dfrac{3x-2y}{2x-2y} - 1 + x$ (4) $\dfrac{2x}{x+1} - \dfrac{3}{x}$

(5) $\dfrac{x+2}{x^2-7} + \dfrac{1-3x}{7-x^2}$ (6) $\dfrac{m+n}{m} - \dfrac{m-n}{n}$

(7) $3a - \dfrac{a^2}{5b}$ (8) $\dfrac{x}{x-4} + \dfrac{-3}{x^2-x-12}$

(9) $\dfrac{2x^2}{x-3} - x - 3$ (10) $\dfrac{x+2}{x^2-x-2} - \dfrac{x}{x^2-5x+6}$

(11) $\dfrac{5}{3p-5} - \dfrac{9}{4p+3}$ (12) $\dfrac{4x+4}{12x^2-x-6} - \dfrac{2x-3}{15x^2-8x-12}$

(13) $\dfrac{-x-3}{25-x^2} + \dfrac{2x}{x+5}$ (14) $\dfrac{3x}{2x+1} + \dfrac{2}{3x-5} - \dfrac{1}{2x-1}$

(15) $\dfrac{1}{x^2-4} + \dfrac{1}{x+2} - \dfrac{1}{x-2}$

(16) $p - \dfrac{p^2-q^2}{p} + \dfrac{p^2+q^2}{q} - q$

(17) $\dfrac{2c}{b^2-c^2} - \dfrac{1}{b+c} + \dfrac{1}{c-b}$

(18) $\dfrac{a^2+1}{a-1} - a + 1$

(19) $\dfrac{b+c}{8bc} + \dfrac{a-4b}{12ab} - \dfrac{5a-c}{6ac}$

(20) $\dfrac{5}{6a^2b} - \dfrac{2}{3ab^2} + \dfrac{3}{4abc}$

(21) $\left(\dfrac{1}{a^3} - \dfrac{1}{a^2} + \dfrac{1}{a}\right)a^4$

(22) $\dfrac{1}{x^2-1} + \dfrac{x}{1-x^2}$

(23) $\dfrac{c}{ab} + \dfrac{a}{bc} + \dfrac{b}{ca}$

(24) $-\dfrac{1}{a} + \dfrac{1}{b} - \dfrac{1}{c}$

(25) $\dfrac{1}{2a(a+b)^2} - \dfrac{1}{3b(a^2-b^2)}$

(26) $\dfrac{a^2-b^2}{ab} - \dfrac{ab-b^2}{ab-a^2}$

(27) $\dfrac{1}{a-b} - \dfrac{3ab}{a^3-b^3} + \dfrac{b-a}{a^2+ab+b^2}$

(28) $-\dfrac{x-1}{x+1} + \dfrac{x}{2x-2} + \dfrac{x-7}{2-2x^2}$

(29) $\dfrac{1}{a-1} + a^2 + a + 1$

(30) $\left(a^4 - \dfrac{1}{a^4}\right) \div \left(a - \dfrac{1}{a}\right)$

4. Simplify.

(1) $\dfrac{3x}{2x^2 + 5x + 3} + \dfrac{6 - x}{4x^2 - 9}$

(2) $\dfrac{x - 1}{x^2 - 5x - 6} - \dfrac{2x - 3}{x^2 - 9x - 10} + \dfrac{5 - x}{x^2 - 16x + 60}$

(3) $\dfrac{x + 2}{x^2 - 2x} - \dfrac{x - 1}{x^2 - 4x + 4} + \dfrac{x + 2}{x - 2}$

(4) $\dfrac{x - 1}{x^2 - x - 6} + \dfrac{2 - x}{x^2 - 9} - \dfrac{x + 1}{x + 2}$

(5) $\dfrac{4a + 12}{3a^2 + 2a - 8} \div \left(a - 2 - \dfrac{5}{a + 2}\right)$

(6) $\left(\dfrac{a + b}{a - b} - \dfrac{a^3 + b^3}{a^3 - b^3}\right)\left(\dfrac{a + b}{a - b} + \dfrac{a^2 + b^2}{a^2 - b^2}\right)$

(7) $\left(\dfrac{x + 2}{x^2 - 2x} - \dfrac{x - 1}{x^2 - 4x + 4}\right) \div \dfrac{4 - x}{x}$

(8) $\left(\dfrac{x^2 + y^2}{xy} - \dfrac{x^2}{xy + y^2} - \dfrac{y^2}{x^2 + xy}\right) \div \dfrac{3}{xy}$

(9) $\dfrac{1}{x^2 - 3x + 2} + \dfrac{1}{x^2 - x} + \dfrac{1}{x^2 + x} + \dfrac{1}{x^2 + 3x + 2}$

(10) $\dfrac{x^3 - x^2 - 4x + 4}{x^2 + x - 2} - \dfrac{x^3 - 2x^2 - 9x + 18}{x^2 + x - 6}$

(11) $\dfrac{x^3 - 1}{x^3 + 3x^2 + 3x + 1} + \dfrac{x^3 + 1}{x^3 - 3x^2 + 3x - 1} - \dfrac{2(x^2 + 1)}{x^2 - 1}$

(12) $\dfrac{1}{1-x} + \dfrac{1}{1+x} + \dfrac{2}{1+x^2} + \dfrac{4}{1+x^4} - \dfrac{5}{1-x^8}$

(13) $\dfrac{1}{1-z} - \dfrac{1}{1+z} - \dfrac{2z}{1+z^2} - \dfrac{4z^3}{1+z^4} + \dfrac{8z^7}{1+z^8}$

(14) $\dfrac{6x}{x^2+x-2} + (1 + \dfrac{1}{x-1}) \div (1 - \dfrac{1}{x+1}) + \dfrac{1}{x+2}$

(15) $(\dfrac{a}{a-b} - \dfrac{b}{a+b} + \dfrac{2ab}{a^2-b^2}) \div (1 + \dfrac{2b}{a-b})$

5. Evaluate.

(1) $[\dfrac{1}{a^2} + \dfrac{1}{b^2} + \dfrac{2}{a+b}(\dfrac{1}{a} + \dfrac{1}{b})] \div \dfrac{(a+b)^2}{ab}$ when $a = -\dfrac{1}{4}, b = \dfrac{4}{3}$

(2) $[\dfrac{4ab}{a-b} + (\dfrac{a}{b} - \dfrac{b}{a}) \div (\dfrac{1}{a} + \dfrac{1}{b})] \div (a^2 + 2ab + b^2)$ when the solutions to $\begin{cases} ax + by = 5 \\ bx + ay = 2 \end{cases}$ are $x = 4, y = 3$.

(3) $\dfrac{x}{xy+x+1} + \dfrac{y}{yz+y+1} + \dfrac{z}{xz+z+1}$ when $xyz = 1$

6. Prove.

(1) If a is an integer and $a^2 \neq 1$, prove $\left(\dfrac{a^9}{a^3 - 1} + \dfrac{a^3}{a^3 + 1}\right) - \left(\dfrac{a^6}{a^3 - 1} - \dfrac{1}{a^3 + 1}\right)$ is an integer.

(2) $\left(\dfrac{a}{b} - \dfrac{b}{a}\right) \div \left(\dfrac{a}{b} + \dfrac{b}{a} - 2\right) \cdot \dfrac{a - b}{a + b} = 1.$

(3) $\dfrac{1}{a^2 b^2} = \left(\dfrac{1}{a + b} - \dfrac{1}{b}\right) \cdot \left(\dfrac{1}{a - b} - \dfrac{1}{a}\right) \cdot \left(\dfrac{1}{a^2} - \dfrac{1}{b^2}\right).$

(4) $\dfrac{a}{b} + \dfrac{b}{a} - 1 = \dfrac{(a - b)^2}{ab} + 1.$

(5) Show $\dfrac{a}{(a - b)(a - c)} + \dfrac{b}{(b - c)(b - a)} + \dfrac{c}{(c - a)(c - b)} = 0.$

6.4   Complex Fractions

A complex fraction is the quotient of two rational expressions (the divisor is not zero). For example,

$$\dfrac{\dfrac{1}{a}}{\dfrac{2}{b}}, \quad \dfrac{\dfrac{1}{x} - \dfrac{2}{y}}{\dfrac{2}{x} + \dfrac{1}{y}}, \quad \dfrac{\dfrac{a}{b} + \dfrac{b}{a}}{3}, \quad \dfrac{-5}{\dfrac{2x^2 y - 3xy^2}{7x^2 y^2}}$$

are all complex fractions. To simplify a complex fraction, we can either multiply both the denominator and the numerator of the complex fraction by the LCD of the top and bottom fractions, or change the quotient into a product according to the definition of division. The two methods are both demonstrated in the following examples.

Examples. Simplify.

(1) $\dfrac{\dfrac{6m}{5ab}}{\dfrac{11m^2}{3a^2}}$

$(2)\ \dfrac{\dfrac{1}{x}+\dfrac{1}{y}}{1-\dfrac{1}{xy}}$

Solutions.

(1) Using the definition of quotient,

$$\frac{\dfrac{6m}{5ab}}{\dfrac{11m^2}{3a^2}} = \frac{6m}{5ab} \div \frac{11m^2}{3a^2}$$

$$= \frac{6m}{5ab} \cdot \frac{3a^2}{11m^2}$$

$$= \frac{6m \cdot 3a^2}{11m^2 \cdot 5ab}$$

$$= \frac{18a}{55mb}.$$

(2) We apply the second method. The least common denominator of $\dfrac{1}{x}+\dfrac{1}{y}$ and $1-\dfrac{1}{xy}$ is xy. Multiplying both the top and the bottom of the complex fraction by xy, we get

$$\frac{\dfrac{1}{x}+\dfrac{1}{y}}{1-\dfrac{1}{xy}} = \frac{\left(\dfrac{1}{x}+\dfrac{1}{y}\right)xy}{\left(1-\dfrac{1}{xy}\right)xy}$$

$$= \frac{y+x}{xy-1}.$$

Both methods work for all complex fractions. We always choose the method that makes the calculations as simple as possible.

Exercises.

1. True or false.

(1) When $a = -b$, $\dfrac{1+\dfrac{b}{a}}{1-\dfrac{b}{a}} = 0\,(a \neq 0)$.

(2) $\dfrac{\dfrac{x+y}{x-y}}{1-\dfrac{x-1}{x-y}} = -\dfrac{x+y}{y+1}.$

(3) $\dfrac{2ac}{\dfrac{4a}{b}} = \dfrac{bc}{2}.$

(4) $\dfrac{\dfrac{c}{4a}}{2ab} = \dfrac{bc}{2}.$

(5) $\dfrac{\dfrac{1}{a}+\dfrac{1}{b}}{\dfrac{1}{ab}} = a + b.$

(6) When $x = \pm 6$, $\dfrac{|x| - 6}{x^2 - 5x - 6} = 0.$

(7) $\dfrac{1}{1 + \dfrac{1}{x}}$ is well-defined when $x \neq 0$ and $x \neq 1$.

(8) When $x \neq 2$ and $x \neq 6$, the fraction $\dfrac{x + 5}{1 - \dfrac{4}{x - 2}}$ is well-defined.

(9) $\dfrac{x^2 + \dfrac{1}{x^2} + 2}{x + \dfrac{1}{x}} = x + \dfrac{1}{x}.$

(10) $\dfrac{1}{1 + \dfrac{1}{1 + \dfrac{1}{1 + x}}}$ simplifies to $\dfrac{x + 2}{2x + 3}.$

2. Multiple choice.

(1) The fraction $\dfrac{5x}{2 + \dfrac{3}{x - 2}}$ is undefined when

    A. $x = 2$ B. $x = 0$

    C. $x = \dfrac{1}{2}$ and $x = 2$ D. $x = \dfrac{1}{2}$ or $x = 2$

(2) The complex fraction $\dfrac{1}{1 + \dfrac{2}{x}}$ is well-defined when

    A. $x \neq 0$ B. $x \neq 2$

    C. $x \neq 0$ and $x \neq -2$ D. $x \neq 0$ and $x \neq 2$

(3) If $a^2 - 2a + b^2 + 6b + 10 = 0$, then $\dfrac{1}{\dfrac{1}{a} + \dfrac{1}{b}}$ equals

    A. $\dfrac{2}{3}$ B. $\dfrac{3}{2}$ C. $\dfrac{3}{4}$ D. $-\dfrac{3}{2}$

(4) If $\dfrac{x}{y} = -\dfrac{1}{2}$, then $\dfrac{1}{\dfrac{y}{x} - \dfrac{x}{y}}$ is

    A. $\dfrac{3}{2}$ B. $\dfrac{2}{3}$ C. $-\dfrac{3}{2}$ D. $-\dfrac{2}{3}$

(5) $\dfrac{x^2 + x}{1 - \dfrac{1}{x}} = 0$ when

A. $x = 0$ B. $x = 0$ or $x = -1$

C. $x = 0$ or $x = 1$ D. $x = -1$

3. Answer the following questions.

(1) Find k such that $\dfrac{1 - |k|}{2 + \dfrac{14}{2k - 5}} = 0$.

(2) Find x such that $\dfrac{x - 3}{1 - \dfrac{1}{x - 2}} = 0$.

(3) Find a such that $\dfrac{a^2 - 5a + 6}{2 - \dfrac{6}{2a - 1}} = 0$.

(4) Find h such that $\dfrac{2}{2 + \dfrac{3}{h - 1}}$ is not well-defined.

(5) Find x such that $\dfrac{x}{1 - \dfrac{1}{1 - \dfrac{1}{1 - \dfrac{2}{x + 1}}}}$ is well-defined.

4. Simplify.

(1) $\dfrac{\dfrac{3ab}{5n}}{\dfrac{24a}{10n}}$

(2) $\dfrac{x-3}{1-\dfrac{1}{x+5}}$

(3) $\dfrac{y-\dfrac{1}{y}}{1+\dfrac{1}{y}}$

(4) $\dfrac{\dfrac{3}{x}-2}{\dfrac{2}{x}}+4$

(5) $\dfrac{\dfrac{2}{x-2}-3}{5+\dfrac{4}{x-2}}$

(6) $\dfrac{a}{1-\dfrac{1}{a}}$

(7) $\dfrac{\dfrac{-1}{x}+\dfrac{3}{x-2}}{\dfrac{2}{x^2-2x}-\dfrac{4}{x}}$

(8) $\dfrac{\dfrac{3}{x+1}-\dfrac{2}{x-1}}{\dfrac{1}{x+1}+\dfrac{2}{x^2-1}}$

(9) $\dfrac{1+\dfrac{b}{x}}{\dfrac{a}{x}+1}$

(10) $\dfrac{5x}{2+\dfrac{3}{x-2}}$

(11) $\dfrac{x - \dfrac{1}{2x}}{x^2 - \dfrac{1}{4x^2}}$

(12) $\dfrac{1}{a - \dfrac{a-1}{1 - \dfrac{1}{a^2}}}$

(13) $\dfrac{x - \dfrac{1}{2x}}{x^3 - \dfrac{1}{8x^3}}$

(14) $\dfrac{\dfrac{1}{(a+b)^2} - \dfrac{1}{(a-b)^2}}{\dfrac{1}{a+b} - \dfrac{1}{a-b}}$

(15) $\dfrac{\dfrac{x^2 - 3xy + 2y^2}{x^2 - 2xy + y^2}}{x - 2y}$

(16) $\dfrac{\dfrac{2(x+y)}{x-y}}{1 - \dfrac{x+y}{2x - 2y}}$

5. Evaluate.

(1) $\dfrac{1 + \dfrac{1}{x}}{1 - \dfrac{1}{1 - \dfrac{1}{x}}}$ when $|x - 3| = 1$

(2) $\dfrac{a}{1 - \dfrac{a}{a + \dfrac{a}{a^2 - 1}}}$ when $a = -\dfrac{1}{3}$

(3) $\dfrac{\dfrac{1}{1+x}+\dfrac{1}{1-x}}{\dfrac{1}{1+x}-\dfrac{1}{1-x}}$ when $x=\dfrac{1}{4}$

(4) $\dfrac{1+\dfrac{1}{x}}{1-\dfrac{1}{1-\dfrac{1}{x}}}$ when $|x-1|=2$

(5) $\dfrac{1+\dfrac{2y^3}{x^3-y^3}}{1+\dfrac{2y}{x-y}}$ when $3x-y=0$ and $xy\neq 0$

6.5 Rational Equations

If we set two rational expressions equal, we get a rational equation(or fractional equation). By moving all fractions to the left side of the equation and then simplify, any rational equation can be transformed into its standard form

$$\frac{P}{Q}=0,$$

where P,Q are polynomials. A fraction is zero if and only if its numerator is zero and its denominator is nonzero. As a result, the above equation is equivalent to the system

$$\begin{cases} P=0 \\ Q\neq 0. \end{cases}$$

Thus, we only need to find solutions to $P=0$ and plug in Q to check if it makes $Q\neq 0$. A solution to $P=0$ such that $Q=0$ is called an extraneous solution. Extraneous solutions should be excluded from the solution set of an equation.

Besides the above method, a rational equation can also be solved by clearing denominators. If we multiply both sides of the equation by the LCD of all fractions, all denominators will be canceled and the equation is transformed into a polynomial equation. Extraneous solutions may appear. We need to make sure that no solutions make the LCD or any of the denominators zero. We demonstrate the methods by the following two examples.

Examples. Solve the equations.

(1) $\dfrac{2}{x+3} = \dfrac{7}{2x+1}$

(2) $\dfrac{3}{x-1} + \dfrac{2}{x} = \dfrac{x+2}{x^2-x}$

Solutions.

(1) Use the first method—move all terms to the left side and simplify.

$$\frac{2}{x+3} - \frac{7}{2x+1} = 0$$

$$\frac{2(2x+1)}{(x+3)(2x+1)} - \frac{7(x+3)}{(x+3)(2x+1)} = 0$$

$$\frac{(4x+2)-(7x+21)}{(x+3)(2x+1)} = 0$$

$$\frac{-3x-19}{(x+3)(2x+1)} = 0.$$

The equation is equivalent to $\begin{cases} -3x-19 = 0 \\ (x+3)(2x+1) \neq 0 \end{cases}$. Solving the first equation we get $x = -\dfrac{19}{3}$, which also satisfies the second inequality, so the solution is $x = -\dfrac{19}{3}$ (or the solution set is $\{-\dfrac{19}{3}\}$).

(2) Use the second method—clear all denominators by multiplying the LCD. The lease common denominator of all three fractions is $x(x-1)$. Multiply both sides by this LCD.

$$\frac{3}{x-1} + \frac{2}{x} = \frac{x+2}{x^2-x}$$

$$x(x-1)\left(\frac{3}{x-1} + \frac{2}{x}\right) = \frac{x+2}{x(x-1)}x(x-1)$$

$$3x + 2(x-1) = x+2$$

$$5x - 2 = x+2$$

$$4x = 4.$$

From the last equation we get a candidate solution $x = 1$; but $x = 1$ makes the LCD $x(x-1)$ equal 0, so it is an extraneous solution. The original equation has no solutions(or the solution set is $\emptyset$).

Essentially, the two methods are the same. One major difference between rational equations and polynomial equations is that rational equations may have extraneous solutions. To check for extraneous solutions, the best way is to plug all candidate solutions into the original equation. Here you may get a better understanding on the importance of rechecking solutions.

Exercises.

1. Solve the equations.

(1) $\dfrac{2}{x-4} = 3$

(2) $\dfrac{3x}{x+1} - 2 = 0$

(3) $\dfrac{5}{x-3} = \dfrac{2}{x}$

(4) $\dfrac{2x-1}{x+2} = \dfrac{2x+5}{x+2}$

(5) $\dfrac{x}{x-3} = \dfrac{x-1}{x-4}$

(6) $\dfrac{1}{x+5} - \dfrac{3}{5-x} = \dfrac{2}{x^2-25}$

(7) $\dfrac{2-x}{x-3} = 2 - \dfrac{2}{x-3}$

(8) $\dfrac{6}{x^2-16} + \dfrac{x-2}{4-x} = -1$

(9) $\dfrac{4}{x-1} + \dfrac{1}{x} = \dfrac{x+3}{x^2-x}$

(10) $\dfrac{|x|-2}{x^2+5} = 0$

2. Solve the equations.

(1) $\dfrac{5x-7}{x^2-3x+2} = \dfrac{2}{x-1} + \dfrac{3}{x-2}$

(2) $\dfrac{x-5}{x-2} - \dfrac{x+2}{x-1} = 0$

(3) $2 + \dfrac{2}{2-x} = \dfrac{1-x}{x-2}$

(4) $\dfrac{3x-4}{3x+4} - \dfrac{x-1}{x+1} = 0$

(5) $1 + \dfrac{4}{4-x} = \dfrac{3-x}{x-4}$

(6) $\dfrac{1-x}{1+x+x^2} - \dfrac{2}{x^3-1} = \dfrac{1}{1-x}$

(7) $\dfrac{7}{x^2-6x+8} + \dfrac{1}{x-4} = \dfrac{2}{2-x}$

(8) $\dfrac{2x}{x^2-2x-3} - \dfrac{x+3}{(x+1)(3-x)} = 0$

(9) $\dfrac{2x-3}{10} - \dfrac{3x-2}{2} = \dfrac{2x+7}{5}$

(10) $\dfrac{3}{x^3+1} + \dfrac{x}{1-x+x^2} = \dfrac{1}{x+1}$

3. Solve the equations.

(1) $\dfrac{2x+3}{2x^2-11x-21} = \dfrac{x+3}{x^2-12x+35}$

(2) $\dfrac{1}{x^2-5x+6} - \dfrac{3}{x^2+2x-8} = \dfrac{6}{x^2+x-12}$

(3) $\dfrac{20}{x^2-25} + 5 = \dfrac{2x-3}{x+5} + \dfrac{3x-5}{x-5}$

(4) $x^2 + 2x + 4 = \dfrac{x^3-3x+1}{x-2}$

(5) $1 - \dfrac{x^2-5}{x^2-8x+7} = \dfrac{3}{x-7}$

(6) $x^2 + x + 1 - \dfrac{5+2x+x^3}{x-1} = 0$

(7) $\dfrac{1 - \dfrac{x}{x+2}}{\dfrac{x}{x+2} + 1} = -\dfrac{1}{2}$

(8) $\dfrac{1}{1 + \dfrac{1}{1 + \dfrac{1}{x+1}}} = -2$

(9) $\dfrac{1}{x^2 - 3x + 2} + \dfrac{3}{x^2 - 5x + 6} = \dfrac{5}{x^2 - 4x + 3}$

(10) $\dfrac{3x + 6}{x^2 + 4x + 4} + \dfrac{x^2 - 4x + 4}{x^2 - 4} = \dfrac{x^2 - 4}{x^2 - 4x + 4}$

4. Tommy, Tina and Christine are working on a math project at the Penn State New Kensington campus. It takes Tommy 15 days to finish the project by himself, while it takes Tina 12 days to finish the project by herself. If Tommy, Tina and Christine work on the project together, it will take them 4 days to finish the project. How long does it take Christine to finish the project by herself?

5. The city of Apple is 50 km away from the city of Pear on Mars. Tommy rides his alien-bike to go from Apple to Pear. After 1 hour and 30 minutes, Tina rides her alien-motorcycle to go from Apple to Pear as well. Given that Tina's speed is 2.5 times faster than Tommy's, and Tina arrives the city of Pear 1 hour earlier than Tommy, find Tina's speed and Tommy's speed.

Coordinate Systems and Lines

Algebra and geometry are two old subjects in mathematics. Although they are quite different in contents, there is a point where they meet — the subject of coordinate geometry. Different from the ordinary geometry, the philosophy of coordinate geometry is to use algebra as tools to study geometry. How can this happen? The secret lies in the invention of coordinate systems. With coordinates, points are modeled by certain patterns of numbers, then we can apply the algebraic tools.

The real number line is a 1-dimension coordinate system, which we have met when we learn real numbers. To study the geometry in planes and spaces, 2-dimension(2-D) and 3-dimension(3-D) coordinate systems are also developed. We are only interested in 2-D geometry in our course. Points and lines are the two basic geometric objects in 2-D geometry. In this chapter we are going to model them as well as to study their positional relations.

Algebra can be used in geometry. Conversely, geometry helps to visualize the theory of algebra. Sometimes, it may even motivates the idea of a proof in algebra, but we enforce the principle that the results of algebra should be proved solely without geometry. Only in this way, algebra keeps as a subject independent and self-complete — once the theory in geometry changes, the validity of algebra won't be hurt.

7.1   2-Dimension Coordinate Systems

A 2-D coordinate system consists of two real number lines perpendicular to each other at their origins.

The two real number lines are called coordinate axes, in which the horizontal axis is referred to as x-axis and the vertical axis is the y-axis by default. The intersection of the two axes is called the origin, denoted by O, which represents zero for both real number lines. The plane where this coordinate system lives is called the coordinate plane or xy-plane. The x and y-axes partition the coordinate plane into four regions, which are called quadrants labeled by I,II,III,IV counterclockwise as in the picture. The points on the axes do not belong to any quadrants. In particular, the origin O does not belong to any quadrants. In the following, a coordinate system always means a 2-D coordinate system. Although the two axes use the same unit length in our picture, it is not a mandatory requirement.

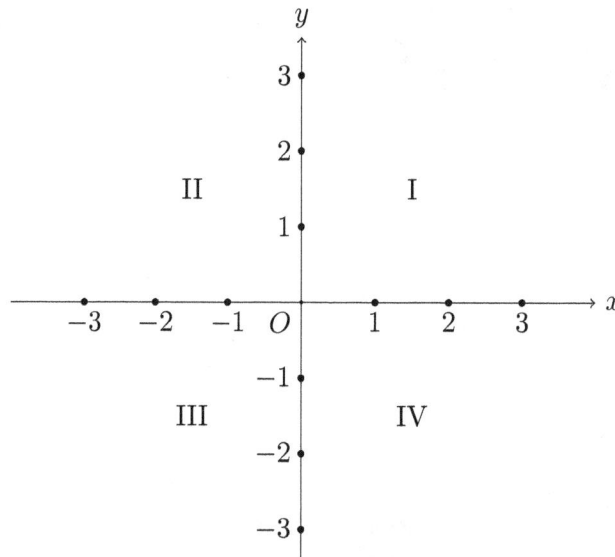

The Cartesian product of the set of real numbers $\mathbb{R}$ with itself, denoted by $\mathbb{R}^2$, is the set of all ordered pairs of real numbers, i.e.

$$\mathbb{R}^2 = \mathbb{R} \times \mathbb{R} = \{(a,b)|a,b \in \mathbb{R}\}.$$

There is a one-one correspondence between the points on a coordinate plane and the ordered pairs of real numbers, shown by the following.

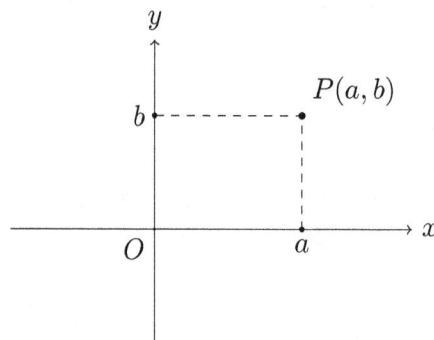

Given any point P in the xy-plane, we project it onto the x and y-axes separately. Suppose the projections are a and b, then P uniquely determines an ordered pair $(a,b) \in \mathbb{R}^2$, called the coordinate of P. Conversely, given any ordered pair (a,b) we can draw a vertical line through a on the x-axis and a horizontal line through b on the y-axis. Their intersection is the only point whose coordinate is (a,b). According to this, it is easy to locate a point by its coordinate.

In the following we identify the points in the coordinate plane with their coordinates and consider $\mathbb{R}^2$ as the set of all points in the coordinate plane.

Given a point $P(a,b)$ (a point P with coordinate (a,b)), a is called the x-coordinate (or x-component) and b is called the y-coordinate (or y-component) of P. The coordinate of a point has the following characteristics of the signs of its x and y-components determined by its location.

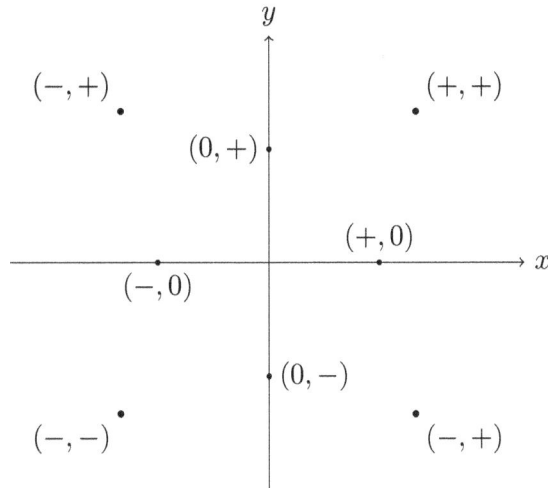

For example, $(-2, 0)$ is a point on the x-axis since its y-coordinate is zero. $(1, -1)$ is a point in Quadrant IV as the sign has the pattern $(+, -)$.

Midpoint Formula. Given any two points $P(x_1, y_1)$ and $Q(x_2, y_2)$, the midpoint R of the line segment PQ has the coordinate (x_0, y_0) given by

$$x_0 = \frac{x_1 + x_2}{2}, \qquad y_0 = \frac{y_1 + y_2}{2}.$$

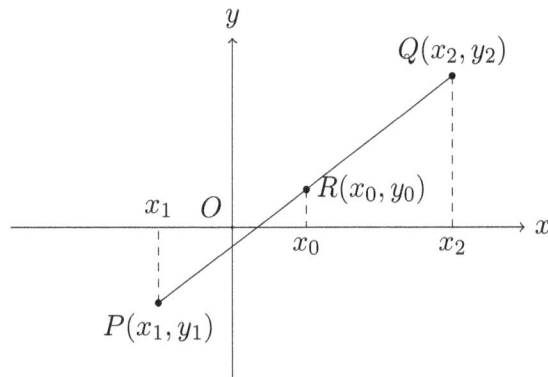

The reason is clear from the picture. If R is the midpoint of PQ, x_0 is the midpoint of $x_1 x_2$, so we must have $x_2 - x_0 = x_0 - x_1$. Solving x_0 from this equation, we get $x_0 = \frac{x_1 + x_2}{2}$. The formula for y-coordinate can be obtained similarly.

Example. B is the midpoint of the line segment AC. If B is $(2, 1)$, C is $(7, 2)$, find A and the quadrant it belongs to.

Solutions. Suppose the coordinate of A is (x, y). Since B is the midpoint of AC, by the midpoint formula

$$2 = \frac{x + 7}{2}, \qquad 1 = \frac{y + 2}{2}.$$

Solving these equations we get $x = -3, y = 0$, so A is $(-3, 0)$. It doest not belong to any quadrant. In fact, A is on the negative x-axis.

Symmetric Points. There are two kinds of symmetries: symmetry about a point and symmetry about a line. Two points P and Q are symmetric about a point R if R is the midpoint of the line segment PQ. P and Q are symmetric about a line l if PQ is perpendicular to l and the midpoint of PQ lies on l.

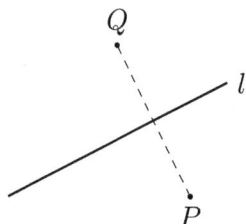

Given any point $P(x, y)$,

- the symmetric point of P about the x-axis is $(x, -y)$;
- the symmetric point of P about the y-axis is $(-x, y)$;
- the symmetric point of P about the origin O is $(-x, -y)$.

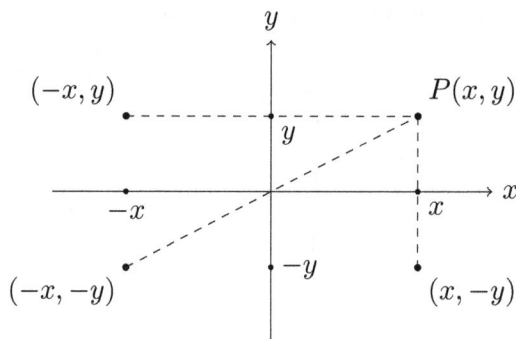

From the picture it is clear that the distance from $P(x, y)$ to the x-axis is $|y|$ and the distance to the y-axis is $|x|$.

Graph of Equations. Given an equation in two variables x and y, the solution set can be treated as a collection of points in the coordinate plane, which is called the graph of the equation.

Examples. Sketch the graphs.

(1) $x^2 + y^2 = 0$

(2) $x = 1$

(3) $y = x$

Solutions. All graphs are sketched at the end of the solutions.

(1) The solution set to $x^2 + y^2 = 0$ is $S_1 = \{(x, y)|x^2 + y^2 = 0\} = \{(0, 0)\}$, i.e. a single point, so the graph of $x^2 + y^2 = 0$ is the origin.

(2) The solution set to $x = 1$ is $S_2 = \{(x, y)|x = 1, y \in \mathbb{R}\} = \{(1, y)|y \in \mathbb{R}\}$, i.e. the collection of all points having the same x-coordinate, 1. The graph is a line through $(1, 0)$ and perpendicular to x-axis.

(3) The solution set to $y = x$ is $S_3 = \{(x,y) \in \mathbb{R}^2 | x = y\} = \{(x,x) | x \in \mathbb{R}\}$. We can sketch the graph by plotting some sample points on the graph, and then connecting them smoothly. The sample points we select are listed in the table below. It is easy to see that the graph should be the angle bisector of the first and third quadrants.

x	y
2	2
1	1
0	0
-1	-1
-2	-2

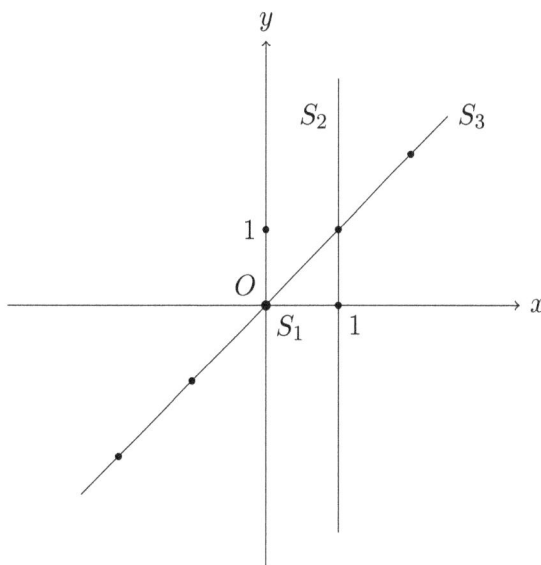

Exercises.

1. True or false.

(1) $(0,0)$ is in all four quadrants.

(2) $(0,0)$ is on both the x and y-axes.

(3) The x-coordinates of points on the x-axis are zero.

(4) The symmetric point of the origin about the x-axis is itself.

(5) The origin is the only point whose symmetric point about the y-axis coincides with itself.

(6) If a line is perpendicular to the y-axis, then it is the x-axis or it is parallel to the x-axis.

(7) $(0,-2)$ is on the y-axis.

(8) The fourth quadrant includes the positive x-axis.

(9) The distance from the origin to $(a,0)$ is a.

(10) The distance from $(c,0)$ to the x-axis is $|c|$.

(11) The symmetric point of a point on the y-axis about the origin is the same as its symmetric point about the x-axis and the symmetric point is again on the y-axis.

(12) The line through $(-2,-2)$ and $(2,-2)$ is perpendicular to x-axis.

2. Multiple choice.

(1) Which of the following is incorrect?

A. The plane where a coordinate system lives is a coordinate plane.

B. The origin in a coordinate system is the intersection of the two axes.

C. There is a one-one correspondence between all points in a coordinate plane and all ordered pairs of real numbers.

D. Any two perpendicular lines form a coordinate system.

(2) If the coordinate of P is $(-4, 6)$, then P is in

    A. Quadrant I   B. Quadrant II   C. Quadrant III   D. Quadrant IV

(3) Which point is in the second quadrant?

    A. $(\pi, -1)$ B. $(-\sqrt{2}, 0.01)$ C. $(-2, 0)$ D. $(-1, -2)$

(4) Which point is on the negative y-axis?

    A. $(0, \sqrt{3} - 2)$ B. $(0, \frac{\pi}{2})$ C. $(-2, 0)$ D. $(0, 0)$

(5) The coordinate of the symmetric point of $P(3, -5)$ about the x-axis is

    A. $(-3, -5)$ B. $(5, 3)$ C. $(-3, 5)$ D. $(3, 5)$

(6) If $P(m + 1, m - 4)$ is in the third quadrant, then

    A. $-1 \leq m \leq 4$ B. $m < -1$ C. $-1 < m < 4$ D. $m \leq -1$

(7) If $P(a, b)$ is in the fourth quadrant, then $Q(-a, b - 1)$ is in

    A. Quadrant I   B. Quadrant II   C. Quadrant III   D. Quadrant IV

(8) The distance from $A(-3, -4)$ to the y-axis is

    A. 3   B. 4   C. -3 D. -4

(9) The point $M(a^2, b^2)$ must be

    A. in Quadrant I   B. not in Quadrant II, III, IV

    C. not in Quadrant I, III, IV   D. not on the y-axis

(10) Given $P(1, 5)$ and $Q(-1, 5)$, the line segment PQ is

    A. parallel to the x-axis   B. parallel to the y-axis

    C. through the origin   D. non of the above

(11) Given $A(5, a)$ and $B(b, -2)$, if AB is parallel to the y-axis, then

    A. $a = -2$ B. $b = 5$ C. $a = 2$ D. $b = -5$

(12) If $ab < 0$, the point (a, b) is in

    A. the first or second quadrant   B. the second or fourth quadrant

    C. the second or third quadrant   D. the first or third quadrant

(13) If $ab \leq 0$, the point (a, b) cannot be

    A. on the positive x-axis   B. on the negative y-axis

    C. in the first or third quadrant   D. the origin

(14) If the mid-point of $(-2, 8)$ and $(2m - 4, 2n)$ is $(-4, 7)$, then

    A. $m = 1, n = 2$ B. $m = -2, n = 0$

    C. $m = -1, n = 3$ D. $m = -1, n = -3$

(15) If $|2x + 1| + |3y - 7| = 0$, then $P(x + 1, y - 3)$ is in

    A. the first quadrant   B. the second quadrant

    C. the third quadrant   D. the fourth quadrant

(16) Which point is on the graph of $2x - 3y^2 = 1$?

    A. $(1, 1)$ B. $(1, -\frac{1}{3})$ C. $(2, -1)$ D. $(3, 2)$

3. Sketch and label the following points in the given coordinate system.

$$(1, -3), \quad (0, 3), \quad (-2, 0), \quad (-3, -2), \quad (0, 0), \quad (2, \frac{1}{2}), \quad (-1, 2), \quad (-1, -2.5)$$

$$(-2, -\frac{3}{2}), \quad (0, -\frac{3}{2}), \quad (\frac{1}{3}, \frac{1}{2}), \quad (3, 0), \quad (-\frac{7}{3}, 1), \quad (-1, -1), \quad (2, -1), \quad (-\frac{5}{2}, \frac{8}{3})$$

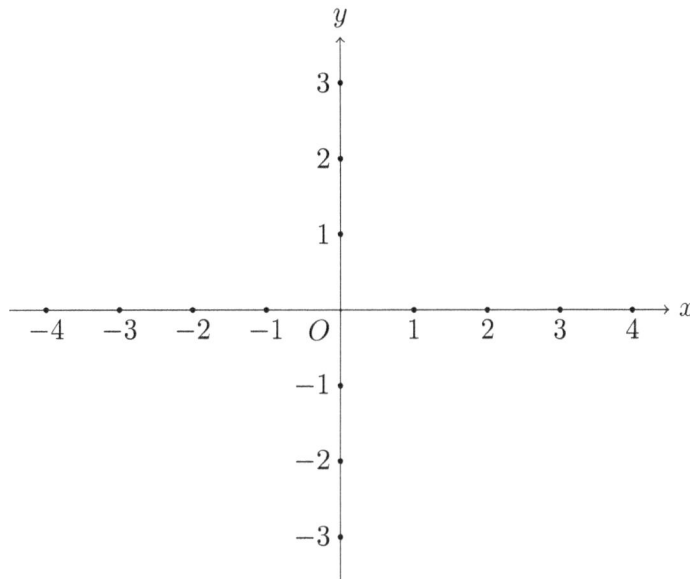

Find all points

(1) on the x-axis:

(2) on the y-axis:

(3) in Quadrant I:

(4) in Quadrant II:

(5) in Quadrant III:

(6) in Quadrant IV:

4. Assume Q is the midpoint of PR. Solve the following problems.

(1) Given $P(-3, 1), R(1, 5)$, find Q.

(2) Given $P(\frac{1}{2}, -\frac{1}{3}), Q(\frac{3}{2}, -\frac{4}{3})$, find R.

(3) Given $Q(-7, -2), R(-7, 3)$, find P.

(4) Given $P(1, -2), R(2, -5)$, find Q.

(5) Given $P(\frac{7}{3}, 0), Q(-2, 3)$, find R.

(6) Given $R(\frac{5}{3}, -4), Q(-\frac{17}{9}, -\frac{9}{7})$, find P.

5. Find the symmetric point about the x-axis, the y-axis and the origin.

(1) $(-1, 0)$

(2) $(-2, -\pi)$

(3) $(-5, -17)$

(4) $(0, 0)$

(5) $(0, -100)$

(6) $(23, -32)$

(7) $(5, 5)$

(8) $(-12, 12)$

6. Find distances to the x-axis and y-axis.

(1) $(-1.7, 17)$

(2) $(-\sqrt{2}, -\sqrt{3})$

(3) $(0, 0)$

(4) $(0, -0.23)$

(5) $(5, -4)$

(6) $(3 - \pi, 7 - 2\pi)$

7. Let A, B, C, D denote the set of points in the first, second, third and fourth quadrant and E, F, G, H denote the set of points on the positive x-axis, negative x-axis, positive y-axis and negative y-axis. Find the following sets in terms of A, B, C, D, E, F, G, H and $\{(0, 0)\}$.

(1) $\{(x, y) | x > 0\}$

(2) $\{(x, y) | y \leq 0\}$

(3) $\{(x, y) | xy \geq 0\}$

(4) $\{(x, y) | xy < 0\}$

(5) $\{(x, y) | x^2 y \geq 0\}$

8. Sketch the graphs of the following equations.

(1) $|x - 1| + (y - 2)^2 = 0$

(2) $y = -2$

(3) $x = 0$

(4) $x + y = 0$

(5) $y = 2x + 1$

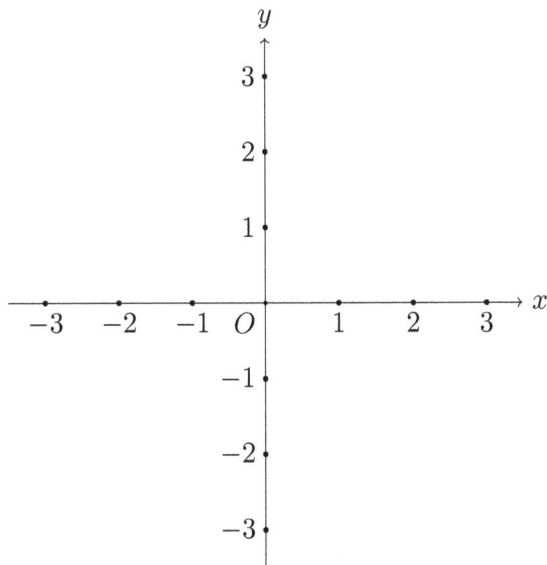

9. Consider the graph of $y - x^3 = 0$.

(1) If (x_0, y_0) is a point on the graph, how about $(-x_0, y_0)$, $(x_0, -y_0)$ and $(-x_0, -y_0)$?

(2) Prove that the graph of $y = x^3$ is symmetric about the origin but not about the x or y-axis.

10. If the points $A(2, 4)$, $B(7, -4)$, $C(2, 1)$ and $D(x_0, y_0)$ in sequence form a parallelogram, find the coordinate of D.

7.2   Slope and Equation of Lines

We have modeled points in a coordinate system as ordered pairs. In this section, we shall model lines, the second class of basic geometric objects.

Slope is used to describe the steepness of a line. Given a line l in the coordinate system, if the angle from the positive x-direction to l is θ, the slope of l is defined to be $m = \tan \theta$.

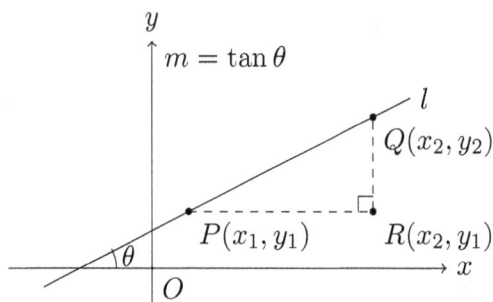

The slope can also be calculated by picking two different points $P(x_1, y_1)$ and $Q(x_2, y_2)$ on l via

$$m = \frac{y_2 - y_1}{x_2 - x_1}.$$

Here $\Delta x = x_2 - x_1$ is the increase in the x-coordinate from P to R, called the run; $\Delta y = y_2 - y_1$ is the increase of the y-coordinate from R to Q, called the rise. Here PQR is the right triangle such that $PR \parallel x$-axis and $QR \parallel y$-axis. As the slope m is determined by θ, it does not depend on the choice of points P and Q. (The formula $\tan \theta = \frac{y_2 - y_1}{x_2 - x_1}$ is from the definition of tangent in trigonometry.) The slope m describes not only the steepness but also the direction of a line l in the coordinate system. They are classified as follows:

- $l \parallel x$-axis if and only if $m = 0$;

- $l \parallel y$-axis if and only if m does not exist;

- l is in the $\nearrow$ direction, if and only if $m > 0$;

- l is in the $\searrow$ direction, if and only if $m < 0$.

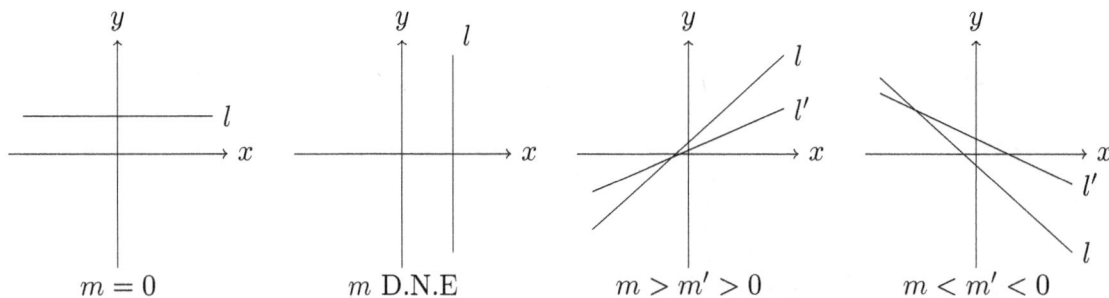

A line l in the coordinate system may intersect the x-axis and the y-axis. The x-intercept of l is the x-coordinate of the x-intersection; similarly, the y-intercept is the y-coordinate of the y-intersection.

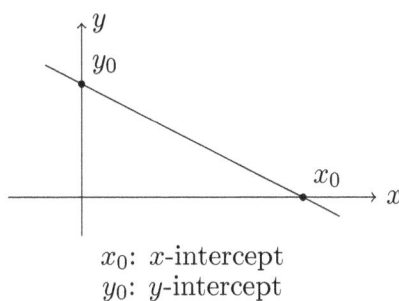

x_0: x-intercept

y_0: y-intercept

Then how can we describe a line in a coordinate plane? Suppose on the xy-plane, there is a line l whose slope is m, and $Q(x_0, y_0)$ is a fixed point on l. Now pick up an arbitrary point $P(x, y)$ on l.

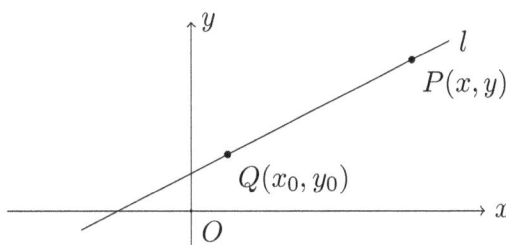

By the slope formula, we should have

$$m = \frac{y - y_0}{x - x_0}.$$

This is an equation in x and y, and we can simplify it to get

$$y - y_0 = m(x - x_0) \quad \text{or} \quad y = m(x - x_0) + y_0.$$

On one hand side, if the point $P(x, y)$ is on l, its coordinate must satisfy this equation. (Notice that the point Q, as a point on l, satisfies this equation if plugging x_0 for x and y_0 for y.) On the other hand side, if an ordered pair (x, y) satisfies the equation, the point (x, y) must be on the line l. Hence, the line l is completely determined by this equation. Moreover, the equation is linear in x and y. We conclude that lines in a coordinate system are modeled by linear equations in two variables.

The equation $y = m(x - x_0) + y_0$ is called in the slope-point form, as the line and its equation is determined by the slope m and a point (x_0, y_0) on it. A line can also be determined by other conditions. All of them are summarized in the following table.

Conditions	Equation of l	Form
slope m y-intercept b	$y = mx + b$	slope-intercept form
slope m a point (x_0, y_0) on l	$y = m(x - x_0) + y_0$	slope-point form
two points on l $(x_1, y_1), (x_2, y_2)$	$\dfrac{y - y_1}{y_2 - y_1} = \dfrac{x - x_1}{x_2 - x_1}$	two-point form
x-intercept $a \neq 0$ y-intercept $b \neq 0$	$\dfrac{x}{a} + \dfrac{y}{b} = 1$	intercept-intercept form

The idea to get the last three equations is again to use the slope formula. If the y-intercept is b, it is equivalent to have a fixed point $Q(0, b)$ on l. Suppose $P(x, y)$ is an arbitrary point on l other than Q, then

$$m = \frac{y - b}{x - 0}.$$

If given two different points $Q(x_1, y_1)$ and $R(x_2, y_2)$ on l, the slope can be calculated by either P, Q or $Q, R (P \neq Q)$, so

$$m = \frac{y - y_1}{x - x_1} = \frac{y_2 - y_1}{x_2 - x_1}.$$

If given nonzero x and y-intercepts a and b, this would be a special case of giving two different points $(a, 0)$ and $(0, b)$ on l, directly

$$m = \frac{y - b}{x - 0} = \frac{0 - b}{a - 0}.$$

Simplifying these equations we get the equations listed in the table.

If the slope of l is 0, i.e. $l \parallel x$-axis, the equation of l is of the form $y = 0 \cdot x + b = b$ where b is the y-intercept of l. This is clear as all points on l have the same y-coordinate, which must be b. This extreme case is included in the slope-intercept and slope-point form.

If the slope of l does not exist, i.e. $l \parallel y$-axis, similar to the above case, all points on l share the same x-coordinate. Supposing the x-intercept is c, the equation of l should be $x = c$. This extreme case is not included in any forms listed in the table.

To remedy this, we need to find the general forms of equations of lines. The equation in two-point form can be rewritten as $(y_2 - y_1)(x - x_1) - (x_2 - x_1)(y - y_1) = 0$ or $(y_2 - y_1)x - (x_2 - x_1)y - (x_1 y_2 - x_2 y_1) = 0$. By denoting the coefficient $(y_2 - y_1)$ of x by a, the coefficient $-(x_2 - x_1)$ of y by b, and the constant term $-(x_1 y_2 - x_2 y_1)$ by c, the equation becomes

$$ax + by + c = 0.$$

Such equations are called the general forms of equations of l. The general form does work in the above extreme cases by making $a = 0$ or $b = 0$. The general form $ax + by = c$ is not unique as we can multiply it by any nonzero number to get an equivalent equation, which determines the same line. Notice here a and b cannot both be 0.

To summarize, the equations of a line in a 2-D coordinate system are always linear equations in two variables, whose general forms are $ax + by + c = 0$ where a, b are not both zero. Suppose $ax + by + c = 0$ is an equation of line l.

- If $b \neq 0$, solve y to get $y = -\frac{a}{b}x + \frac{c}{b}$ (slope-intercept form), so the slope is $-\frac{a}{b}$ and the y-intercept is $\frac{c}{b}$.

- If $b \neq 0, a = 0$, $l \parallel x$-axis. The slope is 0 and the equation is of the form $y = $ constant.

- If $b = 0, a \neq 0$, $l \parallel y$-axis. The slope does not exist and the equation is of the form $x = $ constant.

Examples.

(1) Find an equation of the line passing through $(-3, 9)$ with slope -2;

(2) Find an equation of the line passing through $(-1, -3)$ and $(-1, 7)$;

(3) Given a line $l : 2x + 3x + 3 = 0$, find its slope, x-intercept and y-intercepts and sketch the graph.

Solutions.

(1) Given $m = -2$, $x_0 = -3$, $y_0 = 9$, by the slope-point form, immediately the equation is $y = (-2)[x - (-3)] + 9$, which simplifies to $y = -2x + 3$.

(2) The slope of the line is $\dfrac{-3 - 7}{-1 - (-1)}$, which does not exist, so the line is parallel to the y-axis. The points on the line have the same x-coordinate, -1, so the equation is $x = -1$.

(3) Solving y from the equation $2x + 3y - 3 = 0$, we get $y = -\dfrac{2}{3}x + 1$. Immediately the slope is $-\dfrac{2}{3}$ and the y-intercept is 1. The x-intercept is of the form $(x_0, 0)$, which satisfies the equation of l, i.e. $2x_0 + 3 \cdot 0 - 3 = 0$. Solving the equation we get the x-intercept $\dfrac{3}{2}$. To sketch the graph we only need to find two points on l and draw a line through them. To find a point on l, fix any value of x, say $x = 1$ for example, and solve y from the equation. We can also use the x and y-intercepts which are already found.

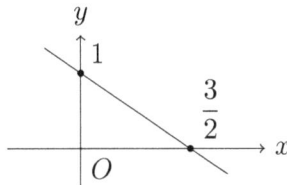

Notice that an equation of the form $ax + by + c = 0$ is a linear equation only when a, b are not both zero.

Exercises.

1. True or false.

(1) The equation of a line in any coordinate system is a linear equation in two variables.

(2) The graph of the equation $ax + by + c = 0$ is always a line.

(3) The line $l : ax + by = c$ passes through the origin if and only if $c = 0$.

(4) $(-4, -1)$ is on the line $-2x + 3y + 5 = 0$.

(5) The slope of the line $2y + 3x - 6 = 0$ is $-\dfrac{2}{3}$.

(6) $y = -\dfrac{1}{2}x + 7$ and $-2x + 4y - 1 = 0$ have the same slope.

(7) The y-intercept of the line $3x - 6y + 2 = 0$ is $-\dfrac{1}{3}$.

(8) The x-intercept of the line $y - 3 = -\dfrac{3}{4}(x - 1)$ is 5.

(9) The slope of a line perpendicular to the x-axis is 0.

(10) The slope of a line can be any real number.

(11) If the slope of a line does not exist, the line must be parallel to the x-axis.

(12) The line $3x + 2y - \dfrac{6}{5} = 0$ passes the first, second and fourth quadrants.

(13) If $x^{2m+1} - 3y^{n^2-n+1} - 7 = 0$ represents a line, then $m \cdot n = 0$.

(14) The graph of $ax + by = c$ is a line only when $ab \neq 0$.

(15) $x = 0$ is the equation of the x-axis; $y = 0$ is the equation of the y-axis.

(16) Any line that does not pass through the origin can be represented by $\dfrac{x}{a} + \dfrac{y}{b} = 1$.

(17) All lines passing through $(0, b)$ can be represented by $y = kx + b$.

(18) The line $l : ax + by = c$ consists of all points whose coordinates satisfy the equation of l.

2. Multiple Choice.

(1) The slope of the line $-3x + 15y - 7 = 0$ is

    A. 5   B. $-\dfrac{1}{5}$ C. $\dfrac{1}{5}$ D. -5

(2) If the line $y = \dfrac{mx}{2} - 1$ passes through the point $(-6, 5)$, then m is

    A. 2   B. -2 C. $-\dfrac{2}{3}$ D. $\dfrac{3}{2}$

(3) The equation of the line passing through $(-1, 3)$ and $(4, -7)$ is

    A. $y + 2x - 1 = 0$ B. $y - 2x + 1 = 0$ C. $2y + x - 1 = 0$ D. $2y - x - 1 = 0$

(4) If the point $(2m + 3, -4)$ is on the line of $-x + 2y - 3 = 0$, then m is

    A. -5 B. -7 C. -9 D. -11

(5) If the y-intercept of the line $2mx - (3m + 5)y + 28 = 0$ is 4, then m is

    A. $-\dfrac{2}{3}$ B. $\dfrac{2}{3}$ C. $-\dfrac{3}{2}$ D. $\dfrac{3}{2}$

(6) Which point is on the line $\dfrac{2y - 1}{-3} = \dfrac{7 - 3x}{-7}$?

    A. $(-1, \dfrac{3}{4})$ B. $(5, -2)$ C. $(\dfrac{7}{3}, \dfrac{1}{2})$ D. $(1, \dfrac{14}{9})$

(7) If the lines $px - 2y + 1 = 0$ and $2qx - y - 1 = 0$ share the same slope, then $5p \div 3q$ equals

    A. $-\dfrac{10}{3}$ B. $\dfrac{5}{3}$ C. $-\dfrac{15}{3}$ D. $\dfrac{20}{3}$

(8) If the points $(-1, 7)$, $(2, -4)$ and $(3m, 4)$ are on the same line, then m is

    A. $-\dfrac{2}{33}$ B. $\dfrac{2}{33}$ C. $\dfrac{33}{2}$ D. $\dfrac{11}{2}$

(9) The slope of the line $x = -1$ is

    A. 0   B. -1 C. 1   D. does not exist

(10) If the x-intercept of a line (not the x or y-axis) is 0, then the y-intercept is

    A. -1 B. 0

    C. 1   D. can't be determined

(11) The graph of $ax + by = c$ is a line if and only if

    A. $ab \neq 0$ B. $c \neq 0$ C. $a^2 + b^2 > 0$ D. $abc \neq 0$

(12) If $k < 0$, the line $y = kx + b$ cannot pass _____ at the same time.

    A. the first, second and third quadrants

    B. the first, second and fourth quadrants

    C. the second, third and fourth quadrants

    D. the second and fourth quadrants

(13) The points $(0, 2)$ and $(-2, 0)$ are symmetric about the line

    A. $y = 0$ B. $x = 0$ C. $y = x$ D. $y = -x$

(14) Given that $ab < 0$ and $bc > 0$, the line $ax - by + 2c = 0$ passes through

    A. Quadrant I, II, III   B. Quadrant II, III, IV

    C. Quadrant I, II, IV   D. Quadrant I, III, IV

(15) If the line $y - 2x + b = 0$ passes the first, third and fourth quadrants, then

    A. $b > 0$ B. $b \leq 0$ C. $b < 0$ D. $b \geq 0$

(16) The intersection of $x = -2$ and $y = 1$ is

    A. $(-2, 1)$ B. $(1, -2)$ C. $(2, -1)$ D. $(-1, 2)$

3. Find equations of the followings lines in the general form.

(1) l_1 passes through $(-3, 7)$ and $(5, 14)$.

(2) l_2 has the y-intercept -1.4 whose slope is the same as $-14x - 20y - 23 = 0$.

(3) l_3 has the x-intercept -3 and the y-intercept -5.

(4) l_4 passes the origin and shares the same slope with $3x - y + 1 = 0$.

(5) l_5 has the slope -7 and passes the midpoint of the line segment whose endpoints are $(3, -1)$ and $(4, -5)$.

(6) l_6 is the x-axis.

(7) l_7 is the y-axis.

(8) l_8 is perpendicular to the x-axis and the distance from the origin to l_8 is 3.

(9) l_9 is parallel to the x-axis whose y-intercept is $-\pi$.

(10) l_{10} is the angle bisector of the second and the fourth quadrants.

4. Find the slope and x, y-intercepts. Sketch the graph.

(1) $l_1 : x = -2$ (2) $l_2 : 2x + y = 0$

(3) $l_3 : 3x - 2y = -6$ (4) $l_4 : 4x + 5y + 7 = 0$

5. If both $(2m^2 - 1, 4)$ and $(3n, -6)$ are on the line $4x - y - 24 = 0$, find $\dfrac{m+n}{m \cdot n}$.

6. If the line $(2b + 1)y - 3x + 1 = 0$ intersects with the y-axis at $(0, -4)$, find b.

7. If the line l intersects with the x-axis at $(-7, 0)$ and shares the same slope with $3x + 5y - 7 = 0$, find the y-intersection of l.

8. If the line passing through $(0, 8)$ and $(3, -10)$ has the same slope with the line passing through $A(m + 1, -2)$ and point $B(2, 4m)$, find the symmetric point of $(m + 1, 4m)$ about the x-axis.

9. If the x-intercept of $3(1 - x) - 7(2 - 3y) - 4 = 0$ is the same as the y-intercept of $x - (2t + 1)y - 4 = 0$, find t.

10. If the line l passes through $(-2.5, 7)$ and its slope is -1.5, find the x-intercept of l.

11. If the points $(-1, 2)$, $(-3, 4)$ and $(a, 6)$ are on the same line, find a.

7.3 Positional Relations between Lines

We have described points and lines in a coordinate system by coordiantes and linear equations in the previous sections. Now we want to study the positional relation between them.

Two points are either distinguished or coincided, depending on if their coordinates are the same.

A point can be on a line or off a line, depending on if the coordinate of the point satisfies the equation of the line. For example, $(-3, 2)$ is on the line $2x + 4y - 2 = 0$, as $2(-3) + 4 \times 2 - 2 = 0$; $(0, 1)$ is off the line $y = 2x$, since $1 \neq 2 \times 0$.

Two lines in a plane can only be parallel or intersecting. Here parallel includes the special case of coincidence; intersecting includes the special case of perpendicular. The relation between two lines can be read from their equations directly. Suppose l_1 and l_2 are two lines with slopes m_1, m_2 (m_1, m_2 may not exist) in the coordinate plane. The following results should be clear from the graphs.

- $l_1 \parallel l_2 \iff m_1 = m_2$ or m_1, m_2 both do not exist;

$l_1 \parallel l_2$ $l_1 \parallel l_2$ $l_1 \parallel l_2$

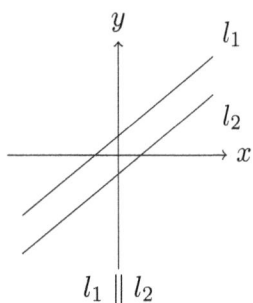

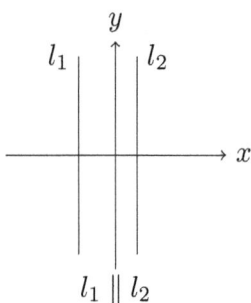

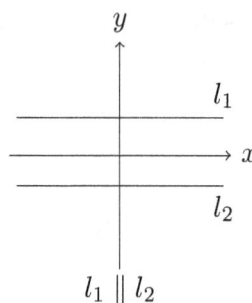

- l_1 coincides with $l_2 \iff l_1 \parallel l_2$ and they have the same x or y-intercepts;

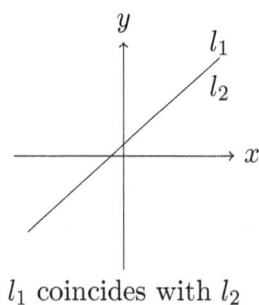

l_1 coincides with l_2

- l_1 intersects $l_2 \iff l_1 \nparallel l_2$, i.e. $m_1 \neq m_2$ when both exist, or only one of m_1, m_2 exists;

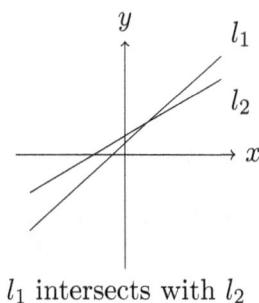

l_1 intersects with l_2

- $l_1 \perp l_2 \iff \begin{cases} m_1 m_2 = -1 \text{ if both } m_1, m_2 \text{ exist,} \\ m_1 = 0, m_2 \text{ does not exist,} \\ m_2 = 0, m_1 \text{ does not exist.} \end{cases}$

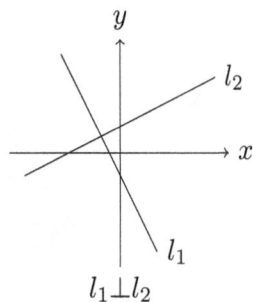

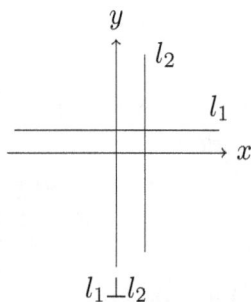

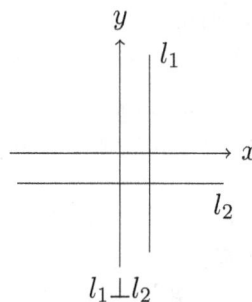

$l_1 \perp l_2$ $l_1 \perp l_2$ $l_1 \perp l_2$

It is easy to see $m_1 = 0$, m_2 does not exist is just the case $l_1 \parallel x$-axis and $l_2 \parallel y$-axis and they are perpendicular. Similarly, $m_2 = 0$, m_1 does not exist encodes the case $l_1 \parallel y$-axis and $l_2 \parallel x$-axis.

Examples. Classify the relation between lines.

(1) $l_1 : y = -1$, $l_2 : y = 3$

(2) $l_1 : y = -2x + 3$, $l_2 : 2x - 4y + 7 = 0$

(3) $l_1 : y = -4x + 5$, $l_2 : 12x + 3y - 15 = 0$

Solutions.

(1) Both l_1 and l_2 have slope 0, so they are parallel. As the y-intercepts are different, they do not coincide.

(2) Solving y from the equation of l_2, we get $y = \dfrac{1}{2}x + \dfrac{7}{4}$. The slope m_1 of l_1 is -2 and the slope m_2 of l_2 is $-\dfrac{1}{2}$. As $m_1 \neq m_2$, l_1 intersects with l_2. Moreover, $m_1 m_2 = -1$, so $l_1 \perp l_2$.

(3) The equation of l_2 can be rewritten, by solving y, as $y = -4x + 5$, which is the same as the equation of l_1. The two lines l_1 and l_2 coincide. In fact, we can get the equation of l_2 by multiplying the both sides of the equation of l_1 by 3. As a conclusion, two lines coincide if and only if their equations are equivalent.

Exercises.

1. True or false.

(1) If the product of the slopes of two lines is -1, the lines are perpendicular.

(2) If the product of the slopes of two lines is 1, the lines are intersecting.

(3) If the product of the slopes of two lines is 0, the lines cannot be parallel.

(4) If two lines intersect, there is only one intersection.

(5) If a line has a common point with the x-axis, its slope is not zero.

(6) If the slope of a line l_1 is 0 and the slope of a line l_2 does not exist, then $l_1 \perp l_2$.

(7) If two lines do not intersect, they are parallel or coincided.

(8) Two lines are parallel only when they have the same slope.

(9) $y = x$ and $y = -x$ are perpendicular at the origin.

(10) If the slope of a line l_1 does not exist but the slope of a line l_2 exists, then l_1 and l_2 must intersect with each other.

(11) If the slopes of two lines l_1 and l_2 both do not exist, then they are parallel to each other.

(12) If the slopes of two lines l_1 and l_2 both exist, then they must intersect.

(13) Two lines coincide if and only if their equations are equivalent.

(14) The line $x + 2y - 1 = 0$ is perpendicular to the line $2x - y + 1 = 0$.

(15) The line $y - 3 = \dfrac{1}{2}(x + 1)$ is parallel to the line $\dfrac{y + 1}{-2} = \dfrac{x - 1}{4}$.

(16) Two lines both perpendicular to another line are parallel.

(17) $x + y = 1$ and $-x - y - 1 = 0$ coincide.

(18) $ax + by + c = 0$ and $bx - ay + d = 0$ are always perpendicular as long as a, b are not both zero.

2. Multiple Choice.

(1) If $(m + 2)x - 2y + n = 0$ and $y = mx + 4$ represent the same line, then

    A. $m = 2, n = -8$ B. $m = -2, n = 8$

    C. $m = 2, n = 8$ D. $m = -2, n = -8$

(2) The line $2x - y + 1 = 0$ and the line $4(x - 1) - 2(3 - y) = -3$ are

    A. parallel but not coincided   B. coincided

    C. intersecting but not perpendicular   D. perpendicular

(3) If the line $y = 4x - 4$ is perpendicular to the line $y = (a - 2)x + 3$, then a is

    A. $\dfrac{7}{4}$ B. $\dfrac{5}{4}$ C. $-\dfrac{7}{4}$ D. $-\dfrac{5}{4}$

(4) The equation of the line passing $(-1, 2)$ and perpendicular to $x - 3y + 1 = 0$ is

    A. $x + 3y - 5 = 0$ B. $x - 3y + 7 = 0$ C. $3x + y + 1 = 0$ D. $3x - y + 5 = 0$

(5) If the line passing $(-1, m)$ and $(m, 3)$ is parallel to $2x + y - 3 = 0$, then m is

    A. 0   B. -5 C. 10   D. -1

(6) The relation between $3x + y + p = 0$ and $x - 3y + q = 0$ is

    A. parallel but not coincided   B. coincided

    C. intersecting but not perpendicular   D. perpendicular

(7) Among the lines $x + 2y - 1 = 0$, $\dfrac{y + 2}{0 - 4} = \dfrac{x - 3}{-5 + 3}$, $\dfrac{y - 1}{-6 + 4} = \dfrac{2x - 1}{-3 + 1}$ and $y = \dfrac{1}{2}x - 7$, how many of them are parallel?

    A. 0   B. 2   C. 3   D. 4

(8) If the lines $-4x - m^2y - 32 = 0$ and $x + 4y - 2m = 0$ are parallel but not coincided, then m is

    A. ± 4 B. 4   C. -4 D. 0

(9) The equation of the line passing through $(-4, 1)$ and parallel to $3x - 6y + 1 = 0$ is

    A. $x - 2y + 6 = 0$ B. $x + 2y - 6 = 0$ C. $2x + y - 6 = 0$ D. $2x - y - 6 = 0$

(10) If the product of the slopes of two lines is negative, then the lines must be

    A. coincided   B. parallel   C. intersecting   D. perpendicular

3. Tell the relation between lines.

(1) $l_1 : 3x + 4y = 7$, $l_2 : 16x - 12y + 15 = 0$

(2) $l_1 : -2x + 5y - 14 = 0$, $l_2 : 6x - 15y = -45$

(3) $l_1 : 9x - 16 = 0$, $l_2 : 2x + 3y = 7$

(4) $l_1 : x - 2y + 9 = 0$, $l_2 : x + 2y - 16 = 0$

(5) $l_1 : 7x - 4y - 23 = 0$, $l_2 : -16x - 28y + 69 = 0$

(6) $l_1 : y = \sqrt{2}$, $l_2 : y = -\sqrt{3}$

(7) $l_1 : 2x + 1 = 0$, $l_2 : 4x = -2$

(8) $l_1 : 6x = 53$, $l_2 : 7y + 37 = 0$

(9) $l_1 : x + y = 2$, $l_2 : x - y = 3$

4. Find an equation in the general form of the line passing through $(4, 3)$ and parallel to $3x + y - 1 = 0$.

5. If the line $2x + (2m + 1)y + 6 = 0$ is parallel to the line $x - (m - 2)y - 1 = 0$, find m.

6. Find an equation in the general form of the line passing through $(5, -2)$ and perpendicular to $x + 2y - 4 = 0$.

7. If the line that passes through $(-7, 4)$ and $(5, 8)$ is also parallel to $(1 + 3m)x + 2y - 1 = 0$, find m.

8. If the line $px + 3y - 4 = 0$ is perpendicular to the line $qx - y + 1 = 0$, $p^2 + q^2 = 10$, find $(p + q)^2$.

9. If the line passing through $(-2, t)$ and $(t, 4)$ is parallel to the line $\frac{7}{3}(6x-1)+2(3y-5) = -9$, find t.

10. If the line passing through $(3, 6)$ and $(-2, 7)$ is perpendicular to $(m-1)x+2y-7 = 0$, find $(m + 1)^2$.

11. Find an equation in the general form of the line perpendicular to $3x - 7y + 12 = 0$ at $(-11, -3)$.

12. If the line $(k^2 - k + 6)x + (k - 3)y - 1 = 0$ is parallel to the y-axis, find k.

13. If neither a_1, b_1 nor a_2, b_2 are both zero, find the conditions for the lines $a_1x+b_1y = c_1$ and $a_2x + b_2y = c_2$ to be (1)parallel, (2)coincided, (3)intersecting.

14. If two lines $a_1x + b_1y = c_1$ and $a_2x + b_2y = c_2$ intersect at (x_0, y_0), what condition should (x_0, y_0) satisfy?

Systems of Linear Equations and Inequalities

When considering two or more equations at the same time, it is convenient to make them into systems. The simplest systems are made of linear equations. The complete theory of systems of linear equations is developed in the course of linear algebra, using the language of matrices. In our course, we only study the systems of linear equations in two or three variables. The standard methods to solve such equations are substitution and elimination. In fact, these methods also work well for linear systems with more variable.

There are also (systems of) linear inequalities in two variables. Armed with the knowledge of coordinate geometry, now we can visualize the solution sets of these (systems of) inequalities.

8.1 Systems of Linear Equations in Two Variables

The standard form of a system of linear equations in two variables is

$$\begin{cases} a_1x + b_1y = c_1 \\ a_2x + b_2y = c_2, \end{cases}$$

where x, y are unknowns and $a_1, b_1, c_1, a_2, b_2, c_2$ are constants. A solution to the equation is an ordered pair $(x_0, y_0) \in \mathbb{R}^2$, which makes both equations hold, i.e. (x_0, y_0) satisfies $a_1x_0 + b_1y_0 = c_1$ and $a_2x_0 + b_2y_0 = c_2$ at the same time. The solution set is the set of all solutions to the equation.

There are basically two methods to solve a system of linear equations: substitution and elimination.

The idea of the substitution method is to solve one variable (say y) by the other (x) using one equation (say the first equation), then plugging the variable solved (y) into the other equation (the second equation). In this way, we get an equation in one variable (x) and can easily solve it.

The idea of the elimination method is to make the two equations have the same coefficient for one variable (say y) using the basic propositions of equations and subtracting one equation from the other. In this way we can also eliminate one variable (y) and get a linear equation in one variable (x).

Both of the methods are demonstrated in the following example.

Example. Solve the system $\begin{cases} 2x + 3y = 13 & ① \\ 5x - 2y = 4 & ② \end{cases}$

Solutions.

(1) (By substitution) Solving y from equation ②, we get

$$y = \frac{5}{2}x - 2 \quad ③$$

Plug ③ into equation ①,

$$2x + 3(\frac{5}{2}x - 2) = 13 \quad ④$$

Solving ④ we get $x = 2$. Plugging $x = 2$ back into ③ we have $y = 3$, so the solution to the equation is $(2, 3)$.

(2) (By elimination) We eliminate x, for example. The coefficients of x in equation ① and ② are 2 and 5 separately. To make them equal, multiply ① by 5 and ② by 2.

$$① \times 5: \quad 5(2x + 3y) = 10x + 15y = 13 \times 5 = 65 \quad ⑤$$
$$② \times 2: \quad 2(5x - 2y) = 10x - 4y = 4 \times 2 = 8 \quad ⑥$$

Then subtract ⑥ from ⑤, i.e. ⑤−⑥, to get

$$(10x + 15y) - (10x - 4y) = 65 - 8 \quad ⑦,$$

which simplifies to $19y = 57$. First we get $y = 3$. Plugging $y = 3$ back into either ① or ②, we can solve for x to get $x = 2$. The solution is $(2, 3)$. Plugging $x = 2, y = 3$ back into the equation we can check that $2 \times 2 + 3 \times 3 = 13$ and $5 \times 2 - 2 \times 3 = 4$ are both identities.

Plugging the solutions back into the equation,check whether the equation hold, helps to discover computation mistakes (if any).

If the equation is not in the standard form, we need to simplify it first. For example, the equation $\begin{cases} 2x - 3y = 4y + 1 \\ 3 - 2x = y + 7 \end{cases}$ is not in the standard form. We can simplify the first equation into $2x - 7y = 1$, and the second equation into $2x + y = -4$, then use the method of substitution or elimination to solve the equivalent system $\begin{cases} 2x - 7y = 1 \\ 2x + y = -4 \end{cases}$.

A linear equation of the form $ax + by = c$, where a, b are both not zero, represents a line in the coordinate plane. (x_0, y_0) is a solution to the system $\begin{cases} a_1x + b_1y = c_1 \\ a_2x + b_2y = c_2 \end{cases}$ if and only if (x_0, y_0) is a common point of the lines $l_1 : a_1x + b_1y = c_1$ and $l_2 : a_2x + b_2y = c_2$. Depending on the relative position of the two lines, we have

- if l_1 coincides with l_2, there are infinitely many solutions to the system;

- if $l_1 \parallel l_2$ but not coincided, there is no solution to the system;

- if l_1 intersects l_2, there is one unique solution to the system.

(Unfortunately, here we break the principle that the results of algebra should not be presented or proved using the language of geometry, but it is not difficult to find the correct algebraic conditions accordingly. Although we omit the proof here, the result below can definitely be proved directly without involving geometry.) If we make the convention that $\dfrac{b}{0}$ equals any number when $b = 0$ and does not equal any number when $b \neq 0$, the classification of the number of solutions to a linear system above can be translated as the following:

- the system has infinitely many solutions $\Longleftrightarrow \dfrac{a_1}{a_2} = \dfrac{b_1}{b_2} = \dfrac{c_1}{c_2}$ (l_1 coincides with l_2);

- the system has no solutions $\Longleftrightarrow \dfrac{a_1}{a_2} = \dfrac{b_1}{b_2} \neq \dfrac{c_1}{c_2}$ ($l_1 \parallel l_2$ but not coincided);

- the system has one unique solution $\Longleftrightarrow \dfrac{a_1}{a_2} \neq \dfrac{b_1}{b_2}$ (l_1 intersects with l_2).

For example, $\begin{cases} 4x - 2y = 7 \\ -12x + 6y = -21 \end{cases}$ has infinitely many solutions as $\dfrac{4}{-12} = \dfrac{-2}{6} = \dfrac{7}{-21}$. The system $\begin{cases} x + y = 1 \\ x + y = 0 \end{cases}$ has no solution as $\dfrac{1}{1} = \dfrac{1}{1} \neq \dfrac{1}{0}$. This is obvious since $x + y$ cannot equal both 0 and 1.

Exercise.

1. True or false.

(1) $\begin{cases} 5x - y = 1 \\ x + 5y = 5 \end{cases}$ has no solutions.

(2) $\begin{cases} 3x - y = -3 \\ x - \dfrac{1}{3}y = -1 \end{cases}$ has infinitely many solutions.

(3) $\begin{cases} 2x + y = 1 \\ \dfrac{1}{2}y + x = -1 \end{cases}$ has a unique solution.

(4) $\begin{cases} 3x - 2y = -6 \\ y - \dfrac{3}{2}x = 5 \end{cases}$ has no solutions.

(5) If the solution to $\begin{cases} 2px + y = q \\ 6x - 3y = 5p + 1 \end{cases}$ is $(1, 0)$, then $p = 1$ and $q = 2$.

(6) If $|x + 1| = 4$ and $x + y = 3$, then $\dfrac{y}{x} = 0$.

(7) If $\begin{cases} ax + ay = 2a - 1 - 2x \\ 2x + 4y = 9 \end{cases}$ has no solution, then $a = -4$.

(8) $\begin{cases} mx + 2y = -4 \\ x - 3y = 6 \end{cases}$ has infinitely many solutions if and only if $m = -\dfrac{2}{3}$.

(9) If $a \neq -\dfrac{1}{15}$, then $\begin{cases} 3a^2x + ay = -1 \\ 5y - x = 3 \end{cases}$ has only one solution.

(10) $3x - 5y = 4$ has infinitely many solutions.

(11) If $\begin{cases} 7x - 2y = 3 \\ x - ky = k \end{cases}$ has a unique solution, then $k \neq \dfrac{2}{7}$.

(12) If a system of linear equations in two variables has more than one solution, it must have infinitely many solutions.

(13) When a system of linear equations in two variables has a unique solution, we can use the method of substitution to solve it.

(14) To solve $\begin{cases} 5x - 3y = 1 \\ 2x + y = 3 \end{cases}$, we can use the method of elimination by multiplying the second equation by -3 and then adding these two equations to eliminate y.

(15) The method of elimination cannot solve a system of linear equations with rational coefficients.

2. Multiple Choice.

(1) If $\begin{cases} x = -1 \\ y = \dfrac{1}{2} \end{cases}$ is a solution to $\begin{cases} mx - 2y = 3 \\ 4x + ny = 1 \end{cases}$, then $m + n$ is

    A. -4 B. 6 C. 10 D. -6

(2) $\begin{cases} 4y - 3x = 1 \\ \dfrac{x-3}{2} - \dfrac{2y+1}{3} = -2 \end{cases}$ has

    A. only one solution   B. infinitely many solutions

    C. no solutions   D. cannot be determined

(3) If $\begin{cases} kx - y = \dfrac{1}{5} \\ 5y = 2 - 3x \end{cases}$ has no solution, then k is

    A. $\dfrac{3}{5}$ B. $-\dfrac{3}{5}$ C. $\dfrac{5}{3}$ D. $-\dfrac{5}{3}$

(4) Solving y from $\dfrac{2x-1}{2} - \dfrac{3-y}{4} = 1 - y$ in terms of x, we get

    A. $y = \dfrac{9-4x}{5}$ B. $y = \dfrac{4-9x}{5}$ C. $y = \dfrac{9-4x}{3}$ D. $y = \dfrac{4-9x}{3}$

(5) Which equation, when combined with $7x - 2y = -1$, makes a system having infinitely many solutions?

    A. $6y - 21x = 3$ B. $14x - 4y = 2$ C. $2y - 7x = -1$ D. $35x - 10y = 5$

(6) If $\begin{cases} kx + 2y = \dfrac{7}{3} \\ x - \dfrac{1}{2}y = \dfrac{5}{3} \end{cases}$ has no solution, then k is

A. -2 B. 6 C. -4 D. 4

(7) If $\dfrac{1}{6}x^3 y^p$ and $4x^{-p+2q}y$ are similar terms, then $p^2 - q^2$ is

   A. -3 B. -2 C. 1 D. -4

(8) How many solutions to $3x + y = 13$ are ordered pair of positive integers?

   A. 1 B. 2 C. 3 D. 4

(9) If $\begin{cases} x = 3 \\ y = \dfrac{1}{2} \end{cases}$ is a solution to $\begin{cases} ax - 4y = 1 \\ 2x + by = -2 \end{cases}$, then $2(8a + b) + a^3$ is

   A. -17 B. -16 C. -15 D. -13

(10) Among the following systems of equations, which one has a unique solution?

A. $\begin{cases} x - y = 2 \\ 3x - 3y = 0 \end{cases}$ B. $\begin{cases} x + y = 0 \\ 2x + 2y = -1 \end{cases}$

C. $\begin{cases} 2x + 2y = 1 \\ x - y = 0 \end{cases}$ D. $\begin{cases} x - y = 3 \\ \dfrac{1}{3}x - \dfrac{1}{3}y = -2 \end{cases}$

(11) If $(a^2 - 1)x^2 + (a+1)x + (a-2)y + 3a = 0$ is a linear equation in two variables, then a is

   A. ± 1 B. 1 C. -1 D. none of above

(12) $\begin{cases} \dfrac{x-y}{3} - \dfrac{x+y}{2} = 1 \\ 3(x-y) - 7(x-y) = -8 \end{cases}$ can be simplified to

A. $\begin{cases} x + 5y = -6 \\ x - y = 2 \end{cases}$ B. $\begin{cases} x - 5y = 6 \\ x + y = 2 \end{cases}$

C. $\begin{cases} -x + 5y = 6 \\ -x + y = 2 \end{cases}$ D. $\begin{cases} x - 5y = 2 \\ x - y = 0 \end{cases}$

(13) The solution to $\begin{cases} 3x - 4y = -55 \\ 8y - 6x = -56 \end{cases}$ is

A. $\begin{cases} x = \dfrac{4}{3} \\ y = \dfrac{49}{4} \end{cases}$ B. $\begin{cases} x = \dfrac{1}{3} \\ y = 14 \end{cases}$

   C. no solution   D. none of the above

(14) If $|(3m - n - 1)x| + |(2m - n + 2)y| = 0$ but $xy \neq 0$, then $2(|m| - |n|)$ is

   A. -10 B. -8 C. -6 D. -4

(15) If the solution (x_0, y_0) to $\begin{cases} 3x + 2y = 5 \\ kx + (k-2)y = 2 \end{cases}$ satisfies $x_0 = y_0$, then k equals

   A. -6 B. 4 C. 2 D. -3

3. Solve the following equations by substitution.

(1) $\begin{cases} x + 3y = 2 \\ 3x - 4y = -1 \end{cases}$
    (2) $\begin{cases} 5x - y = 3 \\ 2x + 3y = -2 \end{cases}$

(3) $\begin{cases} 4x + 3y = 5 \\ 3x - 2y = -5 \end{cases}$
    (4) $\begin{cases} 2a = 7b - 4 \\ 3a - b = 5 \end{cases}$

(5) $\begin{cases} 6x - y = -2 \\ 3y - 18x = 1 \end{cases}$
    (6) $\begin{cases} 4x - 3y = -5 \\ x = 2y + 3 \end{cases}$

4. Solve the following equations by elimination.

(1) $\begin{cases} 4x - 7y = -1 \\ 5x + 2y = 3 \end{cases}$
    (2) $2x - 3y = 4x - 5y = 6$

(3) $\begin{cases} \dfrac{x-2}{2} = \dfrac{3y+1}{4} \\ 2x - 5y = 3 \end{cases}$
    (4) $\begin{cases} 5a + 3b = 6 \\ 2a - 5b = -2 \end{cases}$

(5) $\begin{cases} \dfrac{x-1}{3} - \dfrac{2y-3}{4} = 0 \\ 4x - 3y = 7 \end{cases}$
    (6) $\begin{cases} 7x + 4y = 2 \\ 9x + 5y = -4 \end{cases}$

5. Solve the following equations.

(1) $\begin{cases} \dfrac{x}{2} - \dfrac{y}{3} = \dfrac{1}{6} \\ x - 2y = -2 \end{cases}$

(2) $\begin{cases} x = \dfrac{y-1}{3} - y \\ 2x = \dfrac{y-1}{2} + 4 \end{cases}$

(3) $\begin{cases} \dfrac{x}{2} - \dfrac{y}{3} = 3 \\ \dfrac{x}{4} - \dfrac{y}{5} = -2 \end{cases}$

(4) $\begin{cases} \dfrac{a+b}{3} - \dfrac{a-b}{4} = -1 \\ \dfrac{a+b}{2} - \dfrac{a-b}{3} = 1 \end{cases}$

(5) $\begin{cases} \dfrac{m+n}{2} - \dfrac{2n}{5} = 1 \\ m = 2n - 3 \end{cases}$

(6) $\begin{cases} \dfrac{3x-y}{3} = 1 - 2x \\ 4x - y = -2 \end{cases}$

(7) $\begin{cases} 2(x-5) = 3(1-y) \\ \dfrac{1-2x}{2} - \dfrac{1}{4} = \dfrac{2y}{3} \end{cases}$

(8) $\begin{cases} \dfrac{x}{10} - y = -5 \\ 2 - \dfrac{2x-1}{3} = 1 - \dfrac{2y-1}{2} \end{cases}$

(9) $\begin{cases} \dfrac{2m+3n}{2} + \dfrac{3m+2n}{5} = 1.3 \\ \dfrac{3(2m+3n)}{2} = \dfrac{6(3m+2n)}{5} - 3.3 \end{cases}$

(10) $x + 2y = \dfrac{y-x}{4} = \dfrac{2x+1}{3}$

6. Solve the equations.

$$(1) \begin{cases} \dfrac{1}{s} - \dfrac{8}{t} = 8 \\ \dfrac{5}{s} + \dfrac{4}{t} = 51 \end{cases} \qquad\qquad (2) \begin{cases} \dfrac{2}{x} + \dfrac{4}{y} = 3 \\ \dfrac{4}{x} - \dfrac{8}{y} = 1 \end{cases}$$

$$(3) \begin{cases} \dfrac{5}{2x+3} + \dfrac{2}{y-3} = -4 \\ \dfrac{2}{2x+3} - \dfrac{6}{y-3} = -5 \end{cases} \qquad\qquad (4) \begin{cases} \dfrac{10}{x+y} + \dfrac{3}{x-y} = -5 \\ \dfrac{15}{x+y} - \dfrac{3}{x-y} = 0 \end{cases}$$

7. Tina works at a toy company. It takes Tina 2 hours and 10 minutes to finish making 7 potato-heads and 4 dinosaurs. It takes Tina 2 hours and 15 minutes to finish making 6 potato-heads and 5 dinosaurs. How long does it take Tina to finish one potato-head and one dinosaur?

8. Tommy went shopping for his mom, Ms. Lu. He found that bananas are 52 cents per pound and lady apples are 2 dollars and 98 cents per pound. Tommy spent 16 dollars and 46 cents on 8 pounds of fruit. How many pounds of bananas and how many pounds of apples did Tommy buy?

9. Christina loves animals. She put rabbits and chickens together in a cage. She counted 40 heads and 100 feet of the rabbits and chickens in total. How many rabbits and how many chickens does Christina have?

10. Tell if the lines intersect. If they intersect, find the intersection.

(1) $l_1 : 5x - 7y = 0$, $l_2 : 8x + 11y = 0$

(2) $l_1 : 2x - 3y = 19$, $l_2 : 3x - 2y = 21$

(3) $l_1 : 15x + 24y - 17 = 0$, $l_2 : 16y = 3 - 10x$

8.2   Systems of Linear Equations in Three Variables

The standard form of a system of linear equations in three variables is

$$\begin{cases} a_1x + b_1y + c_1z = d_1 \\ a_2x + b_2y + c_2z = d_2 \\ a_3x + b_3y + c_3z = d_3. \end{cases}$$

Here x, y, z are unknowns and all others are constants. If the system is not in the standard form, we can simplify or rewrite it in the standard form.

Denote by $\mathbb{R}^3$ the Cartesian product $\mathbb{R} \times \mathbb{R} \times \mathbb{R} = \{(x, y, z) | x, y, z \in \mathbb{R}\}$, i.e. the set of all ordered triples (x, y, z) of real numbers. $(x_0, y_0, z_0) \in \mathbb{R}^3$ is a solution to the system if (x_0, y_0, z_0) makes all three equations hold at the same time.

A system of linear equations in three variables can also be solved by substitution or elimination. For substitution, we can solve one variable (say z) from one equation (say the first equation) of the system, and plug it into the other two equations (the second and the third). Then we get a system of linear equations in two variables (x and y), which can be solved by substitution as we showed in the previous section. The unknown z is easy to get as long as we know x and y. For elimination, we demonstrate the method by the following example.

Example. Solve the equation $\begin{cases} 2x + 6y + 3z = 6 & ① \\ 3x + 15y + 7z = 6 & ② \\ 4x - 9y + 4z = 9 & ③ \end{cases}$.

Solutions. The coefficient of x is simpler, so we choose x to eliminate. Performing ①×2−③, we get

$$21y + 2z = 3 \quad ④.$$

Performing ②×2−①×3, we get

$$12y + 5z = -6 \quad ⑤.$$

④ and ⑤ form a new system. ④×5−⑤×2 to eliminate z, we get

$$21y \times 5 - 12y \times 2 = 3 \times 5 - (-6) \times 2,$$

which simplifies to $81y = 27$ and hence $y = \dfrac{1}{3}$ ⑥. Plugging ⑥ into ⑤ we get $12 \times \dfrac{1}{3} + 5z = -6$ and $z = -2$ ⑦. Plugging ⑥ and ⑦ into ①, we get $2x + 6 \times \dfrac{1}{3} + 3 \times (-2) = 6$, so $x = 5$ ⑧. After plugging ⑥, ⑦ and ⑧ back into the original equation, we conclude that $(5, \dfrac{1}{3}, -2)$ is the solution to the system after checking.

Similar to systems of linear equations in two variables, systems of linear equations in three variables can have one unique solution, infinitely many solutions, or no solution at all.

Exercises.

1. Multiple choice.

(1) To solve the system $\begin{cases} x + y = 3 & ① \\ y + z = 4 & ② \\ z + x = 5 & ③ \end{cases}$, which of the following does not work?

A. ①−② to eliminate y and combine it with ③.

B. ①×3+②×4−③×5 to make the constant term 0.

C. Subtract ①,②,③ from $\dfrac{① + ② + ③}{2}$ to get x, y, z directly.

D. $\dfrac{① + ③}{2} - ②$ to eliminate y, z and solve for x.

(2) A system of linear equations in three variables cannot have

A. only one solution   B. only two solutions

C. infinitely many solutions   D. no solutions

(3) Which of the following is the solution to $\begin{cases} 3x - 2y = 8 \\ 2y + 3z = 1 \\ x + 5z = 7 \end{cases}$?

A. $(2, -1, 0)$ B. $(2, 1, 0)$ C. $(0, -1, 1)$ D. $(2, -1, 1)$

(4) To solve $\begin{cases} \dfrac{x}{3} = \dfrac{y}{2} = \dfrac{z}{6} \\ x + 2y + z = 22 \end{cases}$, which of the following is not correct?

A. Get $x = \dfrac{z}{2}, y = \dfrac{z}{3}$ from the first equation and plug them into the second equation to solve z and hence x and y.

B. Let $x = 6$ to solve y and z.

C. Simplify the first equation to $\begin{cases} 2x - 3y = 0 \\ 2x - z = 0 \end{cases}$; combine it with the last equation; then use the elimination method.

D. Let $\dfrac{x}{3} = \dfrac{y}{2} = \dfrac{z}{6} = t$ to get $x = 3t, y = 2t, z = 6t$. Plug them into the second equation to solve t and hence x, y, z.

(5) From the system $\begin{cases} x - 2y + 3z = 0 \\ 2x - 3y + 4z = 0 \end{cases}$, we may get $x : y : z$ equals

    A. $1 : 2 : 1$ B. $1 : (-2) : (-1)$ C. $1 : (-2) : 1$ D. $1 : 2 : (-1)$

(6) The solution to $\begin{cases} 2x = 3y = 6z \\ x + 2y + z = 16 \end{cases}$ is

    A. $(1, 3, 5)$ B. $(6, 3, 2)$ C. $(6, 4, 2)$ D. $(4, 5, 6)$

2. Solve the equations.

(1) $\begin{cases} 3x + 2y + z = 14 \\ x + y + z = 10 \\ 2x + 3y - z = 1 \end{cases}$ (2) $\begin{cases} 2x + 3y - z = 9 \\ x + y + z = 15 \\ 5x - 4y - z = 0 \end{cases}$

(3) $\begin{cases} 3a - 2b + c = -2 \\ 2a - b + c = -1 \\ a + b + c = 2 \end{cases}$ (4) $\begin{cases} x + 2y = 2 \\ 2y + 3z = 4 \\ 3z + x = 6 \end{cases}$

(5) $\begin{cases} 5x - 3y + 4z = 13 \\ 2x + 7y - 3z = 19 \\ 3x + 2y - z = 18 \end{cases}$ (6) $\begin{cases} 2d + e = 10 \\ d - e + f = 4 \\ 3d - e - f = 0 \end{cases}$

(7) $\begin{cases} m + n = 16 \\ n + t = 13 \\ t + m = 21 \end{cases}$ (8) $\begin{cases} 2x + y + z = 4 \\ x + 2y + z = 2 \\ x + y + 2z = 6 \end{cases}$

$(9) \begin{cases} 3x - 5y - 6z = 4 \\ 2x - 2y - 3z = 5 \\ 2x - 3y - 4z = 4 \end{cases}$
$\qquad$
$(10) \begin{cases} 3x + y - 4z = 13 \\ 5x - y + 3z = 5 \\ x + y - z = 3 \end{cases}$

$(11) \begin{cases} \dfrac{a_1 + a_2}{2} = \dfrac{a_2 + a_3}{4} = \dfrac{a_3 + a_1}{3} \\ a_1 + a_2 + a_3 = 27 \end{cases}$
$\qquad$
$(12) \begin{cases} 2p + 3q - 4r = -7 \\ \dfrac{p - 4q}{3} = \dfrac{2q + 3r}{2} = 2 \end{cases}$

3. Solve the equations.

$(1) \begin{cases} \dfrac{2}{x} - \dfrac{1}{y} = 5 \\ \dfrac{3}{y} + \dfrac{1}{z} = 0 \\ \dfrac{5}{x} - \dfrac{2}{z} = 4 \end{cases}$
$\qquad$
$(2) \begin{cases} \dfrac{1}{x} - \dfrac{2}{y} + \dfrac{1}{z} = 1 \\ \dfrac{2}{x} + \dfrac{3}{y} - \dfrac{1}{z} = -\dfrac{7}{2} \\ \dfrac{3}{x} - \dfrac{1}{y} - \dfrac{2}{z} = -\dfrac{19}{2} \end{cases}$

$(3) \begin{cases} \dfrac{x}{y + 1} = 1 \\ \dfrac{y}{z + 2} = 0.25 \\ \dfrac{z}{x + 1} = 1 \end{cases}$
$\qquad$
$(4) \begin{cases} \dfrac{y + 1}{y + z} = 2 \\ \dfrac{y - 2}{2x - z} = 4 \\ \dfrac{z - 1}{4x + y} = 1 \end{cases}$

$$(5) \begin{cases} \dfrac{x+y}{xy} - \dfrac{6}{z} = 9 \\ \dfrac{y-x}{xy} + \dfrac{4}{z} = 5 \\ \dfrac{3x-2y}{xy} - \dfrac{1}{z} = 4 \end{cases} \qquad (6) \begin{cases} \dfrac{x}{2} = \dfrac{y}{3} = \dfrac{z}{4} \\ x+y-z = \dfrac{1}{12} \end{cases}$$

4. If mixing brown(B), white(W) and yellow(Y) rice in the following three ways: (i) B:W:Y= 4 : 3 : 2, (ii) B:W:Y= 3 : 1 : 5, (iii) B:W:Y= 2 : 6 : 1, then how many kilograms of each mixture do we need to buy to make 49 kilograms of a mixture such that B:W:Y=1 : 1 : 1.

5. The speed of a motorcycle is $30m/h$ on the flat road, $35m/h$ down the hills and $28m/h$ up the hills. The distance between A and B is 142km. If it takes 4.5 hours for the motorcycle to go from A to B and 4.7 hours to travel back by the same road, how many miles of flat, up-hill and down-hill roads are there from A to B?

6. On the planet Mars, people eat giant pizzas. It takes Tommy and Tina 36 mars-days, Tommy and Christine 45 mars-days, Tina and Christine 60 mars-days, to finish eating one pizza. How long does it take for each of them to finish eating one pizza? If the three of them eat the pizza together, how long will it take them to finish?

8.3 Linear Inequalities in Two Variables

The standard form of a linear inequality in two variables is

$$ax + by < c,$$

where x, y are variables and a, b, c are constants. We can also use the inequality signs '$>, \geq, \leq$' besides '$<$'. The solution set to the inequality is the set of all ordered pairs of real numbers (x, y) such that the inequality holds, i.e. $\{(x, y) \in \mathbb{R}^2 | ax + by < c\}$. As we identify the ordered pair (x, y) with the point it represents in the coordinate plane, the solution set of the inequality can be visualized as a subset of the coordinate plane. For systems of linear inequalities in two variables, the solution set is the intersection of the solution sets to each inequality of the system.

Examples. Sketch the solution sets.

(1) $x + 2y > 3$

(2) $3x - 2y \leq 6$

(3) $\begin{cases} x + 2y > 3 \\ 3x - 2y \leq 6 \end{cases}$

Solutions.

(1) The line $x + 2y = 3$ divides the coordinate plane into two regions — upper right region and lower left region. The points in the same region make $x + 2y - 3$ have the same sign, so the solution set is one piece of the region and we can plug in an arbitrary point to check. For example, since the y-intercept $\dfrac{3}{2}$ is positive, $(0, 0)$ is a point to the lower left of the line. Since $(0, 0)$ does not make the inequality $x + 2y < 3$ hold, the solution set should be the upper right region. Here the sign is '$>$', so the points on the line $x + 2y = 3$ are not included in the solution set. We denote the solution set by the gray area while the dashed line is not considered as part of the region.

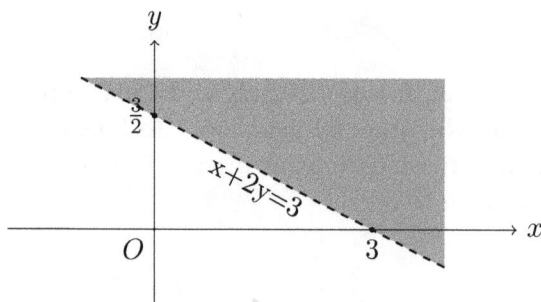

(1)

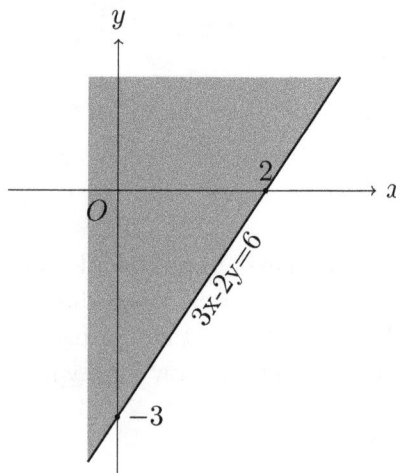

(2)

(2) The line $3x - 2y = 6$ divides the coordinate plane into upper left and lower right

regions. $(0,0)$ is a point in the upper left region which makes the inequality hold, so the solution set is the upper left region. Here '$\leq$' includes '$=$', so the points on the line $3x-2y=6$ are also in the solution set. The solution set is denoted by a gray area, while the solid line indicates it is part of the solution set.

(3) The solution set is the intersection of the solution sets of the two inequalities $x+2y<3$ and $3x-2y\leq 6$. Using the results of the previous two problems, the solution set can be sketched as

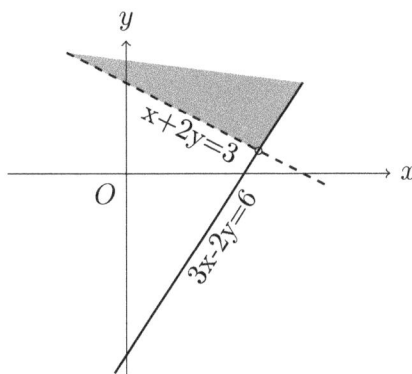

The coordinate of the intersection of the two lines does not satisfy the first inequality, so we delete it by drawing a circle around it.

Exercises.

1. Multiple choice.

(1) Which point is not in the region determined by $3x+2y<6$?

    A. $(0,0)$ B. $(1,1)$ C. $(0,2)$ D. $(2,0)$

(2) The region determined by $x+2y-6<0$ is located at which corner of the line $x+2y-6=0$?

    A. Upper right   B. Upper left   C. Lower right   D. Lower left

(3) The region determined by $x-2y>0$ is

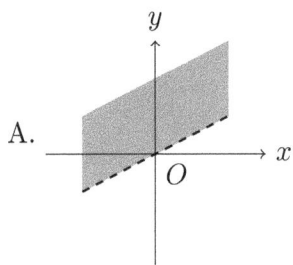

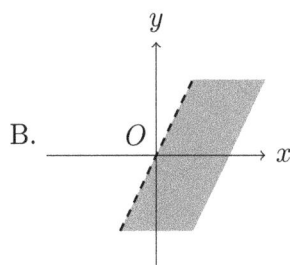

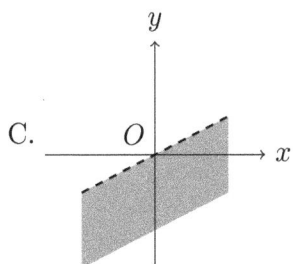

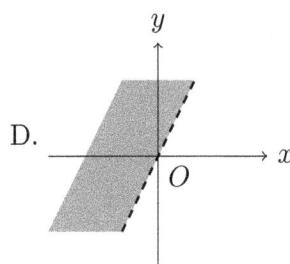

(4) The region determined by $-2x + 3y \geq -1$ is

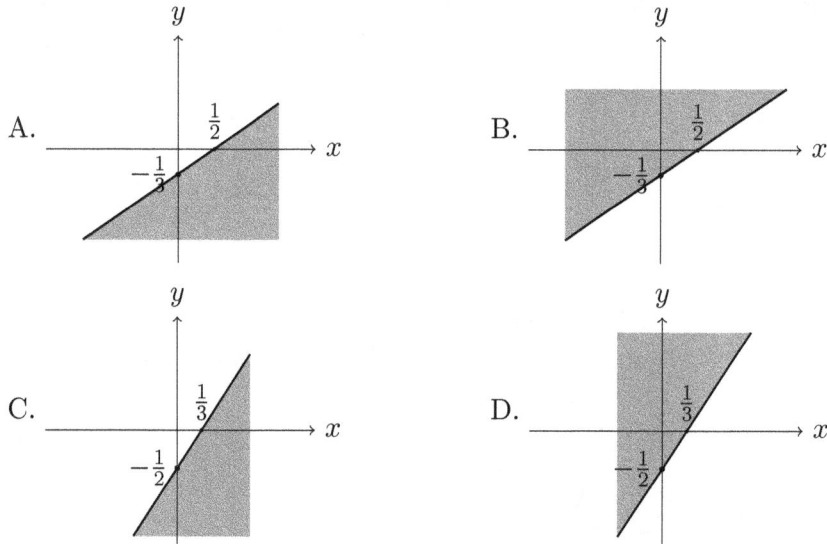

A.

B.

C.

D.

(5) Which system of inequalities determines the following shaded region (boundaries included)?

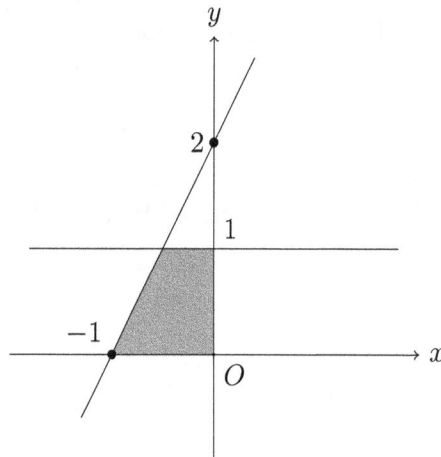

A. $\begin{cases} 0 \leq y \leq 1 \\ 2x - y + 2 \leq 0 \end{cases}$

B. $\begin{cases} y \leq 1 \\ 2x - y + 2 \geq 0 \end{cases}$

C. $\begin{cases} 0 \leq y \leq 1 \\ 2x - y + 2 \geq 0 \\ x \leq 0 \end{cases}$

D. $\begin{cases} y \leq 1 \\ 2x - y + 2 \geq 0 \\ x \leq 0 \end{cases}$

(6) The solution set of $|y| \geq 1$ in the coordinate plane is

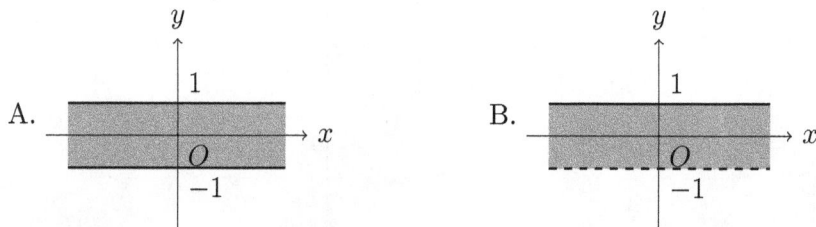

A.

B.

C.

D.

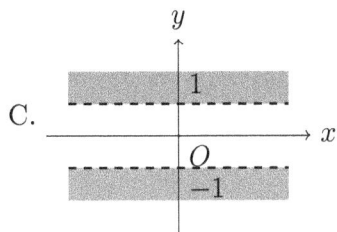

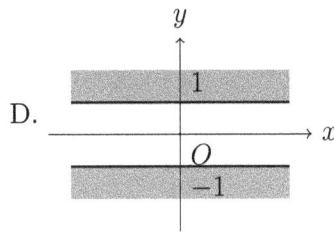

(7) Given a line $l : y = mx + b$, then $\{(x, y) | y \geq mx + b\}$ is the region

A. above l

B. below l

C. to the right of l

D. to the left of l

(8) The region determined by $|y| > |x|$ is

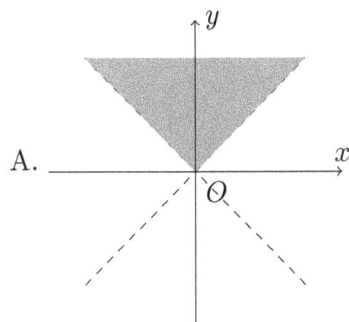

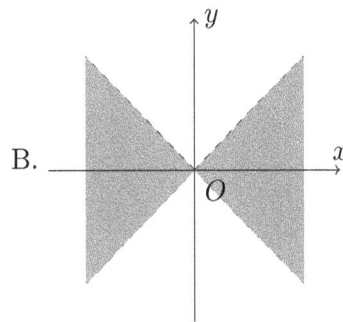

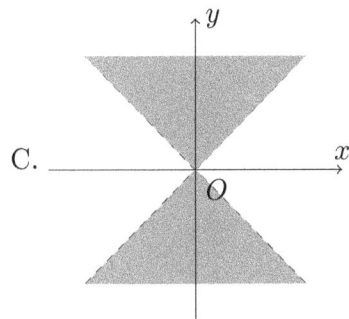

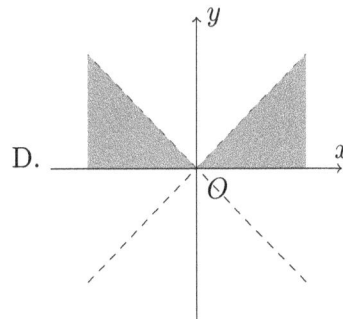

A.

B.

C.

D.

2. Sketch the solution sets.

(1) $4x - 3y < 12$

(2) $3x + 2y > -2$

(3) $x - 7y \geq 5$

(4) $5x - 4y \leq 0$

(5) $\begin{cases} y \geq 2x + 1 \\ x + 2y < 4 \end{cases}$

(6) $\begin{cases} x \geq 2y - 3 \\ y \leq -x - 4 \end{cases}$

(7) $\begin{cases} x + 2y < 8 \\ 9x + 5y > 46 \end{cases}$

(8) $\begin{cases} 3x - 2y > -1 \\ x + 7y \leq 0 \end{cases}$

(9) $\begin{cases} 2x + 3y \leq x + y + 3 \\ -x + y > 1 - 2x - y \end{cases}$

(10) $\begin{cases} 2x - 3y \leq -6 \\ 4x - 6y > 5 \end{cases}$

3. Given three points $A(3, -1)$, $B(-1, 1)$, $C(1, 3)$, find a system of inequalities whose solution set is the region bounded by $\triangle ABC$ with boundaries.

4. Sketch the graph of $|x| + |y| = 1$ and $|x| + |y| \leq 1$.

Radicals and Radical Expressions

In this chapter we shall introduce radicals and complete our study of algebraic expressions. Radicals arise as solutions to equations of the form $x^n = a$. With radicals, we can fulfill the promise to define powers with rational and real exponents. The rules of powers keep valid with these exponents. Radicals also allow us to denote the general solutions to quadratic equations (polynomial equations of degree 2). This is an important topic, and we want to keep it for the next chapter. With radical expressions, it is natural to have radical equations and inequalities. In our course, we concentrate on solving radical equations only.

9.1 Roots and Radicals

Fixing a real number a and a positive integer n, the solutions to the equation

$$x^n = a$$

are called the n-th roots of a.

When $n = 1$, the equation is trivial ($x^1 = a$ is just $x = a$). The situation $n > 1$ would be more interesting. When $n = 2, 3$, we call the solutions square roots and cube roots, respectively. We should distinguish the cases where n is even or odd.

• When n is **even**, i.e. $n = 2k$ for some $k \in \mathbb{N}$, e.g. $n = 2, 4, 6, \cdots$, depending on the sign of a, there are three possibilities.

$a > 0$	$a = 0$	$a < 0$
a has two opposite n-th roots	a has only one n-th root 0	a has no n-th root
e.g. the square roots of 4 are ± 2 as $(\pm 2)^2 = 4$.	as the only solution to $x^n = 0$ is 0.	as $x^n = (x^k)^2 \geq 0$ cannot equal $a < 0$.

It should be noticed, when $n = 2k$ is even, positive numbers always has two opposite n-th roots. The reason is clear: if x_0 is an n-th root of $a > 0$, i.e. $x_0^n = x_0^{2k} = a$, since $(-x_0)^n = (-x_0)^{2k} = x_0^{2k} = a$, $-x_0$ must also be an n-th root. For one more example, the 4-th roots of 64 are ± 4.

• When n is **odd**, i.e. $n = 2k + 1$ for some $k \in \mathbb{N}$, e.g. $n = 1, 3, 5, \cdots$, a always has a unique n-th root. For example, the cube root of -27 is -3 since $(-3)^3 = -27$ and $x^3 = -27$ has no other solutions; the cube root of 0 is 0; the 5th root of 32 is 2.

We apply the following notations for the n-th roots of a.

- When n is even and $a \geq 0$, we denote the nonnegative n-th root of a by $\sqrt[n]{a}$.

- When n is odd, the unique n-th root of a is denoted by $\sqrt[n]{a}$.

Here $\sqrt[n]{}$ is called the radical sign, a is called the radicand and $\sqrt[n]{a}$ is called a radical.

When n is even and $a \geq 0$, the nonnegative n-th root $\sqrt[n]{a}$ of a is called the arithmetic n-th root or the principal n-th root. The other n-th root, as the opposite of $-\sqrt[n]{a}$, can be denoted by $-\sqrt[n]{a}$. When $n = 2$, the symbol $\sqrt[2]{}$ is abbreviated to $\sqrt{}$. For example, the arithmetic square root of 16 is 4, so $\sqrt{16} = 4$ and the square roots of 16 are $\pm\sqrt{16} = \pm 4$. The 4th roots of 2 are $\pm\sqrt[4]{2}$, while the principal one is $\sqrt[4]{2}$. Similarly, the square roots of 5 are $\pm\sqrt{5}$. However, $\sqrt{-2}$ does not exist, as -2 has no square roots. According to our convention, $\sqrt[n]{a} \geq 0$ whenever $a \geq 0$ and n is even.

When n is even, $a \geq 0$ is crucial for $\sqrt[n]{a}$ to exist. For example, $\sqrt{4-t}$ is well-defined only when $4 - t \geq 0$, i.e. $t \leq 4$.

When n is odd, $\sqrt[n]{a}$ is the unique n-th root of a. For example, the cube root of 8 is 2, so $\sqrt[3]{8} = 2$. Similarly $\sqrt[3]{27} = 3$, $\sqrt[5]{-1} = -1$, and the cube root of -3 is $\sqrt[3]{-3}$. We should notice that $\sqrt[n]{a}$ and a always share the same sign in this case (n is odd).

The above analysis are summarized in the following table.

	$a > 0$	$a = 0$	$a < 0$
n is even	$\sqrt[n]{a}$: the positive n-th root of a $-\sqrt[n]{a}$: the negative n-th root of a	$\sqrt[n]{0} = 0$	$\sqrt[n]{a}$ D.N.E
n is odd	$\sqrt[n]{a} > 0$	$\sqrt[n]{0} = 0$	$\sqrt[n]{a} < 0$

From the definition, we always have $\sqrt[n]{0} = 0$ and $\sqrt[n]{1} = 1$ for all $n \in \mathbb{Z}^+$. When n is even, although the n-th root of 0 is 0 only, we also say 0 has two opposite n-th roots (0 and 0). By this convention, any $a \geq 0$ has two opposite roots of even order.

$\sqrt[n]{a}$ (when existing or well-defined) is not always a rational number. For example, $\sqrt{2}$, $\sqrt{3}$, $\sqrt[3]{3}$, $\sqrt[3]{-4}$, $\sqrt[7]{123}$ are all irrational numbers. These numbers form an important class of irrational numbers although a lot of others exist. For example, π is irrational but cannot be written as a radical. As irrational numbers are infinite decimals, we can approximate them by finite decimals: the more digits we take, the more accurate the approximation is. For example, $\sqrt{2} \doteq 1.41421$, $\sqrt[3]{2} \doteq 1.25992$. Radicals themselves represent accurate numbers. Unless required explicitly, we never substitute a radical by its approximations at any time.

From the definition, we have the following propositions.

- $(\sqrt[n]{a})^n = a$ whenever $\sqrt[n]{a}$ exists;

- For any $c \in \mathbb{R}$, $(\sqrt[n]{c^n}) = \begin{cases} |c| & \text{if } n \text{ is even} \\ c & \text{if } n \text{ is odd} \end{cases}$.

These formulas are very important in simplifying radicals. Particularly $\sqrt{c^2} = |c|$ and $\sqrt[3]{c^3} = c$ for all $c \in \mathbb{R}$.

Examples.

(1) $(\sqrt{3})^2 = 3$, $(\sqrt[6]{5})^6 = 6$, $(\sqrt{-2})^2$ D.N.E as $\sqrt{-2}$ does not exist;

(2) $(\sqrt[3]{-5})^3 = -5$, $(\sqrt[5]{100})^5 = 100$;

(3) $\sqrt{(-2)^2} = |-2| = 2$, $\sqrt[6]{3^6} = |3| = 3$, $\sqrt[4]{(-7)^4} = |-4| = 4$;

(4) $\sqrt[4]{a^4 b^8} = \sqrt[4]{(ab^2)^4} = |ab^2| = |a|b^2$, $\sqrt{x^2 + 2x + 1} = \sqrt{(x+1)^2} = |x+1|$;

(5) $\sqrt[3]{x^3 y^6} = \sqrt[3]{(xy^2)^3} = xy^2$.

We can also order radicals. Given $a, b > 0$, $n \in \mathbb{Z}^+$, we know $a > b$ if and only if $a^n > b^n$ (since $a^n - b^n = (a - b)(a^{n-1} + a^{n-2}b + \cdots + b^{n-1})$, $a^n - b^n$ and $a - b$ must have the same sign as the terms in the second parentheses are positive). As a result, $1 < \sqrt{2} < \sqrt{3} < 2$, since $1^2 < (\sqrt{2})^2 = 2 < (\sqrt{3})^2 = 3 < 2^2$. Similarly $1.41 < \sqrt{2} < 1.42$. Using the above result, by making the discussion on the signs of a, b, it is not hard to show: when $n > 0$ is odd, for any $a, b \in \mathbb{R}$, $a < b$ if and only if $a^n < b^n$. For example, $\sqrt[3]{-2} < -1 < 1 < \sqrt[3]{2}$, since $(\sqrt[3]{-2})^3 = -2 < (-1)^3 < 1^3 < (\sqrt[3]{2})^3 = 2$.

Exercises.

1. True or false.

(1) The square roots of a positive number must be a pair of opposite numbers.

(2) The 4th root of 81 is 3.

(3) $\sqrt[2013]{-1^2} = -1$.

(4) The square of a positive number must be greater than its square roots.

(5) -6 is a square root of $(-6)^2$.

(6) The square root of $(-6)^2$ is -6.

(7) $\sqrt[3]{-\dfrac{27}{125}}$ is an irrational number.

(8) The cube roots of -8 are ± 2.

(9) $\sqrt{a} \geq 0$ whenever existing.

(10) $\sqrt[n]{a^n} = a$ for all $n \in \mathbb{N}$ and $a \in \mathbb{R}$.

(11) $\sqrt{-|-1|} = 1$.

(12) $\sqrt[3]{-2} = -\sqrt[3]{2}$.

(13) The n-th root of a^n is a.

(14) If $\sqrt[3]{a} < 0$, then $a < 0$.

(15) The n-th root of x^{2n} is x^2.

(16) If $x^n = a + b$, then $x = \sqrt[n]{a + b}$.

(17) Irrational numbers must be non-repeating infinite decimals.

(18) Irrational numbers must be radicals.

(19) The principal square root of any positive number is positive.

2. Multiple Choice.

(1) Which of the following is not correct?

A. $-\sqrt{2}$ is a square root of 2.

B. $\sqrt{2}$ is a square root of 2.

C. The square root of 2 is $\sqrt{2}$.

D. The arithmetic square root of 2 is $\sqrt{2}$.

(2) Among the real numbers $-2, 0.3, \dfrac{1}{7}, -\sqrt{2}, \pi$, how many of them are irrational numbers?

    A. 2   B. 3   C. 4   D. 5

(3) The arithmetic square root of 9 is

    A. 3   B. -3 C. ± 3 D. $\sqrt{3}$

(4) Which of the following is correct?

    A. The square root of 16 is 4.

    B. The square root of 16 is -4.

    C. 4 is a square root of 16.

    D. -4 is also an arithmetic square root of 16.

(5) If a square has area 5, the length of its edge is

    A. a rational number   B. a finite decimal

    C. an irrational number   D. an integer

(6) Which of the following is correct?

    A. The cube root of a number is either positive or negative.

    B. Negative numbers have no cube roots.

    C. If a number equals its cube root, then this number is 0 or 1.

    D. The cube root has the same sign with its radicand; the cube root of 0 is 0.

(7) If the principal square root of a number equals its cube root, this number must be

    A. $0, 1$ B. $-1, 1$ C. $-1, 0$ D. $-1, 0, 1$

(8) If a is a square root of $(-4)^2$, and 2 is a square root of b, then $a + b$ is

    A. 8   B. 0   C. -4 or 4   D. 8 or 0

(9) Which of the following is correct?

    A. The cube root of $\sqrt{64}$ is 2.   B. -3 is a negative cube root of 27.

    C. The cube roots of $\dfrac{125}{216}$ are $\pm\dfrac{5}{6}$.   D. The cube root of $(-1)^2$ is -1.

(10) If the arithmetic square root of a positive number is a, then the square roots of the number 3 greater is

    A. $\sqrt{a^2 + 3}$ B. $-\sqrt{a^2 + 3}$ C. $\pm\sqrt{a^2 + 3}$ D. $\pm\sqrt{a + 3}$

(11) $-\dfrac{1}{64}$ has

    A. both square and cube roots   B. only square roots

C. only a negative cube root   D. none of the above

(12) Which of the following is correct?

   A. The sum or difference of two irrational numbers is irrational.

   B. All radicals are irrational.

   C. The product or quotient of two irrational numbers is irrational.

   D. A positive irrational number has a positive cube root; a negative irrational number has a negative cube root.

(13) Which of the following is correct?

   A. The reciprocals of rational numbers are rational.

   B. The squares of a rational number and an irrational number can be equal.

   C. The sum of a rational number and an irrational number is irrational.

   D. The product of a rational number and an irrational number is irrational.

(14) Which of the following is not correct?

   A. Positive numbers have roots of any order.

   B. All numbers have roots of odd order.

   C. Only positive numbers have roots of even order.

   D. Negative numbers have no roots of even order.

(15) Which of the following does not have a square root?

   A. $\sqrt{3} - \sqrt{2}$ B. $\sqrt[4]{3}$ C. $\sqrt[3]{-2^2} + 1$ D. $\pi - \sqrt[3]{(-3)^2}$

(16) If $x < 0$, the cube root of x is

   A. $\sqrt[3]{x}$ B. $\sqrt[3]{-x}$ C. $-\sqrt[3]{x}$ D. $\pm\sqrt[3]{x}$

3. Fill in the blanks.

(1) If $x^2 = \dfrac{25}{64}$, then $x = $_____; if $\sqrt{y} = 1.4$, then $y = $_____.

(2) If one square root of a number is 7.12, the other square root of the same number is

_____.

(3) The sum of two square roots of a positive number is _____; the quotient of two square roots of a positive number is _____.

(4) If a number equals its cube root, then the number must be _____; if a number equals its arithmetic square root, the number must be _____; if the arithmetic square root and cube root of a number are equal, then the number is _____.

(5) If the square roots of a number are ± 8, the cube root of the number is _____.

(6) If a square has area a whose side length is b, then b is _____ of a.

(7) Given the approximations $\sqrt{2} \doteq 1.414$, $\sqrt{3} \doteq 1.732$, $2\sqrt{3} - 3\sqrt{2}$ is approximately _____, $\dfrac{2 - \sqrt{3}}{\sqrt{2}}$ is approximately _____.

(8) $\sqrt{3 - m}$ is well-defined when m_____; $\sqrt[3]{3 - m}$ is well-defined when m_____.

(9) If $2a - 1$ and $-a + 2$ are the two square roots of b, then $a = $_____ and $b = $_____.

(10) If $\sqrt{2a - 1} + (b + 3)^2 = 0$, then $\sqrt[3]{\dfrac{2ab}{3}} = $_____.

(11) The minimum value of $\sqrt{a - 1} - 2$ is _____.

(12) If the arithmetic square root of $2a + 1$ is 2, then $a = $_____.

(13) If two legs of a right triangle have lengths 2 and 3, then the hypotenuse has length _____; if the hypotenuse has length 2 and one leg has length 1, then the length of the other leg is _____.

(14) $\sqrt{1 - 2x} = $_____ when $x = -4$; $\sqrt[3]{3x - 2} = $_____ when $x = -2$; $\sqrt{5 - 4x}$ _____ when $x = 2$.

(15) If the cube root of $5x + 19$ is 4, the square roots of $5x + 4$ are _____.

(16) If the square roots of $x - 2$ are ± 2 and the cube root of $2x + y + 7$ is 3, then the square roots of $x^2 + y^2$ are _____.

(17) If $\sqrt[3]{x^3} = 2$, then $x = $_____; if $\sqrt{x^2} = 3$, then $x = $_____.

(18) If $\sqrt{b^2 - 10}$ and $|2a + b^2|$ are opposites, then $a = $_____, $b = $_____.

4. Find all irrational numbers from the following numbers.

$$3.1415, 0.2002002, 2.131131113\cdots, \frac{\pi}{4}, 0.\overline{26}, \sqrt{5},$$

$$\sqrt{1\frac{7}{9}}, \sqrt[3]{-64}, \pi^2, -\sqrt{2}, \sqrt{\pi}, \frac{\pi}{3}, 1.414, 2 + \sqrt{3}, 3.\overline{414}$$

5. Find square roots and arithmetic square roots of the following numbers.

(1) 9

(2) $\dfrac{81}{16}$

(3) 0

(4) -2^2

(5) 15

(6) π

6. Find cube roots of the following numbers.

(1) -8

(2) 0.001

(3) 7

(4) -64

7. Evaluate.

(1) $\sqrt{36}$

(2) $\sqrt{81}$

(3) $\sqrt{0.01}$

(4) $\sqrt{(-41)^2}$

(5) $\sqrt{-(-16)}$

(6) $\sqrt{-|-4|}$

(7) $\sqrt{0.49}$

(8) $\sqrt{1.44}$

(9) $\sqrt[3]{-64}$

(10) $\sqrt[3]{125}$

(11) $\sqrt[3]{0.216}$

(12) $\sqrt[4]{16}$

(13) $\sqrt{2\dfrac{1}{4}}$

(14) $\sqrt{\dfrac{144}{49}}$

(15) $\sqrt{0.04}\sqrt{121}$

(16) $\sqrt{\dfrac{49}{16}} \cdot \sqrt[4]{16}$

(17) $\dfrac{-\sqrt{25}}{\sqrt[3]{27}}$

(18) $\dfrac{\sqrt{196}}{\sqrt{81}}$

(19) $\sqrt[3]{3\dfrac{3}{8}}$

(20) $\sqrt[3]{-\dfrac{125}{27}}$

(21) $-\sqrt{-(7-43)}$

(22) $\sqrt{\dfrac{13^2 - 12^2}{64}}$

(23) $\sqrt{(-3)(-27)}$

(24) $\sqrt{(-\dfrac{1}{6})(1 + \dfrac{1}{24})(-1)}$

(25) $\sqrt[3]{8}\sqrt[3]{-\dfrac{1}{64}}$

(26) $\sqrt[3]{1 - \dfrac{37}{64}}$

(27) $\sqrt[5]{-1024}$

(28) $\sqrt[6]{(-27)^2}$

8. Compare the numbers.

(1) $\sqrt[3]{0.25}$, $\sqrt[3]{0.26}$

(2) $\sqrt[3]{18}$, $\sqrt[3]{\dfrac{1}{18}}$

(3) $\sqrt[3]{-2}$, $\sqrt[3]{-3}$, $\sqrt[3]{0.1}$, $\sqrt[3]{5}$

(4) $-\sqrt[4]{2.9}$, $-\sqrt[4]{3}$, $-\sqrt[4]{3.1}$

(5) $2\sqrt{2}, 3, 2\sqrt{3}$

(6) $3\sqrt{2}, 5, 2\sqrt{5}, \sqrt{15}$

(7) $\sqrt{2}, \sqrt[3]{2}, \sqrt[4]{2}$

(8) $\sqrt{\dfrac{1}{2}}, \sqrt[3]{\dfrac{1}{2}}, \sqrt[4]{\dfrac{1}{2}}$

(9) $\sqrt{x}, \sqrt[3]{x}, \sqrt[4]{x}$ when $x > 1$

(10) $\sqrt{x}, \sqrt[3]{x}, \sqrt[4]{x}$ when $0 < x < 1$

(11) $\sqrt[3]{5}, \sqrt[4]{10}, \sqrt{3}$

(12) $\sqrt[3]{14}, \sqrt{5}, \sqrt{5.5}, \sqrt[3]{12}, \sqrt{6}, \sqrt[3]{13}$

(13) $\sqrt{50}, 7\dfrac{1}{2}$

(14) $-\pi, -\dfrac{22}{7}, \pi, \sqrt{10}$

(15) $-\dfrac{1}{\sqrt{10}}, -\dfrac{1}{\pi}, -\dfrac{1}{3}$

(16) $-\dfrac{\sqrt{2}}{2}, -\dfrac{7}{10}$

9. Find when the following expressions are well-defined.

(1) $\sqrt{2a - 5}$

(2) $\sqrt[3]{x^3 - 9x + 1}$

(3) $\dfrac{\sqrt[4]{3t}}{\sqrt[4]{1 - 2t}}$

(4) $\sqrt{(b + 3)^2}$

(5) $\sqrt[8]{x^2 + 2}$

(6) $\sqrt{-\dfrac{3}{6 - 4a}}$

(7) $\dfrac{\sqrt{2x - 3}}{x - 3}$

(8) $\dfrac{\sqrt{x - 4}}{4 - |x|}$

(9) $\dfrac{\sqrt{8x - 7}}{\sqrt[3]{3x - 4}}$

(10) $\dfrac{\sqrt[4]{3 - 7x}}{\sqrt[3]{-1 - 5x}}$

10. Solve the following equations.

(1) $169x^2 = 100$

(2) $5x^2 - 125 = 0$

(3) $\sqrt{x^2 - 100} = 0$

(4) $2(x-2)^2 - 47 = 3$

(5) $-25(x+1)^2 = (-4)^3$

(6) $25(x^2 - 1) = 24$

(7) $(x-1)^3 = -8$

(8) $\dfrac{(2x+1)^3}{4} = 54$

(9) $\sqrt[3]{-1}(2x-1)^2 = -49$

(10) $(x+4)^3 + 27 = 0$

(11) $(3x-1)(9x^2 + 3x + 1) = 215$

(12) $(x+1)^2 + 4(x+1) = -4$

11. Simplify.

(1) $|\sqrt{2} - 1.\overline{4}|$

(2) $|1.732 - \sqrt{3}|$

(3) $|1 + \sqrt{3}| - |1 - \sqrt{3}|$

(4) $|1 - \sqrt{2}| + |\sqrt{2} - \sqrt{3}| + |2 - \sqrt{3}|$

12. Find the arithmetic square roots.

(1) $1 - 2\pi + \pi^2$

(2) $19 - 4\sqrt{15}$

9.2 Radical Expressions

To manipulate radicals, we also have the following proposition. Whenever $\sqrt[n]{a}$ and $\sqrt[n]{b}$ are well-defined,

- $\sqrt[n]{a}\,\sqrt[n]{b} = \sqrt[n]{ab}$;

- $\dfrac{\sqrt[n]{a}}{\sqrt[n]{b}} = \sqrt[n]{\dfrac{a}{b}}$ if $b \neq 0$.

The proposition can be proved by finding the n-th power on both sides of the identities. We can use this proposition to simplify radicals. An expression of radicals is in simplest form if

- no radicand is a rational fraction;
- no radicand of n-th root contains a perfect n-th power factor;
- no radical appears in the denominator.

When simplifying expressions of radicals, we always simplify it to its simplest form.

Examples. Simplify.

(1) $\sqrt{72}$ (2) $\sqrt[3]{54}$ (3) $\sqrt{\dfrac{3}{8}}$ (4) $\dfrac{1}{\sqrt[3]{25}}$

Solutions.

(1) $72 = 8 \times 9 = 2^3 3^2$, which contains perfect squares 2^2 and 3^2, so we split them from the radicand and simplify by finding the square roots.

$$\sqrt{72} = \sqrt{3^2 2^3} = \sqrt{3^2}\sqrt{2^2 \times 2} = \sqrt{3^2}\sqrt{2^2}\sqrt{2} = 3 \times 2 \times \sqrt{2} = 6\sqrt{2}.$$

(2) $54 = 2 \times 3^3$ contains a perfect cube 3^3, so

$$\sqrt[3]{54} = \sqrt[3]{3^3 \times 2} = \sqrt[3]{3^3}\sqrt[3]{2} = 3\sqrt[3]{2}.$$

(3) $\sqrt{\dfrac{3}{8}} = \dfrac{\sqrt{3}}{\sqrt{8}} = \dfrac{\sqrt{3}}{2\sqrt{2}}$ since $\sqrt{8} = \sqrt{2^2 \times 2} = 2\sqrt{2}$. It has a square root in the denominator. Since $\sqrt{2}\sqrt{2} = 2$, we can multiply both the numerator and denominator by $\sqrt{2}$. Then the expression becomes

$$\frac{\sqrt{3}}{2\sqrt{2}} = \frac{\sqrt{3} \cdot \sqrt{2}}{2\sqrt{2} \cdot \sqrt{2}} = \frac{\sqrt{6}}{2 \times 2} = \frac{\sqrt{6}}{4}.$$

(4) The denominator is a cube root, and a cube root can be simplified when the radicand is a perfect cube. Since $\sqrt[3]{25} = \sqrt[3]{5^2}$, we can transform 5^2 into the cube 5^3 by multiplying both the denominator and numerator by $\sqrt[3]{5}$. As a result

$$\frac{1}{\sqrt[3]{5^2}} = \frac{1 \cdot \sqrt[3]{5}}{\sqrt[3]{5^2} \cdot \sqrt[3]{5}} = \frac{\sqrt[3]{5}}{\sqrt[3]{5^3}} = \frac{\sqrt[3]{5}}{5}.$$

From these examples, to simplify a radical in the denominator, we complete its radicand to a perfect n-th power; to simplify a radical in the numerator, we clear radical signs for factors of perfect n-th power in the radicand.

At this point, all four usual operations on radicals can be done. The proposition at the beginning of this section gives us tools to do multiplication and division. For addition and subtraction, sometimes the result can be simplified by combining similar radicals. Two radicals are similar if they have the same order and the same radicand, i.e. they may only differ by coefficients.

Examples.

(1) $\sqrt{12} - \sqrt{75} = 2\sqrt{3} - 5\sqrt{3} = -3\sqrt{3}$;

(2) $\sqrt{12} + \sqrt{18} - \sqrt{27} = 2\sqrt{3} + 3\sqrt{2} - 3\sqrt{3} = 3\sqrt{2} - \sqrt{3}$;

(3) $(\sqrt{8} - \sqrt{3})(2\sqrt{3} + 3\sqrt{2})$

$= (2\sqrt{2} - \sqrt{3})(2\sqrt{3} + 4\sqrt{2})$

$= 2\sqrt{2}(2\sqrt{3} + 3\sqrt{2}) - \sqrt{3}(2\sqrt{3} + 3\sqrt{2})$

$= (4\sqrt{6} + 12) - (6 + 3\sqrt{6}) = 6 + \sqrt{6}$;

(4) $(\sqrt{6} - 2\sqrt{2})(\sqrt{6} + 2\sqrt{2})$

$= (\sqrt{6})^2 - (2\sqrt{2})^2 (\text{difference of squares})$

$= 6 - 8 = -2$;

(5) $\dfrac{1}{\sqrt{3} + \sqrt{2}} = \dfrac{1 \cdot (\sqrt{3} - \sqrt{2})}{(\sqrt{3} + \sqrt{2})(\sqrt{3} - \sqrt{2})} = \dfrac{(\sqrt{3} - \sqrt{2})}{3 - 2} = \sqrt{3} - \sqrt{2}.$

Here for fractions with denominators of the form $\sqrt{a} \pm \sqrt{b}(a, b \in \mathbb{N}, a \neq b)$, we alway multiply both the denominator and numerator by their conjugate $\sqrt{a} \mp \sqrt{b}$, as in this way the denominator becomes $(\sqrt{a} + \sqrt{b})(\sqrt{a} - \sqrt{b}) = (\sqrt{a})^2 - (\sqrt{b})^2 = a - b$, an integer. Such a procedure is called rationalizing the denominator.

If we allow polynomials as radicands, we get radical expressions. The expressions built from radical expressions by finitely many steps of addition, subtraction, multiplication, division and compositions are again radical expressions. For example, $\sqrt{3}$, $\sqrt[3]{a}$, $1 - \sqrt{2 - x^2}$, $\sqrt[4]{x^y - y^2 + 1}$, $\dfrac{\sqrt[4]{t}}{\sqrt[3]{2t - 1}}$ are all radical expressions. To simplify radical expressions with variables, we recall the formula $(\sqrt[n]{c^n}) = \begin{cases} |c| & \text{if } n \text{ is even} \\ c & \text{if } n \text{ is odd} \end{cases}$. Particularly,

- $\sqrt{a^2} = |a|$ for any $a \in \mathbb{R}$;

- $\sqrt[3]{a^3} = a$ for any $a \in \mathbb{R}$.

Examples.

(1) $\sqrt{x^4} = |x^2| = x^2$ for all $x \in \mathbb{R}$;

(2) If $x < 1$, $x - 1 < 0$, so $\sqrt{(x - 1)^2} = |x - 1| = 1 - x$;

(3) if $x < 0, y > 0$, $\sqrt{75x^2y} = \sqrt{75}\sqrt{x^2}\sqrt{y} = 5\sqrt{3}|x|\sqrt{y} = -5x\sqrt{3y}$;

(4) When $x < 0 < y$, $xy < 0$ and $|xy| = -xy$, hence

$$\sqrt{\frac{1}{x^2y^2}} = \frac{1}{\sqrt{x^2y^2}} = \frac{1}{|xy|} = -\frac{1}{xy};$$

(5) If $y < 0 < x$, $y - x < 0$, $xy < 0$, so

$$\sqrt{\frac{1}{x} - \frac{1}{y}} = \sqrt{\frac{y - x}{xy}} = \sqrt{\frac{x - y}{-xy}} = \frac{\sqrt{x - y}}{\sqrt{-xy}} = \frac{\sqrt{(x - y)(-xy)}}{-xy} = -\frac{\sqrt{xy^2 - x^2y}}{xy};$$

(6) When $x \neq -\dfrac{1}{2}$, $\sqrt[3]{\dfrac{1}{(2x + 1)^3}} = \dfrac{1}{\sqrt[3]{(2x + 1)^3}} = \dfrac{1}{2x + 1}.$

Before applying the formula $\sqrt{ab} = \sqrt{a}\sqrt{b}$ or $\sqrt{\dfrac{a}{b}} = \dfrac{\sqrt{a}}{\sqrt{b}}$, we must make sure $\sqrt{a}, \sqrt{b}$ are well-defined, i.e. $a, b \geq 0$ (b also should not be 0 if it is the denominator). In example (5), as $a = y - x < 0, b = xy < 0$, we change $\dfrac{a}{b}$ into $\dfrac{-a}{-b}$ to make both the denominator and numerator positive.

Exercises.

1. True or false.

(1) $\sqrt{3} + \sqrt{2} = \sqrt{3+2} = \sqrt{5}$.

(2) $\sqrt{5} \times \sqrt{7} = \sqrt{5 \times 7} = \sqrt{35}$.

(3) $\sqrt{3^2 + 4^2} = 3 + 4 = 7$.

(4) $\sqrt{100} - \sqrt{64} = \sqrt{100 - 64} = 6$.

(5) $\sqrt{(-1)(-4)} = \sqrt{-1}\sqrt{-4}$ does not exist.

(6) $\sqrt[3]{25 \times 27 \times 40} = 30$.

(7) $\sqrt{18} + \sqrt{8} - \sqrt{50} = 0$.

(8) $\sqrt{16} - \sqrt{25} = \sqrt[3]{-1}$.

(9) If $(-\sqrt{a})^2 = a$, then $a = 0$.

(10) $\sqrt{4 - 4\pi + \pi^2} = 2 - \pi$.

(11) $\sqrt[3]{-8 \times (-27)} = \sqrt[3]{-8}\sqrt[3]{-27} = 6$.

(12) $\sqrt{\sqrt[3]{-2}} = \sqrt[6]{-2}$.

2. Multiple choice.

(1) Which of the following is not a well-defined radical expression?

  A. $\sqrt{-2\pi}$ B. $\sqrt{x^2 + 4x + 4}$ C. $\sqrt[4]{-x}(x \leq 0)$ D. $\sqrt[3]{-5xy^4}$

(2) Which of the following is in simplest form?

  A. $\sqrt{\dfrac{3xy}{5}}$ B. $\sqrt{4x^2 + 3}$ C. $2\sqrt{\dfrac{3x}{2y}}$ D. $\sqrt{27a^2b}$

(3) $(\sqrt{a} + \sqrt{b})^2 + (\sqrt{a} - \sqrt{b})^2$ simplifies to

  A. $2(a + b)$ B. $2(a - b)$ C. $2a$ D. $2b$

(4) $\dfrac{\sqrt{x} - \sqrt{y}}{\sqrt{x} + \sqrt{y}}$ simplifies to

  A. -1 B. $\dfrac{x + y - 2\sqrt{xy}}{x - y}$ C. $\dfrac{x + y + 2\sqrt{xy}}{x - y}$ D. $\dfrac{x + y}{x - y}$

(5) If $a = \dfrac{4}{1 + \sqrt{5}}, b = \sqrt{5} - 1$, then

  A. $a = -b$ B. $a = \dfrac{1}{b}$ C. $a = b$ D. none of above

(6) If $a > 0, b < 0$, then $\sqrt{a^2} + |b|$ equals

  A. $a + b$ B. $a - b$ C. $b - a$ D. $-a - b$

(7) $\sqrt{1-x} + \dfrac{3x}{3x+1}$ is well-defined when

A. $x \le 1$ and $x \ne -\dfrac{1}{3}$ B. $x \le 1$ and $x \ne \dfrac{1}{3}$

C. $x < 1$ D. $x \le 1$

(8) $\sqrt{|x|-2}$ is well-defined when

A. $x \ge 2$ B. $x \le -2$

C. $x \ge 2$ or $x \le -2$ D. $x \in \mathbb{R}$

(9) Which of the following is true for all $a, b \in \mathbb{R}$?

A. $\sqrt{a^2} + \sqrt{b^2} = a + b$ B. $(\sqrt{a} + \sqrt{b})^2 = a + b$

C. $\sqrt{(a^2+b^2)^2} = a^2 + b^2$ D. $\sqrt{(a+b)^2} = a + b$

(10) $(x-1)\sqrt{-\dfrac{1}{x-1}}$ simplifies to

A. $\sqrt{1-x}$ B. $\sqrt{x-1}$ C. $-\sqrt{1-x}$ D. $-\sqrt{x-1}$

(11) Which of the following is true for all $x \in \mathbb{R}$?

A. $\sqrt{x^2-1} = \sqrt{x+1}\sqrt{x-1}$ B. $\sqrt{x^2} = |x|$

C. $\sqrt{x^2} = (\sqrt{x})^2$ D. $\sqrt[3]{x^3} = \sqrt{x^2}$

(12) If $\sqrt{a^3 + a^2} = -a\sqrt{a+1}$, then

A. $a \le 0$ B. $a \ge -1$ C. $a < 1$ D. $-1 \le a \le 0$

(13) If $x < -4$, $|4 + \sqrt{(x+4)^2}|$ simplifies to

A. $8 + x$ B. $-x$ C. $-8 - x$ D. x

(14) If $y < 0$, $\sqrt{\dfrac{2x}{y}}$ simplifies to

A. $\dfrac{\sqrt{xy}}{y}$ B. $\dfrac{\sqrt{2xy}}{y}$ C. $-\dfrac{\sqrt{2xy}}{y}$ D. $y\sqrt{2xy}$

(15) If $xy \ne 0$, the condition that ensures $\sqrt{4x^2y^3} = -2xy\sqrt{y}$ is

A. $x > 0, y > 0$ B. $x > 0, y < 0$ C. $x < 0, y > 0$ D. $x < 0, y < 0$

(16) When $x < 0, y < 0$, which of the following is true?

A. $\sqrt{x^2y^2} = -xy$ B. $\sqrt{x^4y^2} = x^2y$

C. $\sqrt{16x^3y} = -4x\sqrt{xy}$ D. $\sqrt{4x^4y^4} = -2x^2y^2$

(17) Assume $a, b > 0$, which of the following are similar radicals?

A. $\sqrt{3a^2b}$ and $\sqrt{3ab^2}$ B. $\sqrt{\dfrac{27b}{16a}}$ and $\sqrt{\dfrac{3a^3b^2}{32}}$

C. $\sqrt{\dfrac{4}{3}a^3b^4}$ and $\sqrt{\dfrac{1}{3}a^4b^3}$ D. $\sqrt{\dfrac{2b}{a}}$ and $\sqrt{\dfrac{a}{2b}}$

(18) When $1 \le x \le 2$, $\sqrt{1 - 2x + x^2} + |2 - x|$ simplifies to

A. -1 B. $3 - 2x$ C. $2x - 3$ D. 1

3. Simplify.

(1) $\sqrt{24}$

(2) $\sqrt{27}$

(3) $\sqrt{45}$

(4) $\sqrt{48}$

(5) $-4\sqrt{32}$

(6) $3\sqrt{50}$

(7) $\dfrac{\sqrt{125}}{-5}$

(8) $\dfrac{\sqrt{98}}{14}$

(9) $\dfrac{6}{\sqrt{50}}$

(10) $\dfrac{\sqrt{5}}{5\sqrt{13}}$

(11) $-2\sqrt[3]{48}$

(12) $\dfrac{5\sqrt[3]{-108}}{6}$

(13) $\dfrac{5}{\sqrt{3}}$

(14) $\dfrac{2\sqrt{3}}{\sqrt{8}}$

(15) $\dfrac{2\sqrt{6}}{3\sqrt{17}}$

(16) $\dfrac{\sqrt{3}}{\sqrt{40}}$

(17) $\dfrac{\sqrt{45}}{\sqrt{54}}$

(18) $\dfrac{\sqrt{15}}{\sqrt{5}}$

(19) $\dfrac{-4\sqrt{7}}{\sqrt{60}}$

(20) $\dfrac{\sqrt{11}}{7\sqrt{28}}$

(21) $\dfrac{2}{\sqrt[3]{9}}$

(22) $\dfrac{3}{\sqrt[3]{-2}}$

(23) $\dfrac{\sqrt[3]{6}}{\sqrt[3]{4}}$

(24) $\dfrac{\sqrt[3]{16}}{\sqrt[3]{-5}}$

(25) $\dfrac{\sqrt[3]{25}}{\sqrt[3]{12}}$

(26) $\sqrt{\dfrac{8}{3}}$

(27) $\sqrt{\dfrac{27}{4}}$

(28) $\sqrt{\dfrac{28}{45}}$

(29) $\sqrt{\dfrac{44}{75}}$

(30) $\sqrt{\dfrac{125}{32}}$

(31) $\sqrt[3]{\dfrac{9}{32}}$

(32) $\sqrt[3]{\dfrac{81}{10}}$

4. Simplify.

(1) $\sqrt{6}\sqrt{15}$

(2) $\sqrt{8}\sqrt{6}$

(3) $(-\sqrt{12})(2\sqrt{50})$

(4) $(3\sqrt{5})(5\sqrt{3})$

(5) $\sqrt[3]{12}\sqrt[3]{4}$

(6) $(2\sqrt[3]{6})(3\sqrt[3]{9})$

(7) $\sqrt{30}\sqrt{3\dfrac{2}{3}}\sqrt{\dfrac{2}{5}}$

(8) $\sqrt{5}(\sqrt{15}-\sqrt{7})$

(9) $2\sqrt{6}(3\sqrt{8}-5\sqrt{12})$

(10) $4\sqrt{2}(2\sqrt{12}-5\sqrt{6})$

(11) $(\sqrt{5}+\sqrt{7})(\sqrt{5}-\sqrt{7})$

(12) $(2\sqrt{3}-\sqrt{2})^2$

(13) $(\sqrt{2}+\sqrt{3})(\sqrt{6}-1)$

(14) $(2\sqrt{7}-\sqrt{3})(2\sqrt{5}-3\sqrt{2})$

(15) $(2\sqrt{3}-3\sqrt{2})(5\sqrt{2}+4\sqrt{3})$

(16) $(\dfrac{3}{4}\sqrt{24}+\dfrac{2}{3}\sqrt{6})\sqrt{8}$

(17) $(2\sqrt{3}+5)(\sqrt{3}-2)$

(18) $(2\sqrt{7}-\sqrt{2})(2\sqrt{2}-3\sqrt{7})$

(19) $\sqrt{96}-3\sqrt{28}$

(20) $\sqrt{24}+2\sqrt{54}$

(21) $9\sqrt{3}+12\sqrt{12}-7\sqrt{48}-\sqrt{27}$

(22) $3\sqrt{2}-2\sqrt{\dfrac{1}{2}}+\dfrac{1}{2}\sqrt{8}$

(23) $5\sqrt{8}-\sqrt{\dfrac{25}{2}}$

(24) $8\sqrt{\dfrac{1}{3}}-(\dfrac{4}{\sqrt{3}}+6\sqrt{\dfrac{4}{3}})$

(25) $\sqrt{32}-\sqrt{\dfrac{1}{8}}-\sqrt{0.5}$

(26) $(-2\sqrt{\dfrac{1}{3}}+\sqrt{75})-(\sqrt{18}-2\sqrt{\dfrac{1}{3}})$

(27) $16\sqrt{\dfrac{3}{2}}-5\sqrt{\dfrac{1}{3}}+\dfrac{1}{4}\sqrt{8}-3\sqrt{\dfrac{2}{3}}$

(28) $(\sqrt{18}+\sqrt{24})\div\sqrt{3}$

(29) $\sqrt{2}\div(\dfrac{1}{\sqrt{3}}+\dfrac{1}{\sqrt{2}})$

(30) $\dfrac{5}{4-\sqrt{11}}$

(31) $\dfrac{2}{\sqrt{3}-2}$

(32) $\dfrac{1}{2-3\sqrt{5}}$

(33) $\dfrac{-3}{\sqrt{5} - 2\sqrt{2}}$

(34) $\dfrac{\sqrt{5}}{5 - \sqrt{7}}$

(35) $\dfrac{6}{\sqrt{7} + 2}$

(36) $\dfrac{10}{4 - \sqrt{11}}$

(37) $\dfrac{\sqrt{3} - 1}{\sqrt{3} + 1}$

(38) $\dfrac{\sqrt{2}}{\sqrt{6} - \sqrt{5}}$

(39) $\dfrac{\sqrt{3} - \sqrt{2}}{2\sqrt{2} + \sqrt{3}}$

(40) $\dfrac{11\sqrt{3}}{3 + 2\sqrt{5}}$

(41) $\dfrac{11}{\sqrt{2}(5 - \sqrt{3})}$

(42) $\dfrac{16}{(3\sqrt{2} - \sqrt{6})(\sqrt{7} - \sqrt{11})}$

5. Compare the numbers.

(1) $2\sqrt{15}, 3\sqrt{6}$

(2) $2 + \sqrt{7}, \sqrt{3} + 2\sqrt{2}$

(3) $\sqrt{3} + \sqrt{4}, \sqrt{5} + \sqrt{2}$

(4) $\sqrt{2} - 1, \sqrt{3} - \sqrt{2}, 2 - \sqrt{3}$

(5) $-5\sqrt{\dfrac{1}{5}}, -\dfrac{3}{2}\sqrt{2}$

(6) $\dfrac{1}{2\sqrt{5}}, \dfrac{1}{\sqrt{21}}$

(7) $\sqrt{7}, \dfrac{2}{\sqrt{3} - 1}$

(8) $3 - \sqrt{5}, 2\sqrt{5} - 4$

(9) $2\sqrt{5}, \sqrt{11} + 1$

(10) $\sqrt{2}, \sqrt[3]{3}, \sqrt[4]{4}, \sqrt[5]{5}$

6. Simplify.

(1) $\sqrt{(3 - \pi)^2}$

(2) $\sqrt{(4\sqrt{2} - \sqrt{35})^2}$

(3) $\sqrt{(1 - \sqrt{2})^2} - \sqrt{(2\sqrt{2} - 3)^2}$

(4) $\sqrt{(\sqrt{10} - 3)^2} + \sqrt{(\sqrt{10} - 4)^2}$

(5) $(3\sqrt{\dfrac{1}{2}} - \sqrt{147}) - (9\sqrt{\dfrac{1}{27}} - 4\sqrt{\dfrac{1}{8}})$

(6) $\dfrac{1}{\sqrt{3}(\sqrt{3}+\sqrt{5})+1} - \dfrac{5\sqrt{3}+3\sqrt{5}}{3\sqrt{5}-5\sqrt{3}}$

(7) $(4\sqrt{3}+3\sqrt{2})^2 + [\sqrt{30}(\sqrt{3}-\sqrt{2})]^2$

(8) $\dfrac{\sqrt{3}+\sqrt{7}-2\sqrt{5}}{(\sqrt{7}-\sqrt{5})(\sqrt{3}-\sqrt{5})}$

(9) $\dfrac{\sqrt{5}+\sqrt{7}}{\sqrt{10}+\sqrt{14}+\sqrt{15}+\sqrt{21}}$

(10) $\dfrac{2+\sqrt{30}}{\sqrt{5}+\sqrt{6}-\sqrt{7}}$

(11) $\dfrac{\sqrt{3}-\sqrt{2}+1}{\sqrt{3}+\sqrt{2}-1}$

(12) $(\sqrt{3}+3\sqrt{2}-\sqrt{6})(\sqrt{3}-3\sqrt{2}-\sqrt{6})$

(13) $\dfrac{1}{\sqrt{5}}\left[\left(\dfrac{1+\sqrt{5}}{2}\right)^2 - \left(\dfrac{1-\sqrt{5}}{2}\right)^2\right]$

(14) $\sqrt{\dfrac{165^2 - 124^2}{164}}$

(15) $\dfrac{\sqrt{5}+2\sqrt{7}+3}{(\sqrt{5}+\sqrt{7})(\sqrt{7}+3)}$

(16) $(2\sqrt{2}-\dfrac{1}{3}\sqrt{3}+5\sqrt{6})(3\sqrt{6}-3\sqrt{3})$

(17) $\dfrac{5}{4-\sqrt{11}} - \dfrac{4}{\sqrt{11}-\sqrt{7}} - \dfrac{2}{3+\sqrt{7}}$

(18) $(3\sqrt{5}-5\sqrt{3})^2 - (3\sqrt{5}+5\sqrt{3})^2$

7. Simplify by finding square roots.

(1) $\sqrt{5-2\sqrt{6}}$

(2) $\sqrt{8-2\sqrt{15}}$

(3) $\sqrt{12-2\sqrt{35}}$

(4) $\sqrt{(\sqrt{7}-2)^2} + \sqrt{(\sqrt{7}-3)^2}$

(5) $\sqrt{14-4\sqrt{6}}$

(6) $\sqrt{3-2\sqrt{2}} - \sqrt{4-2\sqrt{3}}$

(7) $\sqrt{3+\sqrt{5}}$

(8) $\sqrt{2-\sqrt{3}}$

(9) $\sqrt{\dfrac{6-2\sqrt{5}}{9}}$

(10) $\sqrt{9+4\sqrt{4+2\sqrt{3}}}$

(11) $\dfrac{\sqrt{2}}{2}\sqrt{\dfrac{\sqrt{5}-\sqrt{3}}{\sqrt{5}+\sqrt{3}}}$

(12) $(\sqrt{9-4\sqrt{5}}-\sqrt{9+4\sqrt{5}})^2$

(13) $\sqrt{5+\sqrt{13-4\sqrt{3}}}$

(14) $\dfrac{1}{\sqrt{11+2\sqrt{30}}}+\dfrac{3}{\sqrt{7+2\sqrt{10}}}$

8. Assume all radicals and fractions below are well-defined. Simplify.

(1) $\sqrt{\dfrac{2x-2y}{x^2}}\div\sqrt{\dfrac{x-y}{2x^2y}}$

(2) $(x+2\sqrt{xy}+y)\div(\sqrt{x}+\sqrt{y})$

(3) $2\sqrt{\dfrac{m}{x}}\left(-3x\sqrt{\dfrac{x}{n}}\right)\dfrac{\sqrt{mn}}{6}$

(4) $x\sqrt{\dfrac{2y}{x}}\div y\sqrt{\dfrac{2x}{y}}$

(5) $-6\sqrt{\dfrac{2m-4n}{x^2}}\div\left(\dfrac{4}{5}\sqrt{\dfrac{m-2n}{2mx^2}}\right)$

(6) $(\sqrt{xy}+2\sqrt{\dfrac{y}{x}}-\sqrt{\dfrac{x}{y}}+\sqrt{\dfrac{1}{xy}})\sqrt{xy}$

(7) $\dfrac{3}{xy}\sqrt{xy^3}\left(\dfrac{2}{9}\sqrt{x^2y}\div\dfrac{1}{3}\sqrt{\dfrac{y}{x}}\right)$

(8) $\dfrac{1}{2}\left(\sqrt{\dfrac{2y}{x}}+\sqrt{\dfrac{2x}{y}}\right)\div\sqrt{x^2+2xy+y^2}\,(x,y>0)$

(9) $\dfrac{\sqrt{x+1}+\sqrt{x}}{\sqrt{x+1}-\sqrt{x}}$

(10) $\dfrac{2}{3}\sqrt{27a^3}-a^2\sqrt{\dfrac{3}{a}}+6a\sqrt{\dfrac{a}{3}}-\dfrac{a}{2}\sqrt{108a}$

(11) $\dfrac{x}{2}\sqrt{4x}+6x\sqrt{\dfrac{x}{9}}-4x^2\sqrt{\dfrac{1}{x}}$

(12) $\sqrt{\dfrac{x}{y}+\dfrac{y}{x}+2}-\sqrt{\dfrac{x}{y}}-\sqrt{\dfrac{y}{x}}\,(x,y>0)$

(13) $\dfrac{1+x^2+x\sqrt{1+x^2}}{x+\sqrt{1+x^2}}$

(14) $\dfrac{\sqrt{a+1}+\sqrt{a-1}}{\sqrt{a+1}-\sqrt{a-1}}+\dfrac{\sqrt{a+1}-\sqrt{a-1}}{\sqrt{a+1}+\sqrt{a-1}}$

(15) $\left(\sqrt{x} + \dfrac{1}{\sqrt{y}}\right)^2 \left(x - \dfrac{2}{y}\sqrt{xy} + \dfrac{1}{y}\right)$ (16) $\left(\sqrt{x} + \dfrac{y - \sqrt{xy}}{\sqrt{x} + \sqrt{y}}\right) \div \left(\dfrac{\sqrt{x}}{\sqrt{x} + \sqrt{y}} + \dfrac{\sqrt{y}}{\sqrt{x} - \sqrt{y}}\right)$

(17) $\dfrac{x - y}{\sqrt{x} + \sqrt{y}} - \dfrac{x + y - 2\sqrt{xy}}{\sqrt{x} - \sqrt{y}}$ (18) $\dfrac{x\sqrt{y} - y\sqrt{x}}{x\sqrt{y} + y\sqrt{x}} - \dfrac{y\sqrt{x} + x\sqrt{y}}{y\sqrt{x} - x\sqrt{y}} \, (x \neq y)$

(19) $\dfrac{1}{m - n}\sqrt{(n - m)^3} - (m - n)\sqrt{\dfrac{1}{n - m}}$

(20) $\left(\dfrac{\sqrt{x}}{\sqrt{x} - \sqrt{y}} + \dfrac{\sqrt{y}}{x - y}\dfrac{x + 2\sqrt{xy} + y}{\sqrt{x} + \sqrt{y}}\right) \div (\sqrt{x} + \sqrt{y})$

(21) $\left(x^2\sqrt{\dfrac{b}{a}} + \dfrac{xy}{a}\sqrt{ab} + \dfrac{b}{a}\sqrt{\dfrac{a}{b}}\right) \div x^2 y^2\sqrt{\dfrac{b}{a}} \, (a, b > 0)$

9. Simplify.

(1) $\sqrt{(x - 2)^2}$ (2) $\sqrt{(x - 2)^4}$

(3) $\sqrt[4]{(x - 2)^4}$ (4) $\sqrt{x^4 + x^2 y^2} \, (x \le 0)$

(5) $\sqrt{\dfrac{-2}{x - y}} \, (x < y)$ (6) $\sqrt{\dfrac{y}{x^2} + \dfrac{x}{y^2}} \, (x, y > 0)$

(7) $(x + y)\sqrt{\dfrac{3(x - y)}{4(x + y)}} \, (x < -|y|)$ (8) $\sqrt{\dfrac{b^2 x^2}{a^2} + 1} \, (a < 0)$

(9) $(x - 1)\sqrt{\dfrac{2x}{x - 1}} \, (x < 0)$ (10) $\sqrt{a + 2\sqrt{a - 1}} + \sqrt{a - 2\sqrt{a - 1}} \, (1 \le a \le 2)$

(11) $\sqrt{x^6 - x^2 y^4} \, (x < y < 0)$ (12) $\sqrt{(x - 2)^2} + \sqrt{(1 - 2x)^2} \, (x > 2)$

(13) $\sqrt{(x-2)^2} + |2x-1|(\dfrac{1}{2} < x < 2)$ (14) $\sqrt{x^2 - 2x + 1} - \sqrt{x^2 + 2x + 1}(x^2 - 1 \geq 0)$

(15) $\sqrt{x-y}\sqrt{x^2 - y^2}(x > y > 0)$ (16) $\dfrac{x}{x-y}\sqrt{x^3 - 2x^2 y + xy^2}(0 < x < y)$

(17) $\dfrac{2}{x-1}\sqrt{x^2 - 2x + 1}(x < 1)$ (18) $\sqrt{(a - \dfrac{1}{a})^2 + 4}(a < 0)$

(19) $\sqrt{a^2 - \dfrac{1}{2}a + \dfrac{1}{16}}(a < \dfrac{1}{4})$ (20) $\sqrt{9 - 12a + 4a^2} - \sqrt{4a^2 + 4a + 1}(2a + 1 < 0)$

(21) $\sqrt{x^2 - 8x + 16} + \sqrt{x^2 - 6x + 9}$ (22) $\dfrac{\sqrt{x^2 - 4x + 4}}{x-2} + \dfrac{\sqrt{x^2 - 2x + 1}}{x-1}$

10. Find the conditions so that the equations are valid.

(1) $\sqrt{4x^2} = -2x$ (2) $\sqrt{(x-2)^2} = 2 - x$

(3) $\sqrt{x^2 - 9} = \sqrt{x+3}\sqrt{x-3}$ (4) $\sqrt{x^2 + 10x + 25} - \sqrt{4 - 4x + x^2} = 7$

11. Evaluate.

(1) $\dfrac{a}{b} - \dfrac{b}{a}$ when $a = \sqrt{2}, b = \sqrt{3}$

(2) $a^3 - b^3$ when $a = \sqrt{2} + 1, b = \sqrt{2} - 1$

(3) $x^2 y - xy^2$ when $x = 3 + 2\sqrt{2}, y = 3 - 2\sqrt{2}$

(4) $\left(\dfrac{b}{a} - \dfrac{a}{b}\right) \div \dfrac{1}{ab}$ when $a = \sqrt{3} - 1, b = \sqrt{3} + 1$

(5) $\dfrac{\sqrt{b}}{\sqrt{a} - \sqrt{b}} - \dfrac{\sqrt{b}}{\sqrt{a} + \sqrt{b}}$ when $a = \dfrac{1}{2}, b = \dfrac{1}{3}$

(6) $\dfrac{1}{x} - \dfrac{1}{y}$ when $x = \dfrac{\sqrt{5} - \sqrt{2}}{3}$, $y = \dfrac{\sqrt{5} + \sqrt{2}}{3}$

(7) $\dfrac{ab + a + b + 1}{a^2 + b^2}$ when $a = \dfrac{\sqrt{3} - \sqrt{2}}{\sqrt{3} + \sqrt{2}}$, $b = \dfrac{\sqrt{3} + \sqrt{2}}{\sqrt{3} - \sqrt{2}}$

(8) $\dfrac{\sqrt{a} - \sqrt{b}}{\sqrt{a} + \sqrt{b}}$ when $a + b = 6$, $ab = 4$ and $a > b$

(9) $\dfrac{2a^2 - 5a + 2}{2a - 1}$ when $a = \dfrac{\sqrt{2}}{\sqrt{2} - 1}$

(10) $\sqrt{a^2 - 4a + 4} - \sqrt{4a^2 - 4a + 1}$ when $|a - 1| = 2$

(11) $\dfrac{a + b}{a^2 + b^2}$ when $a = \dfrac{\sqrt{2} - 1}{\sqrt{2} + 1}$ and $b = \dfrac{\sqrt{2} + 1}{\sqrt{2} - 1}$

(12) $\dfrac{1 - 2x + x^2}{x - 1} - \dfrac{\sqrt{x^2 - 2x + 1}}{x^2 - x}$ when $x = \dfrac{1}{2 + \sqrt{3}}$

(13) $(\sqrt[3]{a^2} - \sqrt[3]{ab} + \sqrt[3]{b^2})(\sqrt[3]{a} + \sqrt[3]{b}) + (\sqrt[4]{a} + \sqrt[4]{b})(\sqrt[4]{a} - \sqrt[4]{b})(\sqrt[4]{a^2} + \sqrt[4]{b^2})$ when $a = 2, b = \dfrac{3}{2}$

(14) $x + \sqrt{4x^2 + 2} - \sqrt{x^2 - 2x + 1}$ when $x = \dfrac{1}{2}$

(15) $\dfrac{\sqrt{\sqrt{x} + 1} - \sqrt{\sqrt{x} - 1}}{\sqrt{\sqrt{x} + 1} + \sqrt{\sqrt{x} - 1}}$ when $x = 8 + 4\sqrt{3}$

(16) $\left(\dfrac{\sqrt{x}}{\sqrt{x} + 1} - \dfrac{1 - \sqrt{x}}{\sqrt{x}}\right) \div \left(\dfrac{\sqrt{x}}{\sqrt{x} + 1} + \dfrac{1 - \sqrt{x}}{\sqrt{x}}\right)$ when $x = \sqrt{9 - 2\sqrt{14}}$

12. Solve the equations.

(1) $\sqrt{3}(x+1) = \sqrt{2}(x-1)$

(2) $(\sqrt{3} - \sqrt{2})^2 x - \dfrac{1}{\sqrt{3} + \sqrt{2}} = 0$

(3) $\dfrac{\sqrt{2}-1}{\sqrt{2}+1} x = (\sqrt{2}+1)^2(x+1)$

(4) $(2 - \sqrt{2} - \sqrt{3})x - \sqrt{3} - \sqrt{2} = -\sqrt{6}x$

13. Factor in the set of real numbers.

(1) $x^2 - 5$

(2) $x^4 - 6x^2 + 5$

(3) $x^2 - 2\sqrt{3}x + 3$

(4) $(x^2 - 1)(x^2 + 2) - 40$

14. Find values.

(1) If the integer and decimal part of $\dfrac{1}{2 - \sqrt{3}}$ are denoted by a, b separately, find $\dfrac{a - b - 2}{a + b}$.

(2) If the decimal parts of $9 + \sqrt{11}$ and $9 - \sqrt{11}$ are denoted by a, b separately, find (i)$a+b$, (ii)$a - b$ and (iii)$\dfrac{2}{a} + b$.

9.3 Rational and Real Exponents

We have defined powers with integer exponents by the following, given $n \in \mathbb{Z}^+$, $a \in \mathbb{R}$,

$$a^n = \overbrace{a \cdots a}^{n \text{ times}},$$
$$a^0 = 1 \text{ if } a \neq 0,$$
$$a^{-n} = \frac{1}{a^n} \text{ if } a \neq 0.$$

For example, $5^0 = 1$, $3^{-1} = \frac{1}{3^1} = \frac{1}{3}$, $4^{-2} = \frac{1}{4^2} = \frac{1}{16}$, $(-2)^{-3} = \frac{1}{(-2)^3} = \frac{1}{-8} = -\frac{1}{8}$. Particularly,

$$a^{-1} = \frac{1}{a}(a \neq 0);$$
$$\left(\frac{a}{b}\right)^{-1} = \frac{b}{a}(ab \neq 0).$$

Exponents can also be rational numbers. Suppose $\frac{m}{n} \in \mathbb{Q}$ with $m, n \in \mathbb{Z}^+$. For any $a \in \mathbb{R}$ such that $\sqrt[n]{a}$ exists, define

$$a^{\frac{m}{n}} = (\sqrt[n]{a})^m = \sqrt[n]{a^m},$$
$$a^{-\frac{m}{n}} = \frac{1}{a^{\frac{m}{n}}} = \frac{1}{\sqrt[n]{a^m}}(a \neq 0).$$

For example, $25^{\frac{1}{2}} = \sqrt{25} = 5$, $27^{\frac{1}{3}} = \sqrt[3]{27} = 3$, $(-27)^{\frac{2}{3}} = (\sqrt[3]{-27})^2 = (-3)^2 = 9$, $9^{-\frac{1}{2}} = \frac{1}{9^{\frac{1}{2}}} = \frac{1}{\sqrt{9}} = \frac{1}{3}$.

Powers with irrational exponents are defined by approximations. For any $b > 0, b \notin \mathbb{Q}$, b can be approximated by rational numbers $\tilde{b}$. As a result, a^b is the unique real number approximated by all such $a^{\tilde{b}}$. Here the base a must be nonnegative as $\tilde{b}$ may have even denominators, which correspond to roots of even orders that require the radicand be nonnegative. For negative irrational exponents, a^{-b} is defined to be the reciprocal $\frac{1}{a^b}$ of a^b. $2^{\sqrt{2}}$, $5^{-2\sqrt{3}}$, 3^π are examples of irrational exponents.

We successfully extended the definition of powers, now with real exponents. The definition of a^b together with the conditions of existence are summarized in the following table.

	$b = n \in \mathbb{Z}^+$	$a^n = \overbrace{a \cdots a}^{n}\ (a \in \mathbb{R})$
$b > 0$	$b = \frac{m}{n}(m, n \in \mathbb{Z}^+)$	$a^{\frac{m}{n}} = \sqrt[n]{a^m} = (\sqrt[n]{a})^m\ (a \in \mathbb{R}, \sqrt[n]{a}\text{ exists})$
	$b \in \mathbb{R}^+ - \mathbb{Q}^+$	$a^b \doteq a^{\tilde{b}}\ (b \doteq \tilde{b} \in \mathbb{Q}^+, a \geq 0)$
$b = 0$		$a^b = 1\ (a \neq 0)$
$b < 0$		$a^b = \frac{1}{a^b}\ (a^b\text{ exists}, a \neq 0)$

The rules of powers are still valid. For any $a, b, \eta, \delta \in \mathbb{R}$, whenever the powers appeared in the formula exist, we have

- $a^\eta b^\eta = (ab)^\eta$; $\frac{a^\eta}{b^\eta} = (\frac{a}{b})^\eta (b \neq 0)$;
- $(a^\eta)^\delta = a^{\eta\delta}$;
- $a^\eta a^\delta = a^{\eta+\delta}$; $\frac{a^\eta}{a^\delta} = a^{\eta-\delta}(a \neq 0)$.

Moreover, $1^\eta = 1$ for all η; $0^\eta = 0$ for all $\eta > 0$; $a^0 = 1$ for all $a \neq 0$. These propositions provides the rules to manipulate the exponents.

Examples. Simplify.

(1) $-2\sqrt{5}\sqrt[3]{10}\sqrt[6]{20}$

(2) $\dfrac{3(a^{\frac{1}{2}}b^{\frac{1}{5}})^3}{2a^{\frac{5}{3}}b^{-\frac{2}{5}}}\ (a,b>0)$

Solutions.

(1) Notice $10 = 2\times 5$ and $20 = 2^2\times 5$, so $\sqrt[3]{10} = \sqrt[3]{5}\sqrt[3]{2}$, $\sqrt[6]{20} = \sqrt[6]{4}\sqrt[6]{5}$. As a result,

$$\begin{aligned}
-2\sqrt{5}\sqrt[3]{10}\sqrt[6]{20} &= -2\times\sqrt{5}(\sqrt[3]{5}\sqrt[3]{2})(\sqrt[6]{2^2}\sqrt[6]{5})\\
&= -2\times 5^{\frac{1}{2}}\times 5^{\frac{1}{3}}\times 2^{\frac{1}{3}}\times (2^2)^{\frac{1}{6}}\times 5^{\frac{1}{6}}\\
&= -2\times 2^{\frac{1}{3}+2\times\frac{1}{6}}\times 5^{\frac{1}{2}+\frac{1}{3}+\frac{1}{6}}\\
&= -2\times 2^{\frac{2}{3}}\times 5^1\\
&= -10\sqrt[3]{4}.
\end{aligned}$$

(2) As $a,b>0$, the fraction and all powers exist.

$$\frac{3(a^{\frac{1}{2}}b^{\frac{1}{5}})^3}{2a^{\frac{5}{3}}b^{-\frac{1}{2}}} = \frac{3a^{\frac{3}{2}}b^{\frac{3}{5}}}{2a^{\frac{5}{3}}b^{-\frac{2}{5}}} = \frac{3}{2}a^{\frac{3}{2}-\frac{5}{3}}b^{\frac{3}{5}-(-\frac{2}{5})} = \frac{3}{2}a^{-\frac{1}{6}}b^1 = \frac{3b}{2\sqrt[6]{a}} = \frac{3b\sqrt[6]{a^5}}{2a}$$

In the last step, to rationalize the denominator, we multiply both the denominator and numerator by $\sqrt[6]{a^5}$.

Exercises.

1. Multiple choice.

(1) When $a > 0$, which of the following is correct?

    A. $3a^{-2} = \dfrac{1}{3a^2}$

    B. $a^{\frac{2}{3}}\div a^{\frac{1}{2}} = \sqrt[6]{a}$

    C. $(a^{-\frac{1}{2}})^2 = a$

    D. $\sqrt[6]{(-8)^6} = \sqrt[3]{-8} = -2$

(2) $\sqrt{2\sqrt{2\sqrt{2}}}$ equals

    A. $2^{\frac{5}{8}}$ B. $2^{\frac{3}{4}}$ C. $2^{\frac{7}{8}}$ D. 2

(3) Which of the following is correct?

    A. $\sqrt[3]{m^2+n^2} = m^{\frac{2}{3}} + n^{\frac{2}{3}}$

    B. $(\frac{2}{3})^{-2} = (\frac{3}{2})^2 = \frac{9}{4}$

    C. $5^{-\frac{2}{3}}\times 5^{\frac{3}{2}} = 5^{-1}$

    D. $(2^0-1)^0 = 1$

(4) $(\dfrac{3}{2}\times 72^{\frac{1}{2}} + 2\times 75^{\frac{1}{2}}) - (162^{\frac{1}{2}} + 147^{\frac{1}{2}})$ is

    A. $2\sqrt{2}+3\sqrt{3}$ B. $-2\sqrt{2}$ C. $3\sqrt{3}$ D. $-2\sqrt{2}+3\sqrt{3}$

(5) Which of the following is correct?

    A. $\sqrt{2}^{\pi}, \pi^{-\sqrt{2}}$ are not numbers.

    B. $(\dfrac{1}{16})^{-\frac{1}{4}}$ is an irrational number.

    C. $0^a = 0$ for all real number a.

    D. $a^0 = 1$ for any $a \neq 0$.

(6) $\left(\dfrac{x}{x-2}\right)^{-\frac{3}{4}}$ is well-defined when

    A. $x \le 0$ or $x \ge 2$ B. $x < 0$ or $x \ge 2$ C. $x < 0$ or $x > 2$ D. $x \le 0$ or $x \ge 2$

(7) If $(-1)^a = 1$, then

    A. $a \in \mathbb{R}$.

    B. $a > 0$.

    C. a is a fraction with even numerator and odd denominator in lowest terms.

    D. $a > 0$ and a is a fraction with even numerator and odd denominator.

(8) If $1^a = 1$, then

    A. $a \in \mathbb{R}$ B. $a > 0$

    C. a is rational   D. $a > 0$ and a is rational

2. Assume all variables are positive. Denote the powers below by radicals.

(1) $a^{-\frac{1}{2}}$ (2) $b^{-\frac{1}{3}}$

(3) $a^{\frac{1}{5}}$ (4) $b^{\frac{3}{4}}$

(5) $c^{-\frac{3}{5}}$ (6) $d^{\frac{2}{3}}$

(7) $2^{-\frac{1}{6}}$ (8) $3^{-\frac{3}{4}}$

(9) $x^{\frac{4}{9}}$ (10) y^{-8}

3. Assume all radicands are positive. Denote the results by powers.

(1) $\sqrt[5]{3y}$ (2) $-4\sqrt{2ab}$

(3) $\sqrt[3]{x^2}$ (4) $\sqrt[4]{(a+b)^3}$

(5) $\sqrt[3]{(m\ \ n)^2}$ (6) $\sqrt{(m-n)^4}$

(7) $\dfrac{1}{\sqrt[4]{a}}$ (8) $\sqrt[3]{(-a)^7}$

(9) $\sqrt{3xy}$ (10) $\sqrt[5]{(5a-4b)^{10}}$

(11) $a\sqrt{a}$ (12) $\sqrt{p^6 q^5}$

(13) $\dfrac{m^3}{\sqrt{m}}$

(14) $(m^{\frac{1}{4}}n^{-\frac{3}{8}})^8$

(15) $\sqrt[3]{-x-y}$

(16) $5x\sqrt{x^3y^5}$

4. Evaluate.

(1) $8^{\frac{2}{3}}$

(2) $100^{-\frac{1}{2}}$

(3) $(\dfrac{1}{4})^{-3}$

(4) $(\dfrac{16}{81})^{-\frac{3}{4}}$

(5) $25^{\frac{3}{2}}$

(6) $(-64)^{-\frac{2}{3}}$

(7) $(\dfrac{36}{49})^{-\frac{3}{2}}$

(8) $(\dfrac{25}{4})^{-\frac{3}{2}}$

(9) $(-125)^{\frac{4}{3}}$

(10) $81^{-\frac{3}{4}}$

(11) $(\dfrac{1}{\sqrt{2}})^{-2}$

(12) $(2\dfrac{1}{4})^{-\frac{1}{2}}$

(13) $0.01^{\frac{1}{2}}$

(14) $(5\dfrac{4}{9})^{\frac{5}{2}}$

(15) $2^{-\frac{1}{2}} \times 128^{\frac{1}{2}}$

(16) $(\dfrac{1}{32})^{-\frac{1}{5}}$

(17) $(\sqrt{5} - \sqrt{125}) \div \sqrt[4]{5}$

(18) $2\sqrt{3} \times \sqrt[3]{1.5} \times \sqrt[6]{12}$

(19) $16^{\frac{1}{2}} - (\dfrac{1}{16})^{\frac{3}{4}} - (\dfrac{1}{2})^{-3}$

(20) $[-5 + 3 \times (\dfrac{4}{15})^0]^{-2}$

(21) $\sqrt[3]{12}\sqrt[4]{18}$

(22) $\sqrt{60}\sqrt[3]{24}\sqrt[4]{75}$

(23) $\dfrac{\sqrt[3]{8}}{\sqrt[4]{4}}$

(24) $\dfrac{\sqrt[7]{27}}{\sqrt[5]{9}}$

(25) $\dfrac{\sqrt{6}}{\sqrt[3]{3}}$

(26) $\dfrac{\sqrt[3]{18}}{\sqrt{12}}$

5. Assume all variables are positive. Simplify. Denote the results by radicals in the simplest forms.

(1) $a \cdot a^{\frac{1}{3}} \cdot a^{-\frac{1}{2}}$

(2) $x^2 \cdot x^{-\frac{1}{3}} \div x^{-\frac{1}{2}}$

(3) $(a^{-\frac{2}{3}})^{\frac{3}{4}}$

(4) $(x^{\frac{1}{2}} y^{-2})^3$

(5) $\sqrt[3]{\dfrac{m^3}{8n^{-4}}}$

(6) $(-2x^{\frac{1}{4}} y^{-\frac{1}{3}})^{-2}$

(7) $\sqrt{x\sqrt{x}}$

(8) $\sqrt[3]{x^2} \div \sqrt{x^3}$

(9) $\sqrt[3]{x\sqrt{x}}$

(10) $\sqrt{ab^3 \sqrt{ab^5}}$

(11) $\dfrac{x}{\sqrt[3]{x^2}}$

(12) $\sqrt{x\sqrt{\dfrac{x}{y}}}$

(13) $(a^{\frac{1}{2}} - b^{\frac{1}{2}})(a^{\frac{1}{2}} + b^{\frac{1}{2}})$

(14) $(x^{\frac{1}{2}} - y^{\frac{1}{2}}) \div (x^{\frac{1}{4}} + y^{\frac{1}{4}})$

(15) $2x^{\frac{3}{4}}(-5x^{\frac{2}{3}})$

(16) $(2a^{\frac{2}{3}} b^{\frac{1}{2}})(-6a^{\frac{1}{2}} b^{\frac{1}{3}}) \div (-3a^{\frac{1}{6}} b^{\frac{5}{6}})$

(17) $\dfrac{a^2}{\sqrt{a}\sqrt[3]{a^2}}$

(18) $(y^{-\frac{1}{2}} z^{\frac{4}{3}})^6$

(19) $(\dfrac{8a^{-3}}{27b^6})^{-\frac{1}{3}}$

(20) $2x^{-\frac{1}{3}}(\dfrac{1}{2}x^{\frac{1}{3}} - 2x^{-\frac{2}{3}})$

(21) $(-14a^{\frac{6}{5}} b^{-\frac{5}{3}})(\dfrac{2}{7}a^{-\frac{3}{10}} b^{-\frac{1}{2}})$

(22) $\dfrac{7m^{\frac{9}{7}}}{6m^{\frac{5}{14}}}$

(23) $\dfrac{18x^{\frac{3}{5}}y^{\frac{1}{3}}}{24x^{\frac{1}{3}}y^{-\frac{1}{2}}}$

(24) $\left(\dfrac{2x^{\frac{2}{5}}}{3y^{\frac{2}{7}}}\right)^{7}$

(25) $\left(\dfrac{6x^{\frac{3}{4}}}{4y^{\frac{1}{9}}}\right)^{\frac{6}{5}}$

(26) $\dfrac{x^{-2}+y^{-2}}{x^{-\frac{2}{3}}+y^{-\frac{2}{3}}} - \dfrac{x^{-2}-y^{-2}}{x^{-\frac{2}{3}}-y^{-\frac{2}{3}}}$ $(x \neq \pm y)$

6. Evaluate.

(1) $[a^{-\frac{2}{3}}b(ab^{-2})^{-\frac{1}{2}}(a^{-1})^{-\frac{2}{3}}]^{-2}$ when $a = \dfrac{1}{\sqrt{2}}$, $b = \dfrac{1}{\sqrt[3]{2}}$

(2) $\dfrac{x^{\frac{1}{2}}+y^{\frac{1}{2}}}{x^{\frac{1}{2}}-y^{\frac{1}{2}}} - \dfrac{x^{\frac{1}{2}}-y^{\frac{1}{2}}}{x^{\frac{1}{2}}+y^{\frac{1}{2}}}$ when $x = \dfrac{1}{2}$, $y = \dfrac{1}{3}$

9.4 Radical Equations

Equations with variables appearing in radicands are radical equations. The basic idea to solve radical equations is to raise radicals to certain powers so that the radicals can be simplified. Like rational equations, radical equations may have extraneous solutions, so we must plug all candidate solutions back into the equation and check the correctness. The methods to solve radical equations are demonstrated in the following examples.

Examples. Solve the equations.

(1) $2\sqrt{x} - 3 = 1$

(2) $\sqrt[3]{4a+5} = -3$

(3) $\sqrt{x} = -3$

(4) $\sqrt{x+4} - \sqrt{x-1} = 1$

Solutions.

(1) x appears only in $\sqrt{x}$, so we keep it on the left side and move all constant terms to the right side to get $2\sqrt{x} = 1 + 3 = 4$, or equivalently $\sqrt{x} = 2$. Square both sides, we get $x = 4$. If we plug $x = 4$ into the equation, we get $2\sqrt{4} - 3 = 2 \times 2 - 3 = 1$, so $x = 4$ is the solution to the original equation.

(2) To simplify the radical sign, finding cubes on both side, the original equation becomes $(\sqrt[3]{4a+5})^3 = (-3)^3$, i.e. $4a + 5 = -27$. Solving the linear equation, we get $a = -8$. It is easy to check that $a = -8$ is the solution to the original equation.

(3) If we square both sides, we get $x = 9$. Plugging $x = 9$ into the equation we get $\sqrt{9} = 3 = -3$, which is false, so $x = 9$ is an extraneous solution. The equation has no solution. This should be clear as $\sqrt{a}$ is always nonnegative whenever it exists.

(4) There are two square roots in the equation. Separating them to two sides first, we get $\sqrt{x+4} = 1 + \sqrt{x-1}$. Then we raise both sides to the 2nd power to get rid of the radical sign on the left side, i.e. $(\sqrt{x+4})^2 = (1 + \sqrt{x-1})^2$. Simplifying the new equation, we get $x + 4 = 1 + 2\sqrt{x-1} + (x-1)$. There is still a radical sign left; we isolate it on the right side and find squares again to get rid of it.

$$x + 4 = 1 + 2\sqrt{x-1} + (x-1)$$
$$x + 4 - 1 - (x-1) = 2\sqrt{x-1}$$
$$4 = 2\sqrt{x-1}$$
$$2 = \sqrt{x-1}$$
$$4 = x - 1$$
$$x = 5.$$

If we plug $x = 5$ into the equation, the left side is $\sqrt{5+4} - \sqrt{5-1} = \sqrt{9} - \sqrt{4} = 3 - 2 = 1$, which equals the right side, so $x = 5$ is the solution to the original equation.

Exercises.

1. Multiple choice.

(1) Which of the following is not a radical equation?

    A. $\sqrt{x} = -1$ B. $\sqrt{x^2} = 0$

    C. $\sqrt{2}x = \sqrt{3}$ D. $\sqrt[3]{x} + \sqrt[5]{x^3} = \sqrt{\pi}$

(2) Which of the following equations has a solution?

    A. $\sqrt{x-2} = 1 - x$ B. $\sqrt{x-2} + \sqrt{2x-1} = 0$

    C. $\sqrt{x+1} + 2 = 0$ D. $\dfrac{1}{x-1} + 1 = \dfrac{x}{x-1}$

(3) The solution set of $\sqrt{x^4} = 4$ is

    A. $\{2\}$ B. $\{\pm 2\}$ C. $\{2^{\frac{1}{2}}\}$ D. $\pm\{2^{\frac{1}{2}}\}$

(4) The solution set of $\sqrt{x} = \sqrt[3]{-2^2}$ is

    A. $\{-2^{\frac{4}{3}}\}$ B. $\emptyset$ C. $\{\sqrt[3]{16}\}$ D. $\{\pm 2^{\frac{4}{3}}\}$

(5) If $\sqrt{2\sqrt{x} + 4} = \sqrt{x} + 1$, then

    A. $x = 1$ B. $x = 2$ C. $x = 3$ D. $x = 4$

(6) If $\sqrt[4]{x^3} = \dfrac{1}{\sqrt[4]{x^3}}$, then

    A. $x = 1$ B. $x = \pm 1$

    C. x does not exist   D. $x = -1$

2. Solve the equations.

(1) $\sqrt{6x} = 3$ (2) $\sqrt{2+y} = -1$

(3) $\sqrt{2x+3} = 5$

(4) $\sqrt{5x-4} - 3 = 0$

(5) $\sqrt{x+1} + 6 = 0$

(6) $\sqrt{4s-3} - 9 = 0$

3. Solve the equations.

(1) $\sqrt{5x-9} = \sqrt{4x+7}$

(2) $\sqrt{1-x} = \sqrt{x-1}$

(3) $\sqrt{5t+3} = \sqrt{2t-9}$

(4) $\sqrt{6y+7} - \sqrt{2y+12} = 0$

(5) $\sqrt{2x-3} - \sqrt{x+1} = 0$

(6) $\sqrt{x-2} - \sqrt{x+1} = 0$

(7) $\sqrt{9a-1} - \sqrt{-5a+20} = 0$

(8) $\sqrt{2x-8} - \sqrt{-3x+9} = 0$

(9) $2\sqrt{3x+5} - \sqrt{9x+17} = 0$

(10) $3\sqrt{7b-2} - 4\sqrt{5b-9} = 0$

4. Solve the equations.

(1) $\sqrt{x^2-7} = 5$

(2) $\sqrt{9-x^2} = 2$

(3) $\sqrt{2x^3-5} = 7$

(4) $\sqrt{4x^2+15} = 11$

(5) $\sqrt{9-5x^2} = 4$

(6) $\sqrt{7-5x^2} = 2$

5. Solve the equations.

(1) $\sqrt{x+1} + \sqrt{x-1} = 1$

(2) $\sqrt{3x+5} - \sqrt{3x-1} = 2$

(3) $\sqrt{4x-5} - \sqrt{4x-3} = 3$

(4) $\sqrt{2x+13} + \sqrt{2x-7} = 5$

(5) $\sqrt{5x-4} + \sqrt{5x+4} = 2\sqrt{2}$

(6) $\sqrt{2x-9} + \sqrt{2x+9} = 6$

(7) $\sqrt{13-6x} + \sqrt{15-6x} = \sqrt{2}$

(8) $\sqrt{3-2x} - \sqrt{-1-2x} = 8$

6. Solve the equations.

(1) $\sqrt[3]{2x-3} = -2$

(2) $\sqrt[3]{4x^2+2} = 3$

(3) $\sqrt[3]{5x+3} = -1$

(4) $\sqrt[3]{25x^2-216} = -7$

(5) $\sqrt[3]{5x-2} = \sqrt[3]{17-4x}$

(6) $\sqrt[3]{7x-5} = \sqrt[3]{2x+4}$

(7) $\sqrt[3]{12x-5} = -\sqrt[3]{3x-16}$

(8) $\sqrt[3]{2x-1} + \sqrt[3]{3x+16} = 0$

7. Solve the equations.

(1) $\sqrt{2x-3} + \sqrt{y+2} = 0$

(2) $\sqrt[5]{x^5} = -\sqrt{2}$

(3) $\sqrt[4]{x^4} = \sqrt[3]{2}$

(4) $\begin{cases} \sqrt{2x+16} - \sqrt{2y+22} = -\sqrt{6} \\ y - x = 12 \end{cases}$

Quadratic Equations

We mentioned that the major task of the elementary algebra is essentially to solve polynomial equations. Our ancestors achieved fruitful results in this field. Linear equations are the simplest polynomial equations, which we have learned in Chapter 5. Among all other polynomial equations, quadratic equations — polynomial equations of degree 2, are comparatively simple. In this chapter we shall learn how to solve such equations. If we keep increasing the degree of polynomial equations, we get cubic, quartic, quintic, $\cdots$ equations. However, we won't cover how to solve these equations in our book. Interested readers may refer to textbooks of abstract algebra.

Quadratic equations can be solved either by completing squares or by the quadratic formula. Besides, there is an interesting relation between the coefficients of a quadratic equation with the sum and product of its solutions, which is summarized in Vieta's formulas.

It is natural that square roots appear in the solutions of quadratic equations. As negative numbers have no square roots, not every quadratic equation has solutions. If we extend the real number system by making all negative numbers have square roots, finally we get the largest number system — the complex number system. The quadratic formula still hold with complex numbers, and every quadratic equation has two complex solutions.

As an application, now we can solve all equations that can be transformed into quadratic equations. Whether an equation has solutions highly depends on the number system we work with. By default, in our book, to solve equations means to solve equations in the set of real numbers, i.e. to find real number solutions. If we want to solve equations in other number systems, we will state it explicitly.

10.1 Solving Quadratic Equations

A quadratic equation is a polynomial equation in one variable in which the highest degree of the variable is 2. Any quadratic equation can be simplified to the standard form

$$ax^2 + bx + c = 0 \quad (a \neq 0).$$

For example, $x^2 = 1$, $x^2 + 2x - 3 = 0$, $2x - x^2 = 0$ are all quadratic equations.

There are four methods to solve quadratic equations: finding square roots, factoring, completing squares and applying the quadratic formula.

Method I: Finding Square Roots. If the quadratic equation is of the form $x^2 = c$ ($c \geq 0$), x must be the square roots of c. Immediately we get $x = \pm\sqrt{c}$. For example, to solve $2x^2 = 3$, first we simplify it to $x^2 = \frac{3}{2}$, then $x = \pm\sqrt{\frac{3}{2}} = \pm\frac{\sqrt{6}}{2}$. Notice that when $c < 0$, $x^2 = c$ has no (real) solutions.

Method II: Factoring. If $a \cdot b = 0$, then $a = 0$ or $b = 0$. According to this, if we can factor $ax^2 + bx + c$ as a product of two linear factors $(p_1 x + q_1)(p_2 x + q_2)$, then it is easy to see that $ax^2 + bx + c = (p_1 x + q_1)(p_2 x + q_2) = 0$ is equivalent to $p_1 x + q_1 = 0$ or $p_2 x + q_2 = 0$, which can be solved easily. Let us solve $2x^2 - 3x - 2 = 0$ as an example. $2x^2 - 3x - 2$ can be factored as $(2x + 1)(x - 2)$, so $2x^2 - 3x - 2 = 0$ is equivalent to $2x + 1 = 0$ or $x - 2 = 0$. Thus, the solution to this equation is $x = 2$ or $x = -\frac{1}{2}$.

Method III: Completing Squares. The polynomial $ax^2 + bx + c$ may not be easy to factor, so we develop the method of completing squares, which works for all quadratic equations. Any polynomial of the form $x^2 \pm px$ can be made into a perfect square by adding a constant term q. In fact,

$$q = (\frac{p}{2})^2,$$

since by the perfect square formula, we have $x^2 \pm px + \frac{p^2}{4} = (x \pm \frac{p}{2})^2$. Shortly speaking, to find q, we divide the coefficient of x by 2 and square the number we get. For example, to make $x^2 + 3x$ into a perfect square, we add $(\frac{3}{2})^2 = \frac{9}{4}$ and $x^2 + 3x + \frac{9}{4} = (x + \frac{3}{2})^2$. The following example demonstrates how to solve quadratic equations by completing squares.

Example. Solve $2x^2 - 3x - 2 = 0$ by completing squares.

Solutions. (i) Divide the coefficient of x^2 on both sides to make x^2 have coefficient 1.

$$x^2 - \frac{3}{2}x - 1 = 0.$$

(ii) Complete the square for the first two terms $x^2 + px$.

$$[x^2 - \frac{3}{2}x] - 1 = 0,$$
$$[x^2 - \frac{3}{2}x + (\frac{3}{4})^2] - (\frac{3}{4})^2 - 1 = 0,$$
$$(x - \frac{3}{4})^2 = \frac{25}{16}.$$

Here we add the term $(\frac{3}{4})^2$ to complete the square. We do not want to modify the equation into an inequivalent one, so we subtract the same number in order to keep the equation unchanged.

(iii) It is easy to see that $x - \frac{3}{4}$ must be the square roots of the right side, thus we have $x - \frac{3}{4} = \pm\sqrt{\frac{25}{16}} = \pm\frac{5}{4}$. Solving $x - \frac{3}{4} = \frac{5}{4}$ and $x - \frac{3}{4} = -\frac{5}{4}$ separately, we get the solutions

$x = -\dfrac{1}{2}$ and 2.

Method IV: Quadratic Formula. If we want to solve the standard quadratic equation $ax^2 + bx + c = 0 (a \neq 0)$ by completing squares, we should proceed as follows:

$$x^2 + \frac{b}{a}x + \frac{c}{a} = 0 \quad \text{(dividing the coefficient } a \text{ of } x^2\text{)};$$

$$\left[x^2 + \frac{b}{a}x + \left(\frac{b}{2a}\right)^2\right] - \left(\frac{b}{2a}\right)^2 + \frac{c}{a} = 0 \quad \text{(completing the square)};$$

$$\left(x + \frac{b}{2a}\right)^2 = \frac{b^2 - 4ac}{4a^2} \quad \text{(simplifying)};$$

$$x + \frac{b}{2a} = \pm\frac{\sqrt{b^2 - 4ac}}{2a} \quad \text{(finding square roots on both sides)}.$$

At last we solve for x, which gives us the quadratic formula

$$x = \frac{-b \pm \sqrt{b^2 - 4ac}}{2a}.$$

For example, we solve the same equation $2x^2 - 3x - 2 = 0$ by quadratic formula. Plugging $a = 2, b = -3, c = -2$ into the quadratic formula, we get $x = \dfrac{-(-3) \pm \sqrt{(-3)^2 - 4 \times 2 \times (-2)}}{2 \times 2}$, which simplifies to $x = \dfrac{3 \pm \sqrt{25}}{4} = \dfrac{3 \pm 5}{4} = -\dfrac{1}{2}$ or 2.

To apply the quadratic formula, we need to find the square roots of $b^2 - 4ac$, whose existence depends on the sign of $b^2 - 4ac$. Only when $b^2 - 4ac \geq 0$, the equation $ax^2 + bx + c = 0$ $(a \neq 0)$ has real number solutions.

$$\Delta = b^2 - 4ac$$

is called the discriminant of the quadratic equation. Depending on the sign of Δ, the nature of the solutions $x = \dfrac{-b \pm \sqrt{\Delta}}{2a}$ are classified as

$\Delta > 0$	$\Delta = 0$	$\Delta < 0$
two different real solutions	two equal real solutions	no real solutions

Notice that when $\Delta = 0$, we still say the quadratic equation has two (real) solutions, although the two solutions are equal. As a result, a quadratic equation always has 2 or 0 solutions. If x_0 is a solution to $ax^2 + bx + c = 0$, we also say x_0 is a root of the trinomial $ax^2 + bx + c$.

Examples. Find the number of real roots to the following trinomials.

(1) $x^2 - 2x + 2$

(2) $9x^2 + 6x + 1$

Solutions.

(1) $x^2 - 2x + 2$ has no real roots as $\Delta = (-2)^2 - 4 \times 1 \times 2 < 0$.

(2) $9x^2 + 6x + 1$ has two equal real roots as $\Delta = 6^2 - 4 \times 9 \times 1 = 0$, which should also be clear from the fact that $9x^2 + 6x + 1 = (3x + 1)^2$ is a perfect square.

Exercises.

1. Multiple choice.

(1) Considering the equation $ax^2 + bx + c = 0$, which of the following is correct?

    A. It is always a quadratic equation.

    B. It always has two real solutions if $a \neq 0$.

    C. It always has two real solutions if $b^2 - 4ac \geq 0$.

    D. It has no solutions if $b^2 - 4ac < 0$.

(2) Which of the following is a quadratic equation in one variable?

    A. $x^2 = -1$ B. $2x^2 - 3xy + 4 = 0$

    C. $x^2 - \dfrac{1}{x} = 4$ D. $x^3 - \dfrac{x^2}{2} + 3 = x^3 - 0.5x^2$

(3) The solutions to $(x - 2)^2 = 9$ are

    A. $x = 5, -1$ B. $x = -5, 1$ C. $x = 11, -7$ D. $x = -11, 7$

(4) The solution(s) to $x^2 = 4x$ is/are

    A. $x = 4$ B. $x = 2$ C. $x = 4, 0$ D. $x = 0$

(5) Which of the following has two different real solutions?

    A. $x^2 = 3x - 8$ B. $x^2 + 5x = -10$

    C. $4x^2 - 12x + 9 = 0$ D. $x^2 - 7x = -5x + 3$

(6) If $(a - 3)x^2 + (b + 1)x + c = 0$ is a quadratic equation, then

    A. $a \neq 0$ B. $a \neq 3$

    C. $a \neq 3, b \neq -1$ D. $a \neq 3, b \neq -1, c \neq 0$

(7) If 2 is a root of $2x^2 - 3x - a^2 + 1$, then a is

    A. 1   B. $\sqrt{3}$ C. $-\sqrt{3}$ D. $\pm\sqrt{3}$

(8) If the constant term in the quadratic equation $(m - 1)x^2 + 5x + m^2 - 3m + 2 = 0$ is 0, then m is

    A. 1   B. 2   C. 1, 2   D. 0

(9) If $k^2 x^2 - (2k + 1)x + 1 = 0$ has two different real solutions, then

    A. $k > -\dfrac{1}{4}$ B. $k > -\dfrac{1}{4}$ and $k \neq 0$

    C. $k < -\dfrac{1}{4}$ D. $k \geq -\dfrac{1}{4}$ and $k \neq 0$

(10) When is $(m^2 - 1)x^2 + (m - 2)x + 4m - 3 = 0$ a quadratic equation?

    A. $m \neq -1$ B. $m \neq -1, 0$ C. $m \neq -1, 0, 1$ D. $m \neq \pm 1$

(11) Which of the following is always a quadratic equation?

    A. $ax^2 + bx + c = 0$ B. $ax^2 + 1 = x^2 - x$

    C. $(a^2 + 1)x^2 - (a^2 - 1)x - 2 = 0$ D. $x - \dfrac{1}{x - 3} = 0$

(12) If $2x + 1$ and $2x - 1$ are reciprocals, then $x =$

A. $\pm\dfrac{1}{2}$　　　　B. ± 1　　　　C. $\pm\dfrac{\sqrt{2}}{2}$　　　　D. $\pm\sqrt{2}$

(13) If one and only one solution to $x^2 + nx + m = 0$ is 0, then

　　A. $m = 0, n = 0$　　B. $m = 0, n \neq 0$　　C. $m \neq 0, n = 0$　　D. $m \neq 0, n \neq 0$

(14) If $x^2 + k = 0$ has real solutions, then

　　A. $k \leq 0$　　　　B. $k < 0$　　　　C. $k \geq 0$　　　　D. $k > 0$

(15) For two numbers a, b, if $ab = 0$, then

　　A. $a = b = 0$　　B. $a = 0$　　　　C. $b = 0$　　　　D. $a = 0$ or $b = 0$

(16) Which of the following are equivalent?

　　A. $ab \neq 0$ and $a^2 + b^2 \neq 0$　　　　　B. $ab = 0$ and $a^2 + b^2 = 0$

　　C. $ab \neq 0$ and $\dfrac{a}{b} \neq 0$　　　　　　D. $ab = 0$ and $\dfrac{a}{b} = 0$

(17) If $a \neq 0$, $a + b + c = 0$, $a - b + c = 0$, then the roots to $ax^2 + bx + c$ are

　　A. $1, 0$　　　　　　　　　　　B. $-1, 0$

　　C. $-1, 1$　　　　　　　　　　D. can't be determined

(18) If m is a solution to $x^2 + nx + m = 0$ and $m \neq 0$, then $m + n$ is

　　A. -1　　　　B. 1　　　　C. $\dfrac{1}{2}$　　　　D. $-\dfrac{1}{2}$

2. Fill in the blanks.

(1) The standard form of the quadratic equation $(1+3x)(1-3x) = 2x^2+1$ is _____.

(2) $(m-1)x^2 + (m+1)x + 3m + 2 = 0$ is a quadratic equation when m_____ and a linear equation when m_____.

(3) If $2x^2 + 1$ and $4x^2 - 2x - 5$ are opposites, then $x =$_____.

(4) The discriminant of $x^2 - px = 1$ is _____.

(5) If -1 is a solution to $(2m - 1)x^2 + 3mx + 5 = 0$, then $m =$_____.

(6) When t_____, $x^2 - 3x + t = 0$ has real solutions.

(7) If $(a + b)(a + b + 2) = 8$, then $a + b =$_____.

(8) If the numbers a, b satisfy $a^2 + ab - b^2 = 0$ and $b \neq 0$, then $\dfrac{a}{b} =$_____.

(9) If the length of three edges in a right triangle are three consecutive natural numbers, then the three numbers are _____.

(10) If $x^2 - 2(m + 1)x + 16$ is a perfect square, then $m =$_____.

(11) If 1 is a solution to $x^2 + kx - 2 = 0$, then $k =$_____ and the other solution is _____.

(12) If 1 is a solution to $(m^2 - 1)x^2 - 2mx + 2 = 0$, then $m =$_____ and the equation is a _____ equation.

(13) If -2 is a solution to the quadratic equation $(m - 1)x^2 + (m + 3)x + m^2 + 7 = 0$, then $m =$_____.

(14) If $mx^2 + 5x = 2$ has at least one real solution, then m_____; if $mx^2 + 5x = 2$

has two real solutions, then m_____.

3. Complete squares and find the perfect squares.

(1) $x^2 - 2x +$_____$= ($_____$)^2$ (2) $x^2 - 6x +$_____$= ($_____$)^2$

(3) $x^2 - 3x +$_____$= ($_____$)^2$ (4) $x^2 + 7x +$_____$= ($_____$)^2$

(5) $x^2 - \dfrac{2}{3}x +$_____$= ($_____$)^2$ (6) $x^2 + \dfrac{1}{2}x +$_____$= ($_____$)^2$

(7) $5x^2 + 6x +$_____$= 5($_____$)^2$ (8) $3x^2 + 4x +$_____$= 3($_____$)^2$

(9) $4x^2 + 3x +$_____$= 4($_____$)^2$ (10) $7x^2 - 6x +$_____$= 7($_____$)^2$

(11) $3x^2 - 5x +$_____$= 3($_____$)^2$ (12) $9x^2 + 12x +$_____$= ($_____$)^2$

(13) $4x^2 - 4x +$_____$= ($_____$)^2$ (14) $49x^2 + 21x +$_____$= ($_____$)^2$

(15) $x^2 - \sqrt{3}x +$_____$= ($_____$)^2$ (16) $3x^2 + 2\sqrt{5}x +$_____$= 3($_____$)^2$

(17) $x^2 + 6x + 21 = ($_____$)^2 +$_____

(18) $x^2 - \dfrac{\sqrt{5}}{2}x + 1 = ($_____$)^2 +$_____

(19) $4x^2 + 6x - 2 = ($_____$)^2 +$_____

(20) $3x^2 + 2x + 3 = 3($_____$)^2 +$_____

4. Solve the following equations by finding square roots or factoring.

(1) $x^2 = 9$ (2) $x^2 + 1 = 0$

(3) $2x^2 - 3 = 0$ (4) $7x^2 - 22 = 32$

(5) $100x^2 = -49$ (6) $2y^2 - 18 = 7 - 13y^2$

(7) $(3t - 2)^2 = 16$ (8) $(7x + 1)^2 - 18 = 0$

(9) $5(5x + 3)^2 = 60$ (10) $5(5z - 4)^2 = 12$

(11) $2x^2 + 3x = 0$ (12) $4x^2 = 15x$

(13) $x^2 - 5x - 6 = 0$ (14) $x^2 - 5x + 6 = 0$

(15) $z^2 - 8z + 12 = 0$ (16) $x^2 + 9x - 36 = 0$

(17) $6x^2 - 5x - 21 = 0$

(18) $16x^2 - 24x + 9 = 0$

(19) $4x^2 + 29x + 30 = 0$

(20) $15x^2 - 17x - 42 = 0$

(21) $25x^2 + 20x + 4 = 0$

(22) $36m^2 + 49m - 72 = 0$

(23) $15x^2 + 23x - 14 = 0$

(24) $(x + 1)^2 - (x + 1) = 0$

(25) $(7n + 4)^2 = 5(7n + 4)$

(26) $2(x - 2)^2 - 7(x - 2) + 3 = 0$

5. Solve the following equations by completing squares.

(1) $x^2 + 2x - 1 = 0$

(2) $x^2 - 6x + 3 = 0$

(3) $x^2 - 8x + 21 = 0$

(4) $x^2 + 4x - 4 = 0$

(5) $x^2 - 3x + 4 = 0$

(6) $x^2 + 5x - 2 = 0$

(7) $x^2 - 4\sqrt{3}x + 9 = 0$

(8) $x^2 + 3\sqrt{5}x - 1 = 0$

(9) $x^2 + \sqrt{2}x - 2 = 0$

(10) $9x^2 + 24x + 10 = 0$

(11) $16x^2 + 8x - 5 = 0$

(12) $25x^2 - 20x - 3 = 0$

(13) $3q^2 - 6q + 1 = 0$

(14) $4t^2 + 4t - 3 = 0$

(15) $3y^2 + 4y - 4 = 0$

(16) $4x^2 + 7x + 4 = 0$

(17) $z^2 - z - 1 = 0$

(18) $n^2 - 7n + 10 = 0$

(19) $3x^2 - 5x - 1 = 0$

(20) $2x^2 + 9x + 8 = 0$

6. Solve the following equations using the quadratic formula.

(1) $x^2 + x - 1 = 0$

(2) $3x^2 + 6x - 1 = 0$

(3) $2x^2 - 5x + 1 = 0$

(4) $x(x + 8) = 16$

(5) $2x^2 - 10x = 3$

(6) $(x + 1)(x - 3) = 6$

(7) $3x^2 - 10x + 6 = 0$

(8) $(x - 3)^2 - 2(x + 1) = x - 8$

(9) $(3 - x)^2 + x^2 = 7$

(10) $x^2 + 2\sqrt{3}x - 3 = 0$

(11) $x^2 + 2\sqrt{2}x - 4 = 0$

(12) $(2x + 1)(4x - 2) = (2x - 1)^2 + 2$

(13) $(x + 1)^2 - 2(x - 1)^2 = 4x + 5$

(14) $6x^2 - x - 3 = 0$

(15) $\sqrt{6}x^2 - x - 2\sqrt{6} = 0$

(16) $\sqrt{2}x^2 + 2x + \sqrt{3} = 0$

7. Find discriminants and classify the nature of the solutions.

(1) $-2a^2 - 3a + 3 = 0$

(2) $5t^2 - 21t + 23 = 0$

(3) $16x^2 + 40x + 25 = 0$

(4) $-6y^2 + 9y - 4 = 0$

(5) $25b^2 - 30b + 9 = 0$

(6) $4m^2 - 7m + 2 = 0$

(7) $-7x^2 + 10x - 3 = 0$

(8) $(x+2)(x-7) = -19$

(9) $2t^2 + 6t + 3 = 0$

(10) $2x^2 + 6x + 5 = 0$

10.2   Vieta's Formulas

Vieta's formulas state the relation between the coefficients of a quadratic equation and the sum and product of its two real roots (when they exist).

Suppose $a \neq 0$ and $\Delta = b^2 - 4ac \geq 0$. Let $x = x_1$ and $x = x_2$ be the two real solutions to the quadratic equation $ax^2 + bx + c = 0$ ($x_1 = x_2$ when $\Delta = 0$). Since x_1 is a root of $ax^2 + bx + c$, the trinomial can be divided by $a(x - x_1)$ and the quotient (by long division) must be of the form $x - x_0$ for some $x_0 \in \mathbb{R}$. As a result, $ax^2 + bx + c = a(x - x_1)(x - x_0)$. We claim that $x_0 = x_2$: (i) if $x_1 \neq x_2$, plugging $x = x_2$ into the identity we get $0 = a(x_2 - x_1)(x_2 - x_0)$. Since $a \neq 0, x_2 - x_1 \neq 0$ we must have $x_2 - x_0 = 0$, i.e. $x_0 = x_2$; (ii) if $x_1 = x_2$, plugging $x = x_0$ into the identity we get $ax_0^2 + bx_0 + c = a(x_0 - x_1)(x_0 - x_0) = 0$, so $x = x_0$ is a solution to $ax^2 + bx + c = 0$, which must be $x_1(= x_2)$ by assumption. At this point we get the first conclusion: if x_1, x_2 are two real roots of $ax^2 + bx + c$, then

$$ax^2 + bx + c = a(x - x_1)(x - x_2)$$

i.e. the trinomial $ax^2 + bx + c$ can be factored by solving the corresponding equation $ax^2 + bx + c = 0$. For example, the roots of $2x^2 - 3x - 1$ are $\dfrac{3 \pm \sqrt{17}}{4}$ (by quadratic formula), so $2x^2 - 3x - 1$ can be factored as $2(x - \dfrac{3 + \sqrt{17}}{4})(x - \dfrac{3 - \sqrt{17}}{4})$.

As a second application, $x^2 + bx + c$ is a perfect square, i.e. $x^2 + bx + c = (x + x_0)^2$ for some $x_0 \in \mathbb{R}$, if and only if $x^2 + bx + c$ has two equal real roots, i.e. $b^2 - 4c = 0$. Thus we can use the discriminant $b^2 - 4c$ to check if a trinomial $x^2 + bx + c$ is a perfect square. For example, $x^2 - \sqrt{3}x + \dfrac{3}{4}$ is a perfect square, since $b^2 - 4c = (\sqrt{3})^2 - 4 \times \dfrac{3}{4} = 3 - 3 = 0$. In fact, $x^2 - \sqrt{3}x + \dfrac{3}{4} = (x - \dfrac{\sqrt{3}}{2})^2$. For the trinomial $ax^2 + bx + c$ ($a \neq 0$) with leading coefficient $a \neq 1$, if $b^2 - 4ac = 0$, then it equals a times a perfect square. For example, in $2x^2 - 4x + 2$, $a = 2, b = -4, c = 2$, which satisfies $b^2 - 4ac = (-4)^2 - 4 \times 2 \times 2 = 0$, so it must be 2 times a perfect square. This is true since $2x^2 - 4x + 2 = 2(x^2 - 2x + 1) = 2(x - 1)^2$.

Simplifying the right side of the identity $ax^2 + bx + c = a(x - x_1)(x - x_2)$ we get $ax^2 + bx + c = ax^2 - a(x_1 + x_2)x + ax_1x_2$, which is true for all real number x. As a result, the two trinomials must be identical, i.e. $b = -a(x_1 + x_2)$ and $c = ax_1x_2$, so the coefficients a, b, c and the roots

x_1, x_2 must satisfy

$$x_1 + x_2 = -\frac{b}{a},$$
$$x_1 x_2 = \frac{c}{a}.$$

These identities are called the Vieta's formulas.

Example. If the sum and product of two numbers are 3 and -1, find these two numbers.

Solutions. Assume the two numbers are x_1 and x_2, then $x_1 + x_2 = 3$ and $x_1 x_2 = -1$. By Vieta's formulas, x_1, x_2 must be the two solutions to the quadratic equation

$$x^2 - 3x + (-1) = 0.$$

The discriminant $\Delta = (-3)^2 - 4 \times 1 \times (-1) > 0$, so the equation has two real solutions. Solving this equation using the quadratic formula, the two numbers must be

$$\frac{-(-3) \pm \sqrt{(-3)^2 - 4 \times 1 \times (-1)}}{2 \times 1}$$

i.e. $\dfrac{3 + \sqrt{13}}{2}$ and $\dfrac{3 - \sqrt{13}}{2}$.

Notice that Vieta's formulas are valid only when the equation is quadratic ($a \neq 0$) and has two real solutions ($\Delta = b^2 - 4ac \geq 0$). To make it clear, let us consider the following example: when does the equation $(b + 2)x^2 - 4x + 1 = 0$ have two positive solutions? It is obvious that the two solutions are positive if and only if both their sum and product are positive, i.e. $-\dfrac{-4}{b+2} > 0$ and $\dfrac{1}{b+2} > 0$. The two conditions both require that $b + 2 > 0$. Immediately we get $b > -2$. However, this is not the correct answer. Since the equation must have two real solutions in advance, first, it must be quadratic, i.e. $b + 2 \neq 0$; and secondly, the discriminant must be nonnegative, i.e. $(-4)^2 - 4(b + 2) \times 1 \geq 0$. Solving these inequalities, b should also satisfy $b \leq 2$ and $b \neq -2$. As a result, we need to combine all these restrictions with $b > -2$. The correct answer is $-2 < b \leq 2$.

Exercises.

1. Multiple choice.

(1) Which of the following may not be a quadratic equation?

    A. $(a^4 + 3)x^2 = 8$ B. $ax^2 + bx + c = 0$

    C. $(x + 3)(x - 2) = x + 5$ D. $\sqrt{3}x^2 + \dfrac{1}{19}x - 2 = 0$

(2) Which of the following equations has constant term 0?

    A. $x^2 + x = 1$ B. $2x^2 - x - 12 = 12$

    C. $2(x^2 - 1) = 3(x - 1)$ D. $2(x^2 + 1) = x + 2$

(3) If $ky^2 - 4y - 3 = 3y + 4$ has real roots, then

    A. $k > -\dfrac{7}{4}$ B. $k \geq -\dfrac{7}{4}$

    C. $k > -\dfrac{7}{4}$ and $k \neq 0$ D. $k \geq -\dfrac{7}{4}$ and $k \neq 0$

(4) If 0 is a root of $(a-1)x^2 + ax + a^2 - 1 = 0$, then

    A. $a = 1$ B. $a = -1$ C. $a = \pm 1$ D. $a = 0$

(5) Given the equation $x^2 + x = 3$, we can assert

    A. the sum of the two solutions is 1.

    B. the sum of the two solutions is -1.

    C. the product of the two solutions is 3.

    D. the equation has no real solutions.

(6) Which of the following makes $x^2 + mx + n = 0$ a perfect square?

    A. $m = 4, n = -4$ B. $m = -3, n = 9$

    C. $m = 8, n = 64$ D. $m = -9, n = \dfrac{81}{4}$

(7) $2x^2 - 6x + 3$ can be factored as

    A. $2\left(x - \dfrac{3+\sqrt{3}}{2}\right)\left(x - \dfrac{3-\sqrt{3}}{2}\right)$ B. $\left(x - \dfrac{3+\sqrt{3}}{2}\right)\left(x - \dfrac{3-\sqrt{3}}{2}\right)$

    C. $2\left(x + \dfrac{3-\sqrt{3}}{2}\right)\left(x + \dfrac{3+\sqrt{3}}{2}\right)$ D. $\left(x + \dfrac{3-\sqrt{3}}{2}\right)\left(x + \dfrac{3+\sqrt{3}}{2}\right)$

(8) If $x^2 + 3x - 3 = (x^2 + 2x - 3)^0$, then

    A. $x = 1$ or -4 B. $x = -4$ C. $x = 1$ D. $x = -3$

(9) If $x^2 + 4x + a^2 - 1$ is a perfect square, then

    A. $a = \pm\sqrt{2}$ B. $a = \pm\sqrt{3}$ C. $a = \pm\sqrt{5}$ D. $a = \pm\sqrt{7}$

(10) If $-\sqrt{2}$ is a root to $x^2 + kx - 4$, then the other root is

    A. $\sqrt{2}$ B. $-\sqrt{2}$ C. $2\sqrt{2}$ D. $-2\sqrt{2}$

2. Fill in the blanks.

(1) If -1 is a solution to both $3ax^2 - bx - 1 = 0$ and $ax^2 + 2bx - 5 = 0$, then $a = $_____, $b = $_____.

(2) The sum of the solutions to $x^2 - 4x - 7 = 0$ is _____; the product of the solutions to $3x^2 + 7x - \pi = 0$ is _____.

(3) The sum of the solutions to $x^2 - x + 3 = 0$ is _____.

(4) If $3 - \sqrt{2}$ is a root to $x^2 + mx - 7 = 0$, then $m = $_____ and the other root is _____.

(5) If 1 is a solution to $x^2 + kx + \sqrt{2} = 0$, then $k = $_____ and the other root is _____.

(6) If the product of two numbers is 12 and the sum of squares of them is 25, then the two numbers are solutions to the quadratic equation _____.

(7) If $m^2 x^2 + (2m - 1)x + 1$ has two different real roots, then m_____.

(8) $mx^2 - mx + 2$ is the multiple of a perfect square when $m = $_____.

(9) If x_1, x_2 are two solutions to $x^2 - 2x - 1 = 0$, then $\dfrac{1}{x_1} + \dfrac{1}{x_2} = $_____.

(10) If a is a solution to $x^2 - 3x + m = 0$ and $-a$ is a solution to $x^2 + 3x - m = 0$, then

$a = \underline{\hspace{2cm}}.$

3. Factor in the set of real numbers.

(1) $x^2 - 12$

(2) $x^2 - 4x - 6$

(3) $x^2 - 8x - 4$

(4) $a^2 + 2\sqrt{3}a - 9$

(5) $6z^2 - 6z + 1$

(6) $t^2 - \sqrt{3}t + \dfrac{1}{2}$

(7) $3x^2 - 2x - 3$

(8) $3x^2 + 9x + 7$

(9) $2x^2 + 20x + 25$

(10) $5x^2 + 6x - 2$

(11) $4x^2 + 7x + 2$

(12) $x^4 - 2x^2 - 3$

4. Solve the following equations for x.

(1) $(x + 3)^2 = (1 - 2x)^2$

(2) $2x^2 - 5kx + 3k^2 = 0$

(3) $16x^2 + 8x - 7 = 0$

(4) $x^2 - 7xy + 12y^2 = 0$

(5) $x^2 - 2|x| - 1 = 0$

(6) $2(x + 1)^2 + 3(x + 1)(2 - x) - 2(x - 2)^2 = 0$

(7) $x^2 + 2\sqrt{2}x - 6 = 0$

(8) $(3x + 2)^2 = 16(x - 3)^2$

(9) $(x - 5)^2 - 17(x - 5) + 30 = 0$

(10) $(x - 3)^2 + 2x(x - 3) = 0$

(11) $x^2 - 9a^2 + 12ab - 4b^2 = 0$

(12) $9x^2 + 12x + 5 = 0$

(13) $x^2 - 2ax + a^2 - b^2 = 0$

(14) $\dfrac{1}{x} + \dfrac{5}{x+2} = 1$

(15) $x^4 - 18x^2 + 72 = 0$

(16) $6x^4 + x^2 - 12 = 0$

(17) $6x^4 + 17x^2 + 5 = 0$

(18) $x + \sqrt{x} - 2 = 0$

5. Find x_1 and x_2 accordingly.

(1) $x_1 + x_2 = 7$, $x_1x_2 = 7$

(2) $x_1 + x_2 = -\dfrac{\sqrt{6}}{2}$, $x_1x_2 = \dfrac{1}{4}$

(3) $x_1 + x_2 = -6$, $x_1x_2 = 11$

(4) $x_1^2 + x_2^2 = 9$, $x_1x_2 = 3$

(5) $x_1 - x_2 = 40$, $x_1x_2 = -144$

(6) $x_1^2 + x_2^2 = 13$, $x_1 - x_2 = -5$

(7) $x_1^2 + x_2^2 = 58$, $x_1 + x_2 = 4$

6. If x_1, x_2 are two solutions to $x^2 - 4\sqrt{2}x - 6 = 0$, find

(1) $(x_1 + 1)(x_2 + 1)$

(2) $\dfrac{1}{x_1} + \dfrac{1}{x_2}$

(3) $x_1^2 + x_2^2$

(4) $|x_1 - x_2|$

(5) $\dfrac{x_1}{x_2} + \dfrac{x_2}{x_1}$

(6) $x_1^3 + x_2^3$

7. If -1 is a root of $ax^2 + bx + c$ $(a \neq 0)$, show that $ax^2 + bx + c = 0$ must have two real solutions.

8. Consider the equation $x^2 - 3x + m - 1 = 0$.

(1) If it has two equal real roots, find m and the roots;

(2) If it has two different real roots, find m.

9. Prove $ax^2 + bx + c = 0$ has two different real roots when a and c have different signs.

10. If $x^2 - (k - 2)x + 3k - 2 = 0$ has two solutions x_1 and x_2, which satisfy $x_1^2 + x_2^2 = 23$, find k.

11. Prove $x^2 - 2mx - 2m - 4 = 0$ has two different real roots.

12. If the two solutions a, b to $x^2 + 2(m - 2)x + m^2 + 4 = 0$ satisfy $(a^2 + b^2) - ab = 21$, find m.

13. If $a^2 + a - 1 = 0$, $b^2 + b - 1 = 0$, $a \neq b$, find $ab + a + b$.

14. Consider the equation $2(m+1)x^2 + 4mx + 3m = 2$. Find m when (1) the equation has two equal solutions; (2) the two solutions are opposites; (3) one of the solution is 0; (4) the two solutions are reciprocal.

15. If the quotient of the two solutions to $x^2 + px + q = 0$ is 2, and the discriminant of the equation is 1, find p, q and solve the equation.

16. If the difference of the two roots of $x^2 + px + q$ equals the difference of the two roots of $x^2 + qx + p$, prove $p = q$ or $p + q + 4 = 0$.

17. Given the equation $x^2 - (5k+1)x + k^2 - 2 = 0$, find k such that the sum of the reciprocals of the two solutions is 4. If such k does not exist, tell the reason.

18. Mike makes a rectangular box without a lid using a piece of paper which measures 40cm$\times$25cm by cutting off four squares of the same size in each corner. If the bottom area of the rectangle box is 450cm^2, find the height of the box.

10.3 Applications to Rational and Quadratic Equations

To solve rational and radical equations, we always simplify them into equivalent systems of polynomial equations and inequalities. We have learned the case when the system only contains linear equations and inequalities. With quadratic formula, we are able to solve more complicated cases. Notice that we should plug the solutions back into the original equation to delete extraneous solutions (if any).

Examples.

(1) $\dfrac{1}{x+2} + \dfrac{4x}{x^2 - 4} - \dfrac{2}{x-2} = 1$

(2) $\sqrt{x+7} - x = 1$

(3) $\sqrt{3x-2} + \sqrt{x+3} = 3$

Solutions.

(1) $x^2 - 4 = (x+2)(x-2)$, so the least common denominator is $(x+2)(x-2)$. Multiplying $(x+2)(x-2)$ on both sides we get

$$(x+2)(x-2)\left(\frac{1}{x+2} + \frac{4x}{x^2-4} - \frac{2}{x-2}\right) = 1 \cdot (x+2)(x-2);$$
$$(x-2) + 4x - 2(x+2) = x^2 - 4;$$
$$3x - 6 = x^2 - 4.$$

The equation simplifies to $x^2 - 3x + 2 = 0$, which is a quadratic equation. Solving this equation by factoring, we get two solutions $x = 1$ and $x = 2$. For fractions, the denominator can never be zero. We plug $x = 1$ and $x = 2$ into all denominators of the original equation to check. $x = 2$ makes two denominators zero, so it is an extraneous solution. The only solution to the equation is $x = 1$.

(2) There is only one square root in the equation. Keep the square root on left side and move all other terms to the right side, we get

$$\sqrt{x+7} = x + 1.$$

Then square both sides, we get

$$x + 7 = (x+1)^2 = x^2 + 2x + 1,$$
$$0 = x^2 + x - 6.$$

Solving the last quadratic equation, we get $x = -3$ or 2. Now we need to plug $x = -3$ and $x = 2$ back into the original equation. When $x = -3$, the equation becomes $\sqrt{-3+7} - (-3) = 2 + 3 = 1$, which is wrong, so $x = -3$ is an extraneous solution. When $x = 2$, $\sqrt{2+7} - 2 = 3 - 2 = 1$, which makes the equation hold, so $x = 2$ is the only solution.

(3) There are two square roots involved in the equation. We need to square it twice to get rid of the square roots. To make the procedure as simple as possible, we isolate one square root on left side and move all other terms to the right side, i.e.

$$\sqrt{3x-2} = 3 - \sqrt{x+3}.$$

Square both sides, we get

$$(\sqrt{3x-2})^2 = (3 - \sqrt{x+3})^2,$$
$$3x - 2 = x + 12 - 6\sqrt{x+3}.$$

Only one square root is left now, so we isolate it on the right side and move all other terms to the left, and then square both sides again.

$$2x - 14 = -6\sqrt{x+3}, \text{ (isolate } \sqrt{x+3} \text{ on right side)}$$
$$x - 7 = -3\sqrt{x+3}, \text{ (divide 2 on both sides)}$$
$$(x-7)^2 = (-3\sqrt{x+3})^2, \text{ (square both sides)}$$
$$x^2 - 14x + 49 = 9x + 27, \text{ (simplify)}$$
$$x^2 - 23x + 22 = 0.$$

Solving the last quadratic equation, we get $x = 1$ or $x = 22$. Plugging these two solutions into the original equation to check: when $x = 22$, the left side becomes $\sqrt{3 \times 22 - 2} + \sqrt{22 + 3} = 8 + 5 = 13$, not equal to the right side 3, so $x = 22$ is an extraneous solution; when $x = 1$, the left side becomes $3 \times 1 - 2 + \sqrt{1 + 3} = 1 + 2 = 3$, equal to the right side, so the only solution is $x = 1$.

We can also solve rational and radical equations by substitution. Here are three examples.

Examples.

(1) $\left(\dfrac{x^2}{x - 1} \right)^2 - \dfrac{3x^2}{x - 1} - 4 = 0$

(2) $\dfrac{8(x^2 + 2x)}{x^2 - 1} + \dfrac{3(x^2 - 1)}{x^2 + 2x} = 11$

(3) $3x^2 + 15x + 2\sqrt{x^2 + 5x + 1} = 2$

Solutions.

(1) If we let $y = \dfrac{x^2}{x - 1}$, the original equation becomes

$$y^2 - 3y - 4 = 0.$$

This equation is easy to solve: $y = 4$ or $y = -1$. By substituting y back, we get two fractional equations

$$\dfrac{x^2}{x - 1} = 4, \qquad \dfrac{x^2}{x - 1} = -1.$$

Solve the first equation: multiply both sides by $x - 1$ to get $x^2 = 4(x - 1)$; simplify it to $x^2 - 4x + 4 = 0$ or $(x - 2)^2 = 0$; so the only possible solution is $x = 2$. If plugging $x = 2$ into $\dfrac{x^2}{x - 1} = 4$, we can check that $x = 2$ is a solution.

Solve the second equation: multiply both sides by $x - 1$ to get $x^2 = 1 - x$, which simplifies to $x^2 + x - 1 = 0$; by the quadratic formula, we get $x = \dfrac{-1 \pm \sqrt{5}}{2}$. We can check that $x = \dfrac{-1 \pm \sqrt{5}}{2}$ are both solutions to $\dfrac{x^2}{x - 1} = -1$.

As a result, the equation has three solutions: $x = 2$, $x = \dfrac{-1 + \sqrt{5}}{2}$ and $x = \dfrac{-1 - \sqrt{5}}{2}$.

(2) We observe that $\dfrac{x^2 + 2x}{x^2 - 1}$ and $\dfrac{x^2 - 1}{x^2 + 2x}$ are reciprocals, so we make the substitution $y = \dfrac{x^2 + 2x}{x^2 - 1}$. The original equation becomes

$$8y + \dfrac{3}{y} = 11.$$

Multiplying y on both sides to clear the denominator, we get $8y^2 + 3 = 11y$ or $8y^2 - 11y + 3 = 0$. Solving this equation, $y = \dfrac{3}{8}$ or $y = 1$. The original equation breaks into the following two equations:

$$\dfrac{x^2 + 2x}{x^2 - 1} = 1 \qquad \text{or} \qquad \dfrac{x^2 + 2x}{x^2 - 1} = \dfrac{3}{8}.$$

Solve the first equation: $x^2+2x = x^2-1$, so $x = -\dfrac{1}{2}$. $x = -\dfrac{1}{2}$ does not make the denominator zero, so it is a solution. Solve the second equation: $8(x^2+2x) = 3(x^2-1)$, $5x^2+16x+3 = 0$, $x = -\dfrac{1}{5}$ or -3. $x = -\dfrac{3}{5}$ and -3 are both solutions, as you can check.

As a result, the solutions to the original equations are $x = -3$, $x = -\dfrac{3}{5}$ and $x = -\dfrac{1}{2}$.

(3) We observe that the terms $3x^2+15x$ can be completed into a multiple of the radicand x^2+5x+1, (i.e. $3x^2+15x = 3(x^2+5x+1)-3$), so we make the substitution $y = \sqrt{x^2+5x+1}$. If we rewrite the equation as $3(x^2+5x+1) + 2\sqrt{x^2+5x+1} = 2+3$, it transforms into

$$3y^2 + 2y - 5 = 0.$$

The equation is easy to solve: $y = 1$ or $y = -\dfrac{5}{3}$. Then the original equation is equivalent to $\sqrt{x^2+5x+1} = 1$ or $\sqrt{x^2+5x+1} = -\dfrac{5}{3}$. The second equation has no solution, since a square root cannot be negative. Solving the first equation we get $x^2+5x+1 = 1$, $x^2+5x = 0$, $x = 0$ or -5. After checking, $x = 0$ and $x = -5$ are both solutions to the original equation.

Exercises.

   1. Multiple choice.

   (1) The solution set of $\dfrac{(x-2)(x+3)}{x^2-4} = 0$ is

   A. $\{2,-3\}$ B. $\{2\}$ C. $\{-3\}$ D. $\{-2,3\}$

   (2) Which of the following equation has a real solution?

   A. $\sqrt{x-8} + \sqrt{5-x} - 2 = 0$ B. $\sqrt{x^2-9} + \sqrt{3x-1} + 2 = 0$

   C. $\sqrt{3-x} + \sqrt{x-3} = 0$ D. $\sqrt{2-x} = x - 3$

   (3) The solution to $\sqrt{x-1}\sqrt{x+5} = 4$ is

   A. $x = 7$ B. $x = 3$ C. $x = -7$ or 3 D. $x = 7$ or -3

   (4) If $\dfrac{3}{x} + \dfrac{ax+3}{x+1} = 2$ has an extraneous solution $x = -1$, then a is

   A. 0 or 1 B. 0 C. 3 D. -1 or 3

   (5) If the solutions to $\sqrt{x^2+1} = m$ are $x = \pm\sqrt{3}$, then m is

   A. 3 B. 2 C. -2 D. ± 2

   (6) How many solutions are there to the equation $\dfrac{2x^2-6x}{x-3} = x+3$?

   A. 0 B. 1

   C. 2 D. infinitely many

   (7) To solve $\dfrac{3x}{x^2-1} + \dfrac{x^2-1}{3x} = 3$ by substitution, if letting $y = \dfrac{3x}{x^2-1}$, the equation transforms into

   A. $y^2 - 3y + 1 = 0$ B. $y^2 + 3y + 1 = 0$

   C. $y^2 + 3y - 1 = 0$ D. $y^2 - y + 3 = 0$

(8) To solve $x^2 + 8x + \sqrt{x^2 + 8x - 11} = 23$ by substitution, if letting $y = \sqrt{x^2 + 8x - 11}$, the equation becomes

    A. $y^2 + y + 12 = 0$ B. $y^2 + y - 23 = 0$

    C. $y^2 + y - 12 = 0$ D. $y^2 + y - 34 = 0$

(9) If $\dfrac{2x}{x+1} - \dfrac{m+1}{x^2 + x} = \dfrac{x+1}{x}$ has extraneous solutions, then m is

    A. -1 or -2 B. -1 or 2 C. 1 or 2 D. 1 or -2

(10) To simplify the denominator in $\dfrac{4}{x} - \dfrac{1}{x-1} = 1$, the least common denominator we multiply by both sides is

    A. x B. $x - 1$ C. $x^2 - x$ D. $x^2(x-1)$

2. Fill in the blanks.

(1) To solve $x + \sqrt{x-2} = 2$: (i) rewrite the equation into $x - 2 + \sqrt{x-2} = 0$; (ii) let $y = \sqrt{x-2}$ and the equation transforms into $y^2 + y = 0$; (iii) solve the quadratic equation to get $y = 0$ or $y = -1$; (iv) when $y = 0$, $\sqrt{x-2} = 0$ and $x = 2$; when $x = -1$, $\sqrt{x-2} = -1$ has no solution; (v) $x = 2$ is the only solution. During the procedure, the method in step (ii) is _____; the reason $\sqrt{x-2} = -1$ has no solution in step (iv) is _____. The procedure is not complete and the step missed is

_____.

(2) The solution to $\sqrt{2x - 3} - \sqrt{x+1} = 0$ is _____.

(3) The solution set to $(x - 5)\sqrt{x - 7} = 0$ is _____.

(4) To make $\sqrt{m(m-1)x + 3} = 2x - 15$ a radical equation, m _____.

(5) If $\dfrac{x - 3}{x - 1} = \dfrac{m}{x - 1}$ has an extraneous solution, then $m =$ _____.

(6) When $m =$ _____, $\dfrac{x}{x - 2} - \dfrac{m+1}{x^2 - 2x} = \dfrac{x+1}{x} + 1$ has an extraneous solution.

3. Solve the equations.

(1) $\dfrac{2x - 1}{(x-1)(x-2)} = \dfrac{x - 5}{(x-2)(x-3)}$ (2) $\dfrac{x}{2x^2 - 11x - 21} = \dfrac{x + 7}{x^2 - 12x + 35}$

(3) $\dfrac{2}{y^2 - 4} = \dfrac{1}{y + 2} - 1$ (4) $\dfrac{15}{x^2 - 4} + \dfrac{2}{2 - x} = 1$

(5) $\dfrac{1}{x + 1} + \dfrac{1}{x - 1} = 2$ (6) $\dfrac{2x - 5}{x^2 - 3x + 2} + \dfrac{4}{x^2 - 4} = \dfrac{1}{x - 2}$

(7) $\dfrac{x-4}{x^2+x-2} = \dfrac{1}{x-1} + \dfrac{x-6}{x^2-4}$

(8) $\dfrac{1}{x+7} = \dfrac{x+1}{(2x-1)(x+7)} + \dfrac{1}{2x^2-3x+1}$

(9) $\dfrac{x-1}{x+1} + \dfrac{2x}{x-1} - \dfrac{4x}{x^2-1} = 0$

(10) $\dfrac{1}{2-x} - 1 = \dfrac{1}{x-2} - \dfrac{6-x}{3x-12}$

(11) $\dfrac{2x}{x^2-4} - \dfrac{1}{x+2} = 1$

(12) $\dfrac{2}{x^2-9} + \dfrac{x-2}{x(x-3)} = \dfrac{1}{x^2+3x}$

(13) $\dfrac{1}{x+3} = \dfrac{x-3}{x^2-9}$

(14) $\dfrac{x+4}{x^2+2x} - \dfrac{1}{x+2} = \dfrac{1}{x} + 1$

4. Solve the equations.

(1) $\sqrt{x+2} = -x$

(2) $\sqrt{x-5} + x = 7$

(3) $\sqrt{x+3} - 2 = x$

(4) $\sqrt{3x+1} = \sqrt{x+4} + 1$

(5) $\sqrt{2x-4} - \sqrt{x+5} = 1$

(6) $2x - \sqrt{2x+1} = 5$

(7) $\sqrt{3x+2} - \sqrt{x-8} = 3\sqrt{2}$

(8) $\sqrt{2-x} + \sqrt{5-4x} = 2$

(9) $\sqrt{3x-2} - \sqrt{x} = \sqrt{x-5}$

(10) $\sqrt{5+x} + \sqrt{x+2} = \sqrt{4x+5}$

(11) $\sqrt{2x-5} + \sqrt{4x-3} = \sqrt{9x+1}$

(12) $\sqrt{9x+1} - \sqrt{4x-8} = \sqrt{x+5}$

5. Solve the equations by substitution.

(1) $2\dfrac{x^2+1}{x+1} + 6\dfrac{x+1}{x^2+1} = 7$

(2) $x^2 + \dfrac{4}{x^2} = 4$

(3) $\dfrac{x^2-5x}{x+1} + \dfrac{24(x+1)}{x(x-5)} + 14 = 0$

(4) $x^2 + 3x + 2 + \dfrac{9}{x} + \dfrac{9}{x^2} = 0$

(5) $\dfrac{x^4+2x^2+1}{x^2} + \dfrac{x^2+1}{x} = 2$

(6) $\dfrac{x^2+2}{2x^2-1} - \dfrac{6x^2-3}{x^2+2} + 2 = 0$

(7) $x - 12 + \sqrt{x} = 0$

(8) $x^2 + 3x + \sqrt{x^2+3x} = 6$

(9) $x^2 + \sqrt{x^2-1} = 3$

(10) $\sqrt{x+10} - \dfrac{6}{\sqrt{x+10}} = 5$

(11) $2x^2 - 4x + 3\sqrt{x^2-2x+6} = 15$

(12) $\sqrt{\dfrac{x-1}{x+2}} - \dfrac{5}{2} = -\sqrt{\dfrac{x+2}{x-1}}$

(13) $6\left(x^2 + \dfrac{1}{x^2}\right) + 5\left(x + \dfrac{1}{x}\right) = 38$

(14) $6\sqrt{x} + \sqrt[4]{x} - 15 = 0$

(15) $x^2 + 2x + 2 = \dfrac{6}{x^2+2x+1}$

(16) $4\left(x^2 + \dfrac{1}{x^2}\right) - 5\left(x - \dfrac{1}{x}\right) - 14 = 0$

(17) $\sqrt{x} + \sqrt{4-x} = \sqrt{2x(4-x)}$

(18) $\sqrt{5-2x} - 2\sqrt{2(2x+5)(5-2x)} = -\sqrt{2x+5}$

6. If real numbers x, y satisfy $2x^2 - 6xy + 9y^2 - 4x + 4 = 0$, find $x - 6y$.

7. Solve the following equations.

(1) $\dfrac{a}{a+x} + \dfrac{a}{a-x} = 6(a \neq 0)$

(2) $\dfrac{\sqrt{a+x}}{\sqrt{a-x}} + \dfrac{\sqrt{a-x}}{\sqrt{a+x}} = \dfrac{10}{3}(a > 0)$

(3) $x^2 - 4x - x\sqrt{x^2 - 2x + 1} = -2$

(4) $\dfrac{a-x}{b+x} = 5 - \dfrac{4(b+x)}{a-x}(a+b \neq 0)$

8. Find a accordingly.

(1) $\dfrac{1}{x} - \dfrac{ax}{x^2 - 4} = \dfrac{3}{x+2}$ has only 1 real solution.

(2) $\dfrac{x}{x+1} + \dfrac{x+1}{x} = -\dfrac{4x+a}{x(x+1)}$ has only 1 real solution.

(3) $\dfrac{3}{x} + \dfrac{6}{x-1} - \dfrac{x+m}{x(x-1)} = 0$ has real solutions.

10.4 Complex Numbers

Negative numbers have no square roots in the set of real numbers, so a quadratic equation has no real solutions when its discriminant Δ is less than 0. To overcome this difficulty, we shall develop the complex number system, in which every quadratic equation has two solutions.

The imaginary unit i is defined to be a square root of -1, i.e.

$$i = \sqrt{-1}.$$

Immediately $i^2 = -1$.

A number of the form $z = a + bi$ where $a, b \in \mathbb{R}$ is called a complex number. a is the real part of z, denoted by $\operatorname{Re} z$; b is the imaginary part of z, denoted by $\operatorname{Im} z$. By convention, we use $\mathbb{C} = \{a + bi | a, b \in \mathbb{R}\}$ to denote the set of all complex numbers. Any real number a can be treated as a complex number $a + 0i$, so the set of real numbers $\mathbb{R}$ is a subset of $\mathbb{C}$. A complex number $z = a + bi$ is real when $b = 0$; when $b \neq 0$, $z = a + bi$ is called imaginary; furthermore, $z = a + bi$ is called pure imaginary if $a = 0, b \neq 0$, i.e. $z = bi$. For example, $-3 + 2i$, $-i$, $\sqrt{2}i$, 4 are all complex numbers in which 4 is real; $-3 + 2i$, $-i$ and $\sqrt{2}i$ are imaginary; $-i$ and $\sqrt{2}i$ are pure imaginary.

Two complex numbers are equal if they have the same real and imaginary parts. Thus a complex number equals 0 if both its real and imaginary parts are 0. Given any complex number $z = a + bi$, define its conjugate $\bar{z}$ to be the complex number $a - bi$, i.e. change the imaginary part to its opposite. It is easy to see, a complex number z is real if and only if $z = \bar{z}$ and pure imaginary if and only if $z = -\bar{z}$.

The operations of complex numbers are defined as follows. Let $z_1 = a_1 + b_1 i$, $z_2 = a_2 + b_2 i$ $(a_1, a_2, b_1, b_2 \in \mathbb{R})$ denote two complex numbers.

• Addition and Subtraction: $z_1 \pm z_2 = (a_1 \pm a_2) + (b_1 \pm b_2)i$, i.e. adding or subtracting the real and imaginary parts separately.

• Multiplication: $z_1 z_2 = (a_1 a_2 - b_1 b_2) + (a_1 b_2 + a_2 b_1)i$.

For any complex number $z = a + bi$,

$$z\bar{z} = a^2 + b^2$$

is a nonnegative real number and we define the module of z to be

$$|z| = \sqrt{|z\bar{z}|} = \sqrt{a^2 + b^2}.$$

The addition and multiplication of complex numbers satisfy the same properties as real numbers (see Section 2.4). By the distributive property, the multiplication can be derived as follows:

$$\begin{aligned}
(a_1 + b_1 i)(a_2 + b_2 i) &= a_1(a_2 + b_2 i) + b_1 i(a_2 + b_2 i) \\
&= a_1 a_2 + a_1 b_2 i + b_1 a_2 i + b_1 b_2 i^2 \\
&= (a_1 a_2 - b_1 b_2) + (a_1 b_2 + a_2 b_1)i.
\end{aligned}$$

In the last second step, we apply the commutative property in the special case: i commutes with any real number; in the last step, we use the fact $i^2 = -1$.

• Division: $z_1 \div z_2 = \dfrac{z_1}{z_2} (z_2 \neq 0)$ can be simplified by multiplying both the numerator and denominator by the conjugate $\bar{z}_2$ of the denominator, i.e.

$$\frac{z_1}{z_2} = \frac{z_1 \bar{z}_2}{z_2 \bar{z}_2}.$$

In this way, we make the denominator a real number and the quotient in the standard form of a complex number.

Examples. Evaluate.

(1) $(2 - 3i) - (5 + 4i)$

(2) $(3 - 2i)(2 + i)$

(3) $\dfrac{-5i}{-2+3i}$

Solutions.

(1) Immediately we can get $(2 - 3i) - (5 - 4i) = (2 - 5) + [-3 - (-4)]i = -3 + i$;

(2) By the distributive property, $(3 - 2i)(2 + i) = 3(2 + i) - 2i(2 + i) = 6 + 3i - 4i - 2i^2 = 6 - i + 2 = 8 - i$;

(3) Multiply both the top and bottom by the conjugate $-2 - 3i$ of the denominator, and simplify:

$$\frac{-5i}{-2+3i} = \frac{(-5i)(-2 - 3i)}{(-2 + 3i)(-2 - 3i)}$$
$$= \frac{-15 + 10i}{(-2)^2 + 3^2}$$
$$= -\frac{15}{13} + \frac{10}{13}i.$$

• Square Roots: For any real number $b > 0$, the square roots of b in $\mathbb{C}$ are still $\pm\sqrt{b}$; the square roots of $-b$ in $\mathbb{C}$ are $\pm\sqrt{b}i$ as $(\pm\sqrt{b}i)^2 = bi^2 = -b$. Particularly the square roots of -1 are $\pm i$. In the future, we will use $\sqrt{-b}$ to denote $\sqrt{b}i$. As a result, in the set of complex numbers, any real number has two opposite square roots (by convention, 0 has two opposite square roots, 0 and 0).

In fact, the fundamental theorem of algebra asserts that any polynomial equation of degree n in $\mathbb{C}$ has n complex roots. Particularly any complex number has n n-th roots. When $n = 2$, any complex number $z = a + bi$ has two square roots. Suppose $x + yi(x, y \in \mathbb{R})$ is a square root of z, then $(x + yi)^2 = (x^2 - y^2) + 2xyi = a + bi$, so

$$\begin{cases} x^2 - y^2 = a \\ 2xy = b. \end{cases}$$

The square roots of z can be found by solving a and b from this equation.

Examples.

(1) Simplify $-3\sqrt{-48}$;

(2) Find the square roots of i.

Solutions.

(1) As $\sqrt{48} = \sqrt{16 \times 3} = 4\sqrt{3}$, from the definition, $\sqrt{-48} = \sqrt{48}i = 4\sqrt{3}i$, so

$$-3\sqrt{-48} = -3(4\sqrt{3})i = -12\sqrt{3}i.$$

(2) From the analysis above, assuming the square root is $x + yi$ $(x, y \in \mathbb{R})$, we only need to solve the equation $\begin{cases} x^2 - y^2 = 0 \\ 2xy = 1 \end{cases}$. From the first equation we get $x = y$ or $x = -y$. Plugging $x = -y$ into the second equation, we get $-2x^2 = 1$, which has no solutions (since x is a real number by assumption, and x^2 cannot be negative). Plugging $x = y$ into the second

equation, we get $2x^2 = 1$ and hence $x = \pm\dfrac{\sqrt{2}}{2}$. The solutions to the system are $x = y = \dfrac{\sqrt{2}}{2}$ or $x = y = -\dfrac{\sqrt{2}}{2}$, so the square roots of i are $\pm\dfrac{\sqrt{2} + \sqrt{2}i}{2}$.

A difference between real numbers and complex numbers is that complex numbers cannot be ordered. If assuming $i > 0$, we should have $i \cdot i > i \cdot 0$, i.e. $-1 > 0$, but this is a contradiction. If assuming $i < 0$, we also get $i \cdot i > i \cdot 0$, the same contradiction.

Exercises.

1. True or false.

(1) Complex numbers are never real numbers.

(2) The equation $x^2 + x + 1 = 0$ has only imaginary solutions.

(3) $\operatorname{Im}(a + bi) = bi$ where $a, b \in \mathbb{R}$.

(4) The square root of -1 is $\sqrt{-1} = i$.

(5) Every real number has two opposite square roots in $\mathbb{C}$.

(6) When $a \geq 0$, the square roots of a are complex.

(7) When $a < 0$, the square roots of a are pure imaginary.

(8) $|1 + i| = 2$

(9) Real numbers and pure imaginary numbers are all the complex numbers.

(10) $\operatorname{Re} z = \dfrac{z + \bar{z}}{2}$.

(11) $\operatorname{Im} z = \dfrac{z - \bar{z}}{2}$.

(12) Given $z_1 = a + bi$, $z_2 = c + di$ where $a, b, c, d \in \mathbb{C}$, if $z_1 = z_2$, then $a = c$, $b = d$.

(13) $\mathbb{R} \subset \mathbb{C}$.

(14) $z \div i = z \times (-i)$.

(15) $i^{50} + i^{100} = 0$.

(16) $\left|\dfrac{1}{3} - \dfrac{1}{2}i\right| = \left|\dfrac{1}{2} - \dfrac{1}{3}i\right|$.

(17) $|z| = z\bar{z}$ for any complex number z.

(18) $|z| = 0$ if and only if $z = 0$.

(19) z^2 is a real number for any $z \in \mathbb{C}$.

(20) When $m = 0$, $2(m^2 - m) + (m - 1)i$ is a pure imaginary number.

2. Multiple Choice.

(1) If $(a^2 + a) + 2(a^2 - 1)i(a \in \mathbb{R})$ is a pure imaginary number, then a is

    A. ± 1 B. -1 or 0 C. -1 D. 0

(2) Given two complex numbers z_1, z_2, which of the following is not correct?

    A. $|z_1 + z_2| = |z_1| + |z_2|$ B. $|z_1 z_2| = |z_1||z_2|$

C. $|\dfrac{z_1}{z_2}| = \dfrac{|z_1|}{|z_2|}$ when $z_2 \neq 0$ 　　　　 D. $\overline{z_1 \bar{z}_2} = \bar{z}_1 z_2$

(3) Given $z \in \mathbb{C}$, which of the following is not correct?

　A. $|\bar{z}| = |z|$ 　　　 B. $|z|^2 = z\bar{z} \geq 0$ 　 C. $z + \bar{z} \in \mathbb{R}$ 　　 D. $z - \bar{z} \in \mathbb{R}$

(4) If $z = 3 + 4i$, then $\dfrac{|z\bar{z}|}{|z| + |\bar{z}|}$ is

　A. 5 　　　　　 B. 10 　　　　 C. $\dfrac{5}{2}$ 　　　　　 D. 25

(5) $\dfrac{5i}{1 - 2i}$ is

　A. $2 - i$ 　　　 B. $i - 2$ 　　　 C. $1 - 2i$ 　　　 D. $-1 - 2i$

(6) If $\dfrac{1 + ai}{2 - i}$ is a pure imaginary number, then a is

　A. ± 2 　　　 B. $-\dfrac{1}{2}$ 　　　 C. 2 　　　　 D. -2

(7) $\left(\dfrac{1 - i}{1 + i}\right)^5$ is

　A. $-i$ 　　　　 B. -1 　　　 C. 1 　　　　 D. i

(8) Which of the following is not correct?

　A. The real part of $-\pi i$ is 0.

　B. The imaginary part of $\sqrt{2}i$ is $\sqrt{2}$.

　C. The product of two imaginary numbers can be real.

　D. The sum of two pure imaginary numbers is pure imaginary.

3. Among the following numbers

$$\sqrt{2}, 5i, -\sqrt[4]{3}, \sqrt{-7}, 1.\bar{1}, \pi, \sqrt{7} + \sqrt[3]{4}i, \sqrt{\dfrac{1}{4}}, 0, 2 - \sqrt{2}i, -1.234, \dfrac{9}{7} - 0.2i,$$

$$-2\sqrt{-\pi}, \dfrac{121}{13i}, 5 + 3i, \sqrt[3]{-9}, \dfrac{37}{41}, \sqrt[4]{5}i, \sqrt{-\pi}i, \sqrt{\sqrt{2} + \sqrt{3}i}, \dfrac{6 - 3i}{i - 2}, \dfrac{3 - 2i}{-9i}$$

using the set notation, find

　the set of rational numbers:

　the set of real numbers:

　the set of imaginary numbers:

　the set of pure imaginary numbers:

4. Add or subtract.

(1) $-8i + (7 - 3i)$ 　　　　　　　 (2) $(6 - 2i) - (10 - 2i)$

(3) $(5 + 2i) - (-i - 1)$

(4) $(-11 + 3i) + (2i + 9)$

(5) $(9 - 4i) - (7 - 3i)$

(6) $(-21 + 5i) + (17 - 6i)$

(7) $(12 - 9i) - (3i - 8)$

(8) $(-2 - 11i) + (-10 - 17i)$

(9) $(\frac{1}{3}i - \frac{2}{5}) + (-\frac{4}{5}i + \frac{3}{2})$

(10) $(\frac{4}{7} + \frac{2}{3}i) - (\frac{3}{7}i - \frac{7}{3})$

(11) $(\frac{2}{11}i - \frac{11}{5}) + (\frac{6}{21} - \frac{5}{2}i)$

(12) $(\frac{1}{6}i - 1) + (\frac{5}{9}i - \frac{7}{4})$

5. Simplify.

(1) $\sqrt{-64}$

(2) $\sqrt{-25}$

(3) $\sqrt{-28}$

(4) $\sqrt{-11}$

(5) $\sqrt{-54}$

(6) $\sqrt{-72}$

(7) $-2\sqrt{-98}$

(8) $-3\sqrt{-108}$

(9) $11\sqrt{-121}$

(10) $\sqrt{-\frac{81}{48}}$

(11) $\sqrt{-\frac{42}{27}}$

(12) $\sqrt{-\frac{35}{18}}$

6. Simplify.

(1) $\sqrt{-2}\sqrt{-7}$

(2) $\sqrt{-3}\sqrt{-10}$

(3) $\sqrt{-12}\sqrt{-9}$

(4) $\sqrt{-20}\sqrt{-75}$

(5) $\sqrt{-32}\sqrt{-8}$

(6) $\sqrt{-4}\sqrt{-10}$

(7) $\sqrt{-27}\sqrt{-6}$

(8) $\sqrt{9}\sqrt{-9}$

(9) $\frac{\sqrt{-2}}{\sqrt{-3}}$

(10) $\frac{\sqrt{-7}}{\sqrt{-5}}$

(11) $\dfrac{\sqrt{-14}}{\sqrt{-24}}$

(12) $\dfrac{\sqrt{-15}}{\sqrt{-35}}$

(13) $\dfrac{\sqrt{-21}}{\sqrt{-6}}$

(14) $\dfrac{\sqrt{-48}}{\sqrt{2}}$

(15) $\dfrac{\sqrt{-72}}{\sqrt{-30}}$

(16) $\dfrac{\sqrt{-20}}{\sqrt{-5}}$

7. Evaluate.

(1) $(3i)(5i)$

(2) $(-2i)(-11i)$

(3) $6i(3-5i)$

(4) $-7i(-i-3)$

(5) $(1-9i)(1+9i)$

(6) $(3-2i)^2$

(7) $(5-7i)(-2i-11)$

(8) $(3i-20)(4-7i)$

(9) $(-9-7i)(-2-13i)$

(10) $(3i-4)(4i-3)$

(11) $(8-4i)(7-2i)$

(12) $(i)^3(3+2i)(2-3i)$

8. Find quotients.

(1) $\dfrac{-5+3i}{-2i}$

(2) $\dfrac{-9-12i}{3i}$

(3) $\dfrac{2i}{1-i}$

(4) $\dfrac{-4i}{2+3i}$

(5) $\dfrac{3-i}{3+i}$

(6) $\dfrac{-8+5i}{9i+1}$

(7) $\dfrac{6+i}{7i-5}$

(8) $\dfrac{12i}{20-15i}$

(9) $\dfrac{4-i}{4+i}$

(10) $\dfrac{17-9i}{21i-3}$

(11) $\dfrac{6-2i}{2i-9}$

(12) $\dfrac{i}{i+1}$

9. Find modules.

(1) $-2+i$

(2) $3+4i$

(3) $\sqrt{2}-\sqrt{6}i$

(4) $2\sqrt{3}-1+(2+\sqrt{3})i$

10. Find square roots.

(1) $-2i$

(2) $-\pi$

(3) $-2+2\sqrt{3}i$

(4) $1+\sqrt{3}i$

11. Given two complex numbers z_1, z_2, show that if $z_1\bar{z}_2 \in \mathbb{R}$ and $z_2 \neq 0$, then $\dfrac{z_1}{z_2} \in \mathbb{R}$.

12. Prove the following identity for any two complex numbers z and w.
$$|z-w|^2 + |z+w|^2 = 2(|z|^2 + |w|^2).$$

10.5   Solving Quadratic Equations in Complex Numbers

If working with complex numbers, any real number has two opposite square roots. As a result, a real-coefficient quadratic equation always has two complex solutions. In this case the quadratic formula still works, i.e. the solutions to $ax^2 + bx + c = 0$ $(a, b, c \in \mathbb{R}, a \neq 0)$ are given by

$$x = \frac{-b \pm \sqrt{b^2 - 4ac}}{2a}.$$

Depending on the sign of the discriminant $\Delta = b^2 - 4ac$, we have the following classifications of the nature of the solutions:

$\Delta > 0$	$\Delta = 0$	$\Delta < 0$
two different real solutions	two equal real solutions	two conjugate imaginary solutions

Vieta's formulas remain true. They can also be proved by the quadratic formula directly. As the two solutions to the quadratic equation $ax^2 + bx + c = 0$ are $x_1 = \dfrac{-b + \sqrt{b^2 - 4ac}}{2a}$ and $x_2 = \dfrac{-b - \sqrt{b^2 - 4ac}}{2a}$, immediately we have

$$
\begin{aligned}
x_1 + x_2 &= \frac{-b + \sqrt{b^2 - 4ac}}{2a} + \frac{-b - \sqrt{b^2 - 4ac}}{2a} \\
&= \frac{-2b}{2a} = -\frac{b}{a} \\
x_1 x_2 &= \frac{-b + \sqrt{b^2 - 4ac}}{2a} \cdot \frac{-b - \sqrt{b^2 - 4ac}}{2a} \\
&= \frac{(-b)^2 - (\sqrt{b^2 - 4ac})^2}{(2a)^2} \\
&= \frac{b^2 - (b^2 - 4ac)}{4a^2} = \frac{c}{a}.
\end{aligned}
$$

Notice. In this book, numbers mean real numbers by default. When we work with complex numbers, we will state it explicitly. This section is the only section in which we solve quadratic equations in the set of complex numbers. Outside of this section, to solve equations always means to solve equations in the set of real numbers, and in case we want to apply Vieta's formulas, we need to check the existence of solutions (as real numbers) in advance.

Examples. Solve $3x^2 - 2x + 1 = 0$ in complex numbers.

Solutions. The discriminant $(-2)^2 - 4 \times 3 \times 1 = -8$ is negative, so there are two imaginary solutions. Plugging $a = 3, b = -2, c = 1$ into the quadratic formula, the solutions are

$$x = \frac{-b \pm \sqrt{b^2 - 4ac}}{2a} = \frac{-(-2) \pm \sqrt{(-2)^2 - 4 \times 3 \times 1}}{2 \times 3} = \frac{2 \pm \sqrt{-8}}{6}$$

Notice that $\sqrt{-8} = 2\sqrt{2}i$, so $x = \dfrac{2 \pm 2\sqrt{2}i}{6} = \dfrac{1 \pm \sqrt{2}i}{3}$. By Vieta's formula, the sum and product of the two solutions are $-\dfrac{b}{a} = \dfrac{2}{3}$ and $\dfrac{c}{a} = \dfrac{1}{3}$, as you can check by the solutions directly.

By the quadratic formula, the complex solutions to any real-coefficient quadratic equation are always conjugate to each other. This is also clear in the example above.

Exercises.

1. Multiple choice.

(1) Given the equation $ax^2 + bx + c = 0$ $(a, b, c \in \mathbb{R})$, which of following is not correct?

A. The equation always has two complex solutions as long as $a \neq 0$

B. If the equation has two imaginary solutions, the two imaginary solutions are conjugated to each other.

C. The equation has real solutions only when $b^2 - 4ac \geq 0$ and $a \neq 0$.

D. The equation has two equal solutions when $b^2 - 4ac = 0$ and $a \neq 0$ and the solutions must be real.

(2) Which of the following is not correct?

A. The sum and product of two imaginary numbers may both be real.

B. Any trinomial can be factored in $\mathbb{C}$.

C. The solutions to $x^2 = c$ $(c \in \mathbb{C} - \mathbb{R})$ cannot be real.

D. Vieta's formulas only work when a quadratic equation has real solutions.

(3) The equation $7x^2 - 13x + 12 = 0$ has

A. two conjugate imaginary solutions   B. two equal real solutions

C. two different real solutions   D. none of the above

(4) The solutions to $x^2 + 2x + 2 = 0$ in $\mathbb{C}$ are

A. $x = 1 \pm i$ B. $x = -1 \pm i$ C. $x = -1, 2$ D. $x = 2, -1$

(5) Denote the real solutions to $3x^2 + 2x + 3 = 0$ by x_1, x_2. $x_1 + x_2$ is

A. $\dfrac{2}{3}$ B. 1   C. $-\dfrac{2}{3}$ D. D.N.E

(6) Suppose a, b are the complex solutions to $2x^2 + 3x + 2 = 0$; then $|a - b| =$

A. $\dfrac{1}{2}$ B. $\dfrac{\sqrt{3}}{2}$ C. $\dfrac{\sqrt{5}}{2}$ D. $\dfrac{\sqrt{7}}{2}$

2. Solve the following equations in the set of complex numbers.

(1) $-5x^2 + 3 = 18$ \hspace{2cm} (2) $7x^2 + 2 = 0$

(3) $x^2 + x + 2 = 0$ \hspace{2cm} (4) $x^2 + 9x + 18 = 0$

(5) $2x^2 - 5x + 3 = 0$ \hspace{2cm} (6) $3t^2 + 4t + 2 = 0$

(7) $\dfrac{y^2}{4} + \sqrt{2}y + 1 = 0$

(8) $x^2 - \sqrt{6}x + 2 = 0$

(9) $x^2 - 3x + 4 = 0$

(10) $6x^2 - 3x + 1 = 0$

(11) $x^2 - 2ix + 1 = 0$

(12) $x^2 - 3ix - 2 = 0$

(13) $z^2 - 2\sqrt{2}i - 1 = 0$

(14) $x^2 - 4\sqrt{5}i + 1 = 0$

3. Factor in the set of complex numbers.

(1) $3x^2 + 6$

(2) $2t^2 + 5$

(3) $m^2 + 2m + 5$

(4) $n^2 + 4m + 20$

(5) $3x^2 + 12x + 13$

(6) $9n^2 - 6n + 26$

(7) $z^2 - 12z + 85$

(8) $x^2 + 10x - 74$

4. Classify the nature of solutions by discriminants.

(1) $3x^2 - 7x + 9 = 0$

(2) $5x^2 + 4x + 3 = 0$

(3) $49z^2 - 28z + 4 = 0$

(4) $(2x + 3)(3x - 2) = 1$

(5) $15x^2 + 9x + 4 = 0$ (6) $6b^2 - 7b + 2 = 0$

(7) $(5a + 1)(4 - 7a) = 5$ (8) $8t^2 + 24t + 8 = 0$

5. Denote the complex solutions to the equation $ax^2 + bx + c = 0$ $(a \neq 0)$ by x_1 and x_2. Given $a, b, c \in \mathbb{R}$, find $x_1^2 + x_2^2$ and $|x_1 - x_2|$.

(1) $a = 1, b = -2, c = 3$ (2) $a = 3, b = 2, c = 1$

(3) $a = 4, b = 5, c = -3$ (4) $a = 6, b = 9, c = 12$

6. If the sum of two complex numbers is 2 and their product is 4, find these two complex numbers.

10.6   Polynomial Equations and Inequalities

We are back working with real numbers again. In the following contents, to solve equations and inequalities means to solve them in the set of real numbers. All coefficients and parameters are understood as real numbers as well.

A polynomial equation in one variable is of the form $P(x) = 0$ where $P(x)$ is a polynomial in x. Linear and quadratic equations are two special cases, and we have formulas to solve them. When the degree of $P(x)$ is 3 or 4, we still have formulas, but when the degree is greater than 4, there are no formulas anymore, and the equation becomes hard to solve. Luckily, there is a special case in which a polynomial equation of high degree can be solved easily — when $P(x)$ can be factored as a product of linear and quadratic factors. The idea is demonstrated by the following example.

Example. Solve $(x - 2)(2x + 1)^2(2x^2 - 3x - 1)(x^2 + 2x + 2) = 0$.

Solutions. The product $(x - 2)(2x + 1)^2(2x^2 - 3x + 1)(x^2 + 2x + 2)$ is 0 only when one of the factors is 0, i.e. $x - 2 = 0$ or $(2x + 1)^2 = 0$ or $2x^2 - 3x - 1 = 0$ or $x^2 + 2x + 2 = 0$. Solve these four equations separately:

- $x - 2 = 0$ when $x = 2$;

- $(2x + 1)^2 = 0$ when $2x + 1 = 0$, so $x = -\dfrac{1}{2}$;

- $2x^2 - 3x - 1 = 0$ has two real solutions; by the quadratic formula $x = \dfrac{3 \pm \sqrt{17}}{4}$;

- $x^2 + 2x + 2 = 0$ has no real solutions, since its discriminant is negative, which means $x^2 + 2x + 2$ can never be 0. This is also clear from completing the square: $x^2 + 2x + 2 = (x + 1)^2 + 1 \geq 1 > 0$.

Collecting all roots of the four factors, the solutions to the original equation are $x = 2, -\dfrac{1}{2}, \dfrac{3 + \sqrt{17}}{4}, \dfrac{3 - \sqrt{17}}{4}$.

We say x_0 is a root of multiplicity k of the polynomial $P(x)$ if $(x - x_0)^k$ is a factor of $P(x)$ but $(x - x_0)^{k+1}$ is not a factor of $P(x)$.

A polynomial inequality in one variable is of the form $P(x) > 0$. We can also use '$\geq$', '$\leq$', '$<$' instead of '$>$'. Such inequalities can be solved when $P(x)$ is factored as a product of linear factors. Suppose $P(x)$ can be factored as

$$P(x) = a(x - x_1)^{k_1}(x - x_2)^{k_2} \cdots (x - x_n)^{k_n},$$

where $a \neq 0$, $x_1, x_2, \cdots, x_n$ are roots of $P(x)$ with multiplicity $k_1, k_2, \cdots, k_n$ and $x_1 > x_2 > \cdots > x_n$. Let $Q(x) = (x - x_1)^{k_1}(x - x_2)^{k_2} \cdots (x - x_n)^{k_n}$. First, we divide a on both sides of the inequality and reduce the inequality into one of the following: $Q(x) > 0$, $Q(x) < 0$, $Q(x) \geq 0$, $Q(x) \leq 0$. Secondly, we sketch the graph of $y = Q(x)$ in the coordinate system by the following rule:

- Label all points $x_1, x_2, \cdots, x_n$ on the real number line (keep them in order).

- Draw a wave curve starting from the right part above the x-axis.

- The curve goes to the left direction and touches the first point x_1. If k_1 is odd, cross the x-axis; if k_1 is even, do not cross the x-axis and stay on the same side.

- Repeat the above step for the points $x_2, \cdots, x_n$ in order.

At last we can read the solution set from the graph easily. For example, the solution set to $Q(x) > 0$ is the union of all intervals over which the curve is above the x-axis; the solution set to $Q(x) \leq 0$ is the union of intervals and points over which the curve is below and on the x-axis.

Examples. Solve the inequalities.

(1) $(x - 1)(x - 2)(x - 3)(x - 4) \geq 0$

(2) $2(x + 1)(x^2 - 3x - 4)(-x^2 - 1) < 0$

(3) $(-2x - 2)(x^2 - 3x - 4) \leq 0$

Solutions.

(1) Obviously, the roots to $(x - 1)(x - 2)(x - 3)(x - 4)$ are $x = 1, 2, 3, 4$. The polynomial already has leading coefficient 1 and the coefficients of x in each factor are positive, so we can start to sketch the graph of $y = (x - 1)(x - 2)(x - 3)(x - 4)$. According to our rules, the graph is like

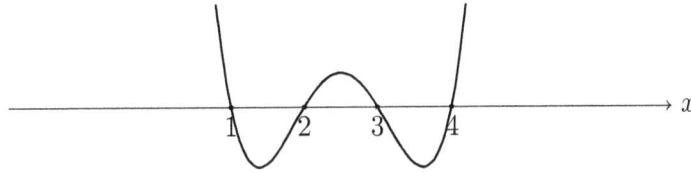

Reading from the graph, the solution set to $(x-1)(x-2)(x-3)(x-4) \geq 0$ is the union of all intervals and points over which the curve is above and on x-axis, i.e. $(-\infty, -1] \cup [2, 3] \cup [4, +\infty)$.

(2) First, $-x^2 - 1 < 0$; we can cancel this item by dividing it on both sides to get $2(x+1)(x^2 - 3x - 4) > 0$. Secondly, $x^2 - 3x - 4 = (x-4)(x+1)$ and the inequality becomes $2(x+1)^2(x-4) > 0$. The roots of $(x+1)^2(x-4)$ are $x = -1, 4$. As the leading coefficient and both the coefficients of x in each factor are already positive, we can sketch the graph.

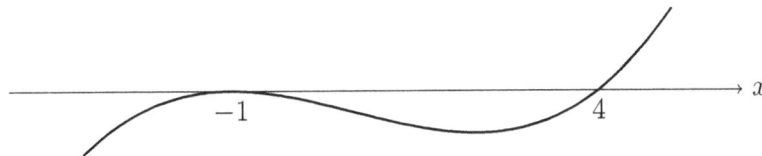

As a result, the solution set should be the interval over which the curve is above the x-axis, i.e. $(4, +\infty)$.

(3) The coefficient of x in $-2x - 2$ is negative; we should divide -2 to make it 1. As $x^2 - 3x - 4 = (x-4)(x+1)$, the inequality simplifies to $(x+1)^2(x-4) \geq 0$. Using the same graph in the previous problem, besides the interval $(4, +\infty)$ over which the curve is above the x-axis, the points $x = -1$ and 4 make the equality hold, so the solution set should be their union $\{-1\} \cup [4, +\infty)$.

No matter whether we are solving a polynomial equation or an inequality, to deal with a quadratic factor, we can factor it in $\mathbb{R}$ as a product of two linear factors by solving the corresponding quadratic equation when the discriminant is nonnegative, but if the discriminant is negative, how do we handle it? In fact, a quadratic polynomial with negative discriminant is always positive or negative (nonzero), depending on the sign of the leading coefficient, and we can cancel such terms in the equations or inequalities. To be specific, consider the trinomial $ax^2 + bx + c$ $(a \neq 0)$. If $\Delta = b^2 - 4ac < 0$ (the trinomial has no real roots), then

- when $a > 0$, $ax^2 + bx + c > 0$ for all $x \in \mathbb{R}$;
- when $a < 0$, $ax^2 + bx + c < 0$ for all $x \in \mathbb{R}$.

This can be proved by completing squares. For example, since $ax^2 + bx + c = a(x - \frac{b}{2a})^2 - \frac{\Delta}{4a}$, when $a > 0$ and $\Delta < 0$, $ax^2 + bx + c = a(x - \frac{b}{2a})^2 - \frac{\Delta}{4a} \geq -\frac{\Delta}{4a} > 0$. The case $a < 0$ and $\Delta < 0$ is proved similarly. By these results, we have $2x^2 - 3x + 2 > 0$, since the discriminant is $(-3)^2 - 4 \times 2 \times 2 < 0$ and the leading coefficient is $2 > 0$.

Exercises.

1. True or false.

(1) A polynomial equation always has real solutions.

(2) 1 is the only real solution to $x^3 - 1 = 0$.

(3) The solution set to $x^2 + x + 2 > 0$ is $\mathbb{R}$.

(4) The solution set to $x^2 > 1$ is $(1, +\infty)$.

(5) $9x^2 + 6x + 1 \leq 0$ has no solutions.

2. Multiple choice.

(1) Which of the following is always positive?

    A. $2x^2 + 4x + 1$ B. $x^2 + 6x + 9$ C. $4x^2 - 8x + 4$ D. $2x^2 - 8x + 9$

(2) Which of the following has 2 as a root of multiplicity 3?

    A. $x^2 - 4x + 4$ B. $(x^3 - 8)(x^2 - 4)$

    C. $(x^2 - 4)(x^4 - 16)(x^3 - 8)$ D. $(x^2 - 4)(x - 2)^3$

(3) Which of the following is not true for all x?

    A. $-x^2 + 2x - 3 < 0$ B. $-4x^2 - 4x - 1 \leq 0$

    C. $2x^2 - 4\sqrt{3}x + 6 > 0$ D. $6x^2 + 3x + 1 \geq 0$

(4) The following are always nonnegative except

    A. the module of a complex number

    B. the square root of a nonnegative real number

    C. the square of a real number

    D. the absolute value of a real number

(5) If $A = \{x | x^2 - x - 6 < 0\}$, $B = \{x | x^2 + 2x - 8 \geq 0\}$, then $A \cap B =$

    A. $[2, 3)$ B. $(-2, 2]$ C. $(-2, 3)$ D. $(-3, -2]$

3. Solve the following equations.

(1) $x^3 + 1 = 0$ (2) $(x^2 - 2)(x^2 + \sqrt{2}x - 4) = 0$

(3) $(x^2 - 3x + 2)x(3x - 6)^2 = 0$ (4) $(3x + 4)(x - 1)^2(x^2 - 3)(x^3 - 4x) = 0$

(5) $(3x^2 - 6x + 2)(3x^2 - 7x + 2) = 0$ (6) $(-4x^2 + 8x - 4)(4x^2 + 7x - 2) = 0$

(7) $(-x^2 - 3x - 6)(x^2 - 4x + 4)(2x^2 - 5x + 2) = 0$

(8) $(2x + 1)^3 x(x^2 - 2x + 3)(x^2 + 4x + 2)(3x - 2)^2 = 0$

4. Find solution sets to the following inequalities.

(1) $x^2 - x - 6 < 0$

(2) $3x^2 - 7x + 2 \leq 0$

(3) $4x^2 + 4x + 1 < 0$

(4) $(x - 1)(x - 2) \geq -3$

(5) $(3x - 1)^2 > -12x$

(6) $6x^2 + x - 2 \geq 0$

(7) $-x^2 - 2x - 1 \geq 0$

(8) $x(x - 1) \leq 1$

(9) $-2x^2 + x + 2 < 0$

(10) $3x^2 - 6x + 2 > 0$

(11) $(x + 1)(x - 2)(x + 3) > 0$

(12) $(x^2 - 1)(x + 1)x(2 - 2x) \leq 0$

(13) $(x^2 - 5x + 6)(x^2 - 2x - 3) > 0$

(14) $(x+2)^3(x+1)^2(2x-1)(x-1)(2x^2+x-1) < 0$

(15) $(x^2 - 7)x^2(x^2 + x - 1) \leq 0$

(16) $(x + 3)^4(3x + 5)^3(2x - 3)^5(x - \sqrt{7})^2 \geq 0$

5. If the sum of the squares of the two solutions to $x^2 - 2mx + (m + 2) = 0$ is greater than 2, find m.

6. Find k to make the system $\begin{cases} xy = -2k + 1 \\ x + y = 3k - 2 \end{cases}$ have real solution.

7. Prove the inequalities by completing squares.

(1) $x^2 + 4x + 4 \geq 0$ (2) $-2x^2 + 12x - 21 < 0$

(3) $3x^2 + 4x + 5 > 0$ (4) $-4x^2 - 8x - 4 \leq 0$

(5) $-5t^2 + 30t - 51 < 0$ (6) $2x^2 - 6x + 5 > 0$

10.7   Rational Inequalities

If connecting two rational expressions by an inequality sign, we get rational or fractional inequalities. Rational inequalities can be transformed into an equivalent system of polynomial inequalities. The idea is for any fraction $\dfrac{a}{b}$ $(a, b \in \mathbb{R}, b \neq 0)$,

- $\dfrac{a}{b} > 0$ when a, b have the same sign, i.e. $ab > 0$;

- $\dfrac{a}{b} < 0$ when a, b have different signs, i.e. $ab < 0$;

- since $b \neq 0$, $\dfrac{a}{b} = 0$ only when $a = 0$, so $\dfrac{a}{b} \geq 0$ is equivalent to $ab \geq 0$ and $b \neq 0$;

- similarly, $\dfrac{a}{b} \leq 0$ is equivalent to $ab \leq 0$ or $b \neq 0$.

Consider the rational expression $\dfrac{P(x)}{Q(x)}$ where $P(x)$ and $Q(x)$ are polynomials in x and $Q(x)$ is not the zero polynomial. From the analysis above,

- $\dfrac{P(x)}{Q(x)} > 0 \iff P(x)Q(x) > 0;$

- $\dfrac{P(x)}{Q(x)} < 0 \iff P(x)Q(x) < 0;$

- $\dfrac{P(x)}{Q(x)} \geq 0 \iff P(x)Q(x) \geq 0$ and $Q(x) \neq 0;$

- $\dfrac{P(x)}{Q(x)} \leq 0 \iff P(x)Q(x) \leq 0$ and $Q(x) \neq 0.$

As long as we can factor $P(x)$ and $Q(x)$ as products of linear and quadratic factors, the rational inequality can be solved by solving the equivalent system of polynomial inequalities.

Examples.

(1) $\dfrac{x-5}{x^2-9} \geq 0$

(2) $\dfrac{2x^2+x-1}{x^2-3x+2} < 0$

Solutions.

(1) The inequality $\dfrac{x-5}{x^2-9} \geq 0$ is equivalent to

$$(x-5)(x^2-9) \geq 0 \text{ and } x^2 - 9 \neq 0.$$

As $x^2 - 9$ can be factored as $(x+3)(x-3)$, it is easy to see that the solution set to $(x-5)(x^2-9) = (x-5)(x-3)(x+3) \geq 0$ is $[-3,3] \cup [5,+\infty)$.

The solution set to $x^2 - 9 \neq 0$ is $x \neq \pm 3$. As a result, the solution set to the fractional inequality is the difference $[-3,3] \cup [5,+\infty) - \{\pm 3\} = (-3,3) \cup [5,+\infty)$.

(2) Since $2x^2+x-1 = (2x+1)(x-1)$, $x^2-3x+2 = (x-1)(x-2)$, the original inequality is equivalent to

$$(2x^2 + x - 1)(x^2 - 3x + 2) = (2x+1)(x-1) \cdot (x-1)(x-2)$$
$$= (2x+1)(x-1)^2(x-2) < 0.$$

The graph of $(2x+1)(x-1)^2(x-2)$ is like

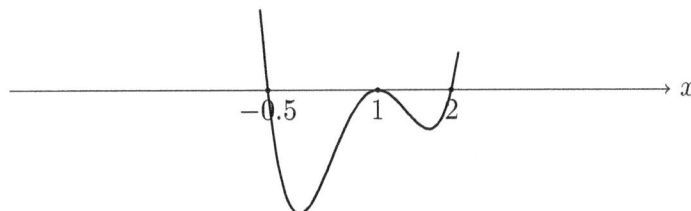

Immediately, the solution set is $(-0.5, 1) \cup (1, 2)$.

Exercises.

1. Multiple choice.

(1) $\dfrac{x-3}{2-x} \geq 0$ is equivalent to

  A. $(x-3)(x-2) \geq 0$ 　　　　　　B. $\dfrac{x-3}{x-2} \leq 0$

  C. $\dfrac{x-2}{x-3} \geq 0$ 　　　　　　D. $(x-3)(x-2) > 0$

(2) If the solution set to $\dfrac{x+a}{x^2+4x+3} > 0$ is $(-3,-1) \cap (2,+\infty)$, then $a =$

  A. 2 　　　　　B. -2 　　　　　C. $\dfrac{1}{2}$ 　　　　　D. $-\dfrac{1}{2}$

(3) If the solution set to $x^2 + qx + p < 0$ is $(1,2)$, then the solution set to $\dfrac{x^2+qx+p}{x^2-5x-6} > 0$

is

  A. $(1,2)$ 　　　　　　　　　　　B. $(-\infty,-1) \cup (6,+\infty)$

  C. $(-1,1) \cup (2,6)$ 　　　　　　D. $(-\infty,-1) \cup (1,2) \cup (6,+\infty)$

(4) Given polynomials $P(x), Q(x)$, which of the following is not correct?

  A. $\dfrac{P(x)}{Q(x)} \geq 0 \iff P(x)Q(x) > 0$ or $P(x) = 0$

  B. $\dfrac{P(x)}{Q(x)} > 0 \iff P(x)Q(x) \geq 0$ and $P(x)Q(x) \neq 0$

  C. $\dfrac{P(x)}{Q(x)} \leq 0 \iff P(x)Q(x) \leq 0$ and $Q(x) \neq 0$

  D. $\dfrac{P(x)}{Q(x)} < 0 \iff P(x)Q(x) < 0$

(5) If $A = \{x | x^2 \leq 9\}$, $B = \{x | \dfrac{x-7}{x+1} \leq 0\}$, then $A \cap B$ is

  A. $[-3,-1)$ 　　　　　　　　　B. $[3,7]$

  C. $(-1,3]$ 　　　　　　　　　D. $(-\infty,-3] \cup [3,7]$

2. Find the solution sets of the following inequalities.

(1) $\dfrac{x-1}{x} \geq 2$ 　　　　　　　　　(2) $x + \dfrac{2}{x+1} > 2$

(3) $\dfrac{x(x+2)}{x-3} < 0$ 　　　　　　　　(4) $\dfrac{x^2-1}{x^2-3x+2} \geq 0$

(5) $\dfrac{5x^2 - 8x + 3}{3x^2 - 5x + 2} \geq 1$

(6) $x - 2 \leq \dfrac{3}{x}$

(7) $\dfrac{x - 1}{x^2 - 5x + 6} \geq 0$

(8) $\dfrac{x^2 + 2x - 3}{-x^2 + x + 6} < 0$

(9) $\dfrac{(x - 1)^2(x + 1)(x - 2)}{x + 4} < 0$

(10) $\dfrac{x^2 + x - 2}{x^3 + 7x^2 - 8x} \geq 0$

(11) $\dfrac{4x^2 - 20x + 18}{x^2 - 5x + 4} \geq 3$

(12) $\dfrac{(x^2 + 3x + 2)(x^2 - 1)}{x - 2} \geq 0$

(13) $\dfrac{1}{x} < \dfrac{1}{2}$

3. If a is an arbitrary real number, solve the inequality for x.

(1) $\dfrac{x}{x - 1} < 1 - a$

(2) $\dfrac{x - a}{x^2 - a^2} \geq 0$

4. If the solution set to $ax^2 + bx + c > 0$ is (m, n) where $0 < m < n$, find the solution set to $cx^2 + bx + a < 0$.

5. If the solution set to $ax^2 + bx + 1 > 0$ is $(-1, \frac{1}{3})$, find the solution set to $x^2 + bx + a \geq 0$.

6. If $ax^2 - x + a > 0$ is true for all $x \in \mathbb{R}$, find a.

7. If $\dfrac{3x^2 + 2x + 2}{x^2 + x + 1} > m$ is true for all x, find m.

8. If $\dfrac{2x^2 + 2kx + k}{4x^2 + 6x + 3} < 1$ is true for all x, find k.

Conic Sections

We shall complete the study of 2-D coordinate geometry in this chapter. Besides the two basic geometric objects — points and lines, in 2-D, there are also advanced objects, such as circles, ellipses, parabolas and hyperbolas. These four types of objects are characterized by polynomial equations of degree two in two variables. Thus, they are also referred to as quadratic curves. The classification theorem asserts that these four types of curves together with their degenerates exhaust all possible quadratic curves.

Another interesting phenomenon, which also indicates the close relation between these curves, is that all of them can be implemented as intersections of planes with a cone in 3-D. As a result, these curves are assigned another name — conic sections.

Although circles are special ellipses, they can be defined independently, and we will investigate them first.

11.1 Distance Formula and Circles

Given two points $P(x_1, y_1)$ and $Q(x_2, y_2)$ in a coordinate system, we can complete the following right triangle where R is the point (x_2, y_1).

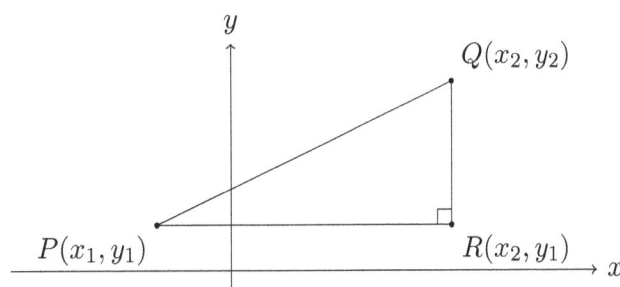

The distance between P and R is $|PR| = |x_2 - x_1|$ and the distance between R and Q is $|RQ| = |y_2 - y_1|$. By the Pythagorean theorem, the length of PQ satisfies $|PQ|^2 = |PR|^2 + |RQ|^2 = (x_2 - x_1)^2 + (y_2 - y_1)^2$. Finding square roots on both side (only take nonnegative square roots as a distance is never negative), we get the distance formula

$$\mathrm{dist}(P, Q) = |PQ| = \sqrt{(x_2 - x_1)^2 + (y_2 - y_1)^2}.$$

For example, the distance between $(-1, 2)$ and $(3, -2)$ is $\sqrt{(-1-3)^2 + [2-(-2)]^2} = 4\sqrt{2}$.

For a special case, the distance between any point $P(x, y)$ and the origin $O(0, 0)$ is $|OP| = \sqrt{x^2 + y^2}$ by plugging the coordinates of P and O into the distance formula.

A circle is determined by its center and length of radius. A circle of center C and length of radius r is the trajectory of points in the plane whose distance to C is r. Suppose the circle lies in a coordinate plane and the coordinate of C is (x_0, y_0). Picking an arbitrary point $P(x, y)$ on the circle, as $|PC| = r$, by the distance formula, we have $\sqrt{(x-x_0)^2 + (y-y_0)^2} = r$, or equivalently

$$(x - x_0)^2 + (y - y_0)^2 = r^2.$$

Thus, if a point is on the circle, its coordinate satisfies the above equation. On the other hand, if the coordinate of a point satisfies this equation, its distance to the center $C(x_0, y_0)$ is r, which means the point must be on the circle. As a result, the solutions to the above equation give exactly all points on the circle with center (x_0, y_0) and length of radius r. It is called the standard equation of the circle. From the standard equation, it is easy to find the center and length of radius.

For example, the equation of the circle with center $(1, -2)$ and length of radius 3 is $(x-1)^2 + [y-(-2)]^2 = 3^3$, or $(x-1)^2 + (y+2)^2 = 9$. $(4, -2)$ is a point on the circle as it satisfies the equation, i.e. $(4-1)^3 + [(-2)-(-2)]^2 = 3^2 + 0^2 = 9$. On the contrary, $(3, 2)$ is not on the circle, since $(3-1)^2 + [2-(-2)]^2 = 20 > 9$. In fact, it is outside the circle. In general, for a fixed point $Q(a, b)$,

- if $(a - x_0)^2 + (b - y_0)^2 > r^2$, Q is outside the circle, since its distance to the center is bigger than r;

- if $(a - x_0)^2 + (b - y_0)^2 < r^2$, Q is inside the circle, since its distance to the center is smaller than r.

If the center of a circle is the origin, then its equation is of the form $x^2 + y^2 = r^2$ by plugging $x_0 = 0, y_0 = 0$ into the standard equation. The circle centered at the origin with length of radius 1 is called the unit circle, whose equation is $x^2 + y^2 = 1$.

If expanding the equation $(x - x_0)^2 + (y - y_0)^2 = R^2$, it looks like $x^2 + y^2 - 2x_0x - 2y_0y + x_0^2 + y_0^2 - R^2 = 0$ (x_0, y_0, R are constants) and is of the form

$$x^2 + y^2 + ax + by + c = 0.$$

This is called the general equation of a circle. Given the equation of a circle in general form, we can rewrite it into the standard form by completing squares, to get the center and length of radius.

Example. Find the center and length of radius of the circle $x^2 + y^2 + 2x - 4y - 7 = 0$.

Solutions. Complete squares for x and y separately:

$$(x^2 + 2x) + (y^2 - 4y) - 7 = 0,$$
$$[(x^2 + 2x + 1) - 1] + [(y^2 - 4y + 4) - 4] - 7 = 0,$$
$$(x + 1)^2 + (y - 2)^2 - 12 = 0,$$
$$(x + 1)^2 + (y - 2)^2 = 12.$$

From the last equation in standard form, the center is $(-1, 2)$ while the length of radius is $r = \sqrt{12} = 2\sqrt{3}$.

Each equation in the standard form of a circle can be transformed into the general form. However, not every equation in the general form defines a circle. For example, $x^2 + 2x + y^2 + 2 = 0$ is in the general form, but after completing squares we get $(x + 1)^2 + y^2 = -1$. This a circle with negative length of radius, so it does not exist. (We can also derive the nonexistence from the fact that squares are nonnegative.) Another special case is when the length of radius is zero and the circle degenerates to a point.

Exercises.

   1. Multiple choice.

   (1) Choose a point whose distance to the origin is 1.

      A. $(1, 1)$ B. $(\frac{1}{2}, \frac{1}{2})$ C. $(-\frac{1}{2}, \frac{\sqrt{3}}{2})$ D. $(\frac{1}{3}, \frac{2}{3})$

   (2) Which point is outside the circle $(x + 3)^2 + (y - 3)^2 = 20$?

      A. $(-1, 3)$ B. $(1, 1)$ C. $(1, 2)$ D. $(-2, 9)$

   (3) The set of points inside the unit circle (excluding the points on the circle) can be denoted by

      A. $\{(x, y) | x^2 + y^2 \leq 1\}$ B. $\{(x, y) | x^2 + y^2 > 1\}$

      C. $\{(x, y) | x^2 + y^2 < 1\}$ D. $\{(x, y) | x^2 + y^2 \geq 1\}$

   (4) Which of the following makes $\triangle ABC$ an isosceles triangle?

      A. $A(0, 0), B(2, 1), C(3, 3)$ B. $A(-3, 5), B(-2, 3), C(-1, 2)$

      C. $A(-3, -4), B(2, 0), C(-4, 3)$ D. $A(0, 1), B(1, 0), C(-2, -1)$

   (5) Whose graph is a point?

      A. $(x - 1)^2 + (y - 2)^2 = 1$ B. $x^2 + y^2 - 2y + 1 = 0$

      C. $x^2 + y^2 = -1$ D. $2x - y + 3 = 0$

   (6) If $Ax^2 + 5y^2 + 2y - 1 = 0$ is the equation of a circle, then A equals

      A. 0 B. 1 C. 5 D. -5

   (7) If the origin is on the circle $x^2 + y^2 + ax + by + c = 0$ with positive length of radius, then

      A. $a = b = 0, c \neq 0$ B. $a = b = c = 0$

      C. $a^2 + b^2 \neq 0$ but $c = 0$ D. $c = 0$

   (8) If the center of $x^2 + y^2 + 2ax + 2by + c = 0$ is on x-axis, then

      A. $a = 0$ B. $b = 0$

      C. $a = 0$ and $b^2 - c \geq 0$ D. $b = 0$ and $a^2 - c \geq 0$

   (9) Which equation describes the upper half of the circle centered at the origin with length of radius 4?

      A. $x = \sqrt{4 - y^2}$ B. $x = \sqrt{16 - y^2}$ C. $y = \sqrt{4 - x^2}$ D. $y = \sqrt{16 - x^2}$

(10) The symmetric circle of $(x-2)^2 + (y+4)^2 = 8$ about the origin is

    A. $(x-2)^2 + (y+4)^2 = 8$ B. $(x+2)^2 + (y+4)^2 = 8$

    C. $(x-2)^2 + (y-4)^2 = 8$ D. $(x+2)^2 + (y-4)^2 = 8$

2. Given A, B as follows, find the distance between A and B.

(1) $A(-1,3), B(2,7)$ (2) $A(5,-2), B(17,3)$

(3) $A(1,1), B(-1,4)$ (4) $A(-2\sqrt{2}, -9), B(\sqrt{2}, , -3)$

(5) $A(-\sqrt{3}, \sqrt{2}), B(\sqrt{2}, \sqrt{3})$ (6) $A(\frac{1}{2}, 1), B(-1, 1)$

3. In $\triangle ABC$, D is the midpoint of BC. Given A, B, C as follows, find the length of $|AD|$.

(1) $A(-2,1), B(5,3), C(-1,-1)$

(2) $A(-1,1), B(3,-2), C(9,6)$

(3) $A(-2\sqrt{3}, -1), B(2\sqrt{3}-5, 7), C(5,-1)$

(4) $A(0, -3\sqrt{8}), B(2\sqrt{15}, -9\sqrt{2}), C(-4\sqrt{15}, \sqrt{18})$

4. Determine if $\triangle ABC$ is a right triangle, an acute triangle or an obtuse triangle.

(1) $A(-1,2), B(4,-8), C(-3,1)$ (2) $A(-3,4), B(-2\sqrt{3}-3, -3), C(2\sqrt{3}-3, -2)$

(3) $A(-3,-1), B(1,-2), C(3,1)$ (4) $A(5,-4), B(-3,4), C(-5,2)$

5. Find an equation in the general form for the circle S given below.

(1) S is centered at $(-2,5)$ with length of radius 7.

(2) S is centered at the origin and passing through $(5, -4)$.

(3) S is centered at $(-3, -2)$ and passing through $(1, 1)$.

(4) S has a diameter whose endpoints are $(-7, 4)$ and $(-5, -10)$.

(5) S passes through $(3, 4)$, $(4, 3)$ and $(-2, -1)$.

6. Find the center and length of radius of the circles.

(1) $x^2 + y^2 - 8x + 6y - 1 = 0$

(2) $x^2 + y^2 + 5x - 3y + 8 = 0$

(3) $x^2 + y^2 - 10\sqrt{2}x + 6\sqrt{3}y = 0$

(4) $3x^2 + 3y^2 - 4x - 2 = 0$

(5) $x^2 + 2x + y^2 + 4y + 5 = 0$

7. Given points $A(4, 3), B(0, -2), C(-3, -1)$, if D, E are midpoints of AB and AC separately, (1)find $|DE|$; (2)show $|DE| = \dfrac{1}{2}|BC|$.

8. Determine if the line l intersects with the circle S. If intersecting, find the intersection(s); if not intersecting, show the reasons.

(1) $l : 2x + y - 7 = 0$, $S : x^2 + y^2 + 2y - 8 = 0$

(2) $l : x - y + 10 = 0$, $S : x^2 + y^2 + 6x - 16 = 0$

(3) $l : y = -3x + 1$, $S : x^2 + y^2 + 4x - 2y - 5 = 0$

9. Find the equations of the symmetric circle of $x^2 + y^2 - 4x + 8y - 1 = 0$ about (1) the x-axis, (2) the y-axis, (3) the origin.

10. Match the circle pairs with their relative positions.

A. $\begin{cases} S_1 : (x + 3)^2 + (y - 2)^2 = 9 \\ S_2 : (x + 3)^2 + (y + 6)^2 = 25 \end{cases}$

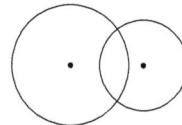

(1) Intersecting

B. $\begin{cases} S_1 : (x + 3)^2 + (y - 1)^2 = 16 \\ S_2 : (x - 2)^2 + (y - 9)^2 = 25 \end{cases}$

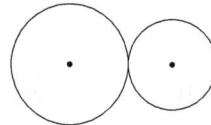

(2) Tangent Externally

C. $\begin{cases} S_1 : x^2 + y^2 + 4x + 12y + 31 = 0 \\ S_2 : x^2 + y^2 + 2x + 14y + 46 = 0 \end{cases}$

(3) Tangent Internally

D. $\begin{cases} S_1 : x^2 + y^2 + 2x - 6y - 26 = 0 \\ S_2 : x^2 + y^2 - 2x - 10y + 22 = 0 \end{cases}$

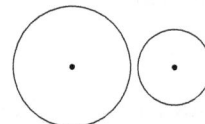

(4) Separate

E. $\begin{cases} S_1 : x^2 + y^2 = 81 \\ S_2 : x^2 + y^2 - 2x - 2\sqrt{3}y = 45 \end{cases}$

(5) Inside

11. Find equations and/or inequalities to describe the following.

(1) Lower half of the unit circle.

(2) Right half of the circle centered at the origin with radius 4.

(3) The disk centered at the origin whose radius is 2.

(4) The outside of the unit circle.

(5) The upper half of the disk centered at $(1, -1)$ with radius 1.

12. Tell if the two circles $x^2 + y^2 + 4x - 6y + 9 = 0$ and $x^2 + y^2 - 2x = 9$ intersect or not. If intersecting, find the intersections; if not intersecting, tell the reason.

11.2 Parabolas

The graph of an equation of the form $y = ax^2 + bx + c$ $(a, b, c \in \mathbb{R}, a \neq 0)$ is a parabola.

Special Case: Basic Parabola $y = ax^2$ $(a \neq 0)$

The basic parabolas $y = ax^2$ $(a \neq 0)$ are symmetric about the y-axis, which is called the line of symmetry, as if plugging both x and $-x$ into the equation $y = ax^2$, we get the same y-value. The line of symmetry intersects with the parabola at its vertex. The vertex of a basic parabola $y = ax^2$ is always $(0, 0)$. The parabola $y = ax^2$ opens upward when $a > 0$ and downward if $a < 0$. The graphs of basic parabolas are sketched below.

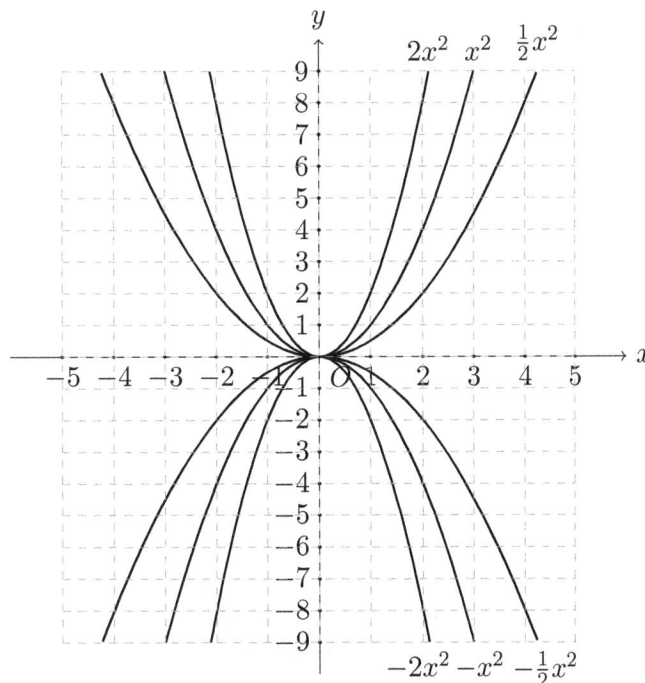

It can be concluded from the graphs: (1) the greater $|a|$ is, the narrower the graph of $y = ax^2$ is; (2) the graphs of $y = ax^2$ and $y = -ax^2$ are symmetric about the x-axis. This is true since if (x_0, y_0) is a point on $y = ax^2$, then $(x_0, -y_0)$ is on $y = -ax^2$.

General Case: $y = ax^2 + bx + c \ (a \neq 0)$

Given any equation in two variables $F(x, y) = 0$ whose graph is S (i.e. $S = \{(a, b) \in \mathbb{R}^2 | F(a, b) = 0\}$), if S' is the graph by shifting S to the right by a units and upward by b units, then the equation of S' is $F(x - a, y - b) = 0$. The reason is as follows:

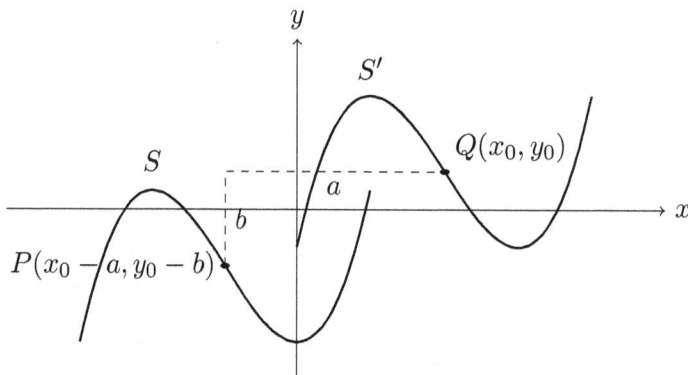

Pick an arbitrary point $Q(x_0, y_0)$ on S'. Shifting Q to the left by a units and downward by b units, we get a point $P(x_0 - a, y_0 - b)$ on S. The coordinate of P should satisfy the equation $F(x, y) = 0$, so $F(x_0 - a, y_0 - b) = 0$. As (x_0, y_0) is an arbitrary point on S', the equation of S' is $F(x - a, y - b) = 0$.

The graph of $y = ax^2 + bx + c(a \neq 0)$ can be treated as the shifting graph of $y = ax^2$. The tool we need here is again to complete squares.

Example. Find the vertex and line of symmetry of $y = -\dfrac{1}{2}x^2 - 2x - 1$. Sketch the graph.

Solutions. Completing the square for x, we get

$$y = -\frac{1}{2}(x^2 + 4x) - 1,$$

$$y = -\frac{1}{2}[(x^2 + 4x + 4) - 4] - 1,$$

$$y = [-\frac{1}{2}(x^2 + 4x + 4) + 2] - 1,$$

$$y = -\frac{1}{2}(x + 2)^2 + 1,$$

$$(y - 1) = -\frac{1}{2}(x + 2)^2.$$

So the parabola $y = -\dfrac{1}{2}x^2 + 2x - 1$ can be obtained by shifting $y = -\dfrac{1}{2}x^2$ to the left by 2 units and upward by 1 unit. As a result, the vertex is $(-2, 1)$ and the line of symmetry is $x = -2$. As $a < 0$, the parabola opens downward. We can sketch the graph by finding several points on the graph symmetric about the line of symmetry $x = -2$.

x	y
-4	-1
-3	0.5
-2	1
-1	0.5
0	-1

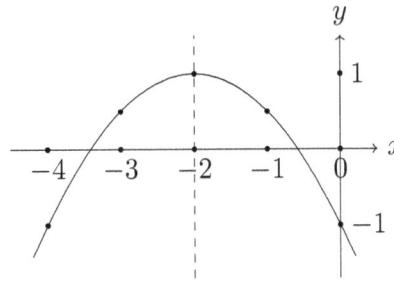

Any parabola can be obtained by shifting basic parabolas. In order to get the vertex and line of symmetry, we make the equation of a parabola into the square-completed form $y = a(x - x_0)^2 + y_0$. Its graph is the same as the parabola $y = ax^2$, but moved to the right by x_0 units and upward by y_0 units. As a result, the vertex of the parabola is (x_0, y_0) and the line of symmetry is $x = x_0$.

The y-intercept of the parabola $y = ax^2 + bx + c$ ($a \neq 0$) is c, the constant term, by plugging $x = 0$ into the equation. To find the x-intercepts, we set $y = 0$ and get an equation $ax^2 + bx + c = 0$. The x-intercepts are just solutions to this quadratic equation. Depending on the sign of $\Delta = b^2 - 4ac$, there are three possibilities:

- $\Delta > 0$, there are two x-intercepts;

- $\Delta = 0$, there is only one x-intercept, which means the parabola is tangent to the x-axis;

- $\Delta < 0$, there are no x-intercepts.

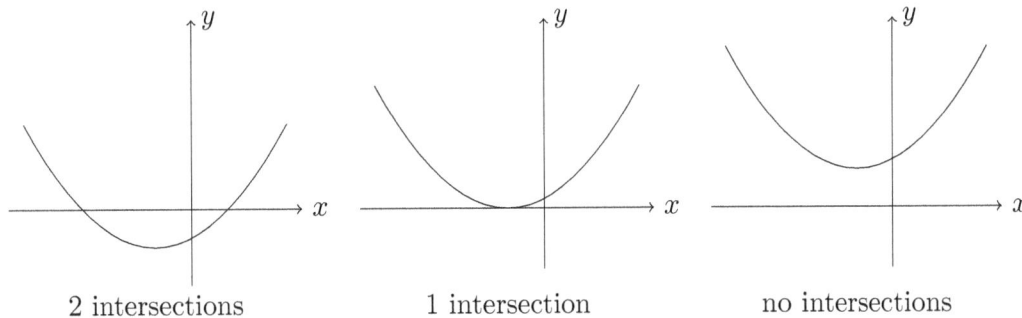

2 intersections   1 intersection   no intersections

For example, the y-intercept of $y = -x^2 - 2x + 3$ is 3. To find x-intercepts, solving the equation $-x^2 - 2x + 3 = 0$ we get $x = -3, 1$.

When $\Delta = b^2 - 4ac < 0$, the parabola $y = ax^2 + bx + c$ has no intersections with the x-axis. As a result, $ax^2 + bx + c$ is greater than 0 for all $x \in \mathbb{R}$ when $a > 0$, and less than 0 for all x when $a < 0$. This is already proved by completing squares in Section 10.6, but parabolas help us to visualize the results.

Exercises.

1. Multiple choice.

(1) The vertex of $y = 3(x - 2)^2 + 4$ is

    A. $(-2, 4)$ B. $(-2, -4)$ C. $(2, -4)$ D. $(2, 4)$

(2) The number of intersections between $y = -2x^2 + x - 1$ and the x-axis is

    A. 0   B. 1   C. 2   D. 3

(3) Shift the graph of $y = -3x^2$ towards right by 2 units and upward by $\dfrac{1}{5}$ unit, the equation of the new parabola is

A. $y = -3(x-2)^2 - \dfrac{1}{5}$ B. $y = -3(x+2)^2 + \dfrac{1}{5}$

C. $y = -3(x-2)^2 + \dfrac{1}{5}$ D. $y = 3(x-2)^2 + \dfrac{1}{5}$

(4) The number of intersections between $y = -x + 1$ and $y = -2x^2 + 4x - 1$ is

A. 0   B. 1

C. 2   D. cannot be determined

(5) The axis of symmetry of $y = -3x^2 + x$ is

A. $y = \dfrac{1}{6}$ B. $x = \dfrac{1}{6}$ C. $y = -\dfrac{1}{6}$ D. $x = -\dfrac{1}{6}$

2. Find the opening direction, vertices and lines of symmetry of the following parabolas.

(1) $y = 2(x+3)^2 + 2$ (2) $y = -x^2 - 2x$

(3) $y = -5x^2 - 7$ (4) $y = 6x^2 - 24x + 17$

(5) $y = -3x^2 + x - 1$ (6) $y = 4x^2 + 2x + 5$

(7) $y = \dfrac{1}{2}x^2 - \dfrac{5}{2}x + \dfrac{7}{8}$ (8) $y = -\dfrac{4}{3}x^2 - 2x + \dfrac{1}{4}$

3. State the procedure to get the second parabola from the first.

(1) $y = (x-1)^2 - 2 \Rightarrow y = x^2$ (2) $y = 2(x+5)^2 - 6 \Rightarrow y = 2(x-7)^2 - 1$

(3) $y = x^2 \Rightarrow y = x^2 + 2x + 2$ (4) $y = 3x^2 + 1 \Rightarrow y = 3x^2 - 12x + 7$

(5) $y = 6x^2 - 36x + 35 \Rightarrow y = 6x^2 - 12x$ (6) $y = -x^2 - 6x - 16 \Rightarrow y = -x^2 + x - 2$

(7) $y = x^2 \Rightarrow y = -x^2$ (8) $y = -2x^2 + 1 \Rightarrow y = 2x^2 - 4x + 1$

4. Find equations of the parabolas.

(1) The parabola with y-intercept $-\sqrt{2}$ and equation of the form $y = 7x^2 - 6x + c$.

(2) The parabola with vertex $(4, 2)$ which passes through the point $(3, -1)$.

(3) The parabola with vertex $(-2, 4)$ and y-intercept 6.

(4) The parabola with x-intercepts $5, -4$ and y-intercept -20.

(5) The parabola with x-intercepts $0, 1$ and equation of the form $y = 4x^2 + bx + c$.

(6) The parabola passing $(3, -9)$, $(5, -33)$ with equation of the form $y = -2x^2 + bx + c$.

(7) The parabola passing three points $(-1, -2)$, $(1, 0)$ and $(2, -8)$.

5. Find the x and y-intercepts of the parabolas.

(1) $y = 3x^2 + 5x + 2$ (2) $y = -2x^2 + 6x - 7$

(3) $y = -4x^2 - 16x - 16$ (4) $y = x^2 + 3x + 1$

6. Sketch the parabolas.

(1) $y = 2x^2 - 1$ (2) $y = -x^2 - 2x + 3$

(3) $y = 2x^2 + 5x - 3$ (4) $y = -4x^2 + 3x + 7$

7. Tell if the line and the parabola intersect. If intersecting, find the intersections and tell if they are tangent to each other.

(1) $y = 2x + 1$, $y = -x^2 - 2x + 1$

(2) $y = -5x - 17$, $y = 2x^2 + 7x + 1$

(3) $y = 4x + 3$, $y = 2x^2 - 6x + 5$

(4) $y = 7x - 1$, $y = 5x^2 - 3x + 4$

(5) $y = -3x - 5$, $y = 3x^2 - x + 6$

11.3 Ellipses

An ellipse is the trajectory of points in the plane the sum of whose distances to two fixed points is a constant. The two fixed points are called foci (focus points).

If we denote the two foci by F_1, F_2 and assume the constant is $2a$ ($a > 0$), then a point P is on the ellipse if and only if

$$|PF_1| + |PF_2| = 2a.$$

Assume the distance between F_1 and F_2 is $2c(0 < c < a)$, called the focal length, and set up a coordinate system such that the coordinates of F_1 and F_2 are $(-c, 0)$ and $(c, 0)$ separately. Applying the distance formula to the above equation, $P(x, y)$ is a point on the ellipse if and only if $\sqrt{(x + c)^2 + y^2} + \sqrt{(x - c)^2 + y^2} = 2a$. Simplifying the equation, let $b = \sqrt{a^2 - c^2}$ (i.e. $b > 0, a^2 = b^2 + c^2$), we get the standard equation of an ellipse

$$|PF_1| + |PF_2| = 2a$$

$$\frac{x^2}{a^2} + \frac{y^2}{b^2} = 1(a > b > 0).$$

The graph of the ellipse is given below.

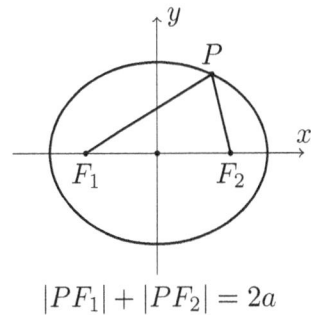

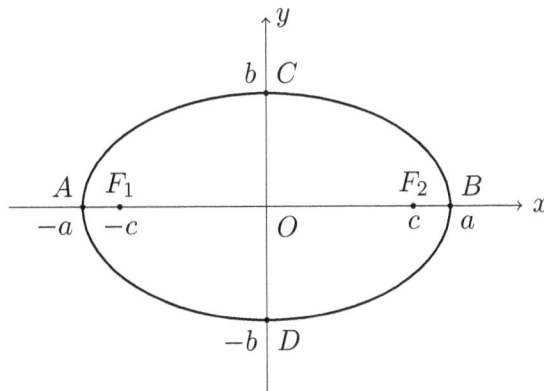

The points A, B, C, D are called vertices; the line segment AB is called the major axis and CD is called the minor axis. As a result, $|AB| = 2a$ and $|CD| = 2b$ are the lengths of the major and minor axis separately. $|OA| = |OB| = a$ is the length of the semi-major axis and $|OC| = |OD| = b$ is the length of the semi-minor axis.

Any ellipse in the standard equation is symmetric about the origin, x-axis and y-axis. In this case $O(0,0)$, the midpoint of the major axis, is called the center; the x and y-axes, the lines where the major and minor axes reside, are called the axes of symmetry.

$\dfrac{x^2}{a^2} + \dfrac{y^2}{b^2} = 1 \ (b > a > 0)$ is also the equation of an ellipse, but in this case the major axis is on the y-axis. If $a = b > 0$, $\dfrac{x^2}{a^2} + \dfrac{y^2}{b^2} = 1$ degenerates to an equation of a circle and the two foci coincide at the origin. The three cases are summarized in the table below.

$\dfrac{x^2}{a^2} + \dfrac{y^2}{b^2} = 1 \ (a > b > 0)$	$\dfrac{x^2}{a^2} + \dfrac{y^2}{b^2} = 1 \ (a = b > 0)$	$\dfrac{x^2}{b^2} + \dfrac{y^2}{a^2} = 1 \ (a > b > 0)$
$c = \sqrt{a^2 - b^2}$	$c = 0$	$c = \sqrt{b^2 - a^2}$

To summarize, for an ellipse with equation of the form $\dfrac{x^2}{a^2} + \dfrac{y^2}{b^2} = 1 \ (a, b > 0)$, the major axis is on the x-axis if $a > b$, and on the y-axis if $b > a$. The length of the semi-major axis is always the greater one between a and b, while the length of semi-minor axis is always the smaller one.

If we shift the ellipse in the standard equation $\dfrac{x^2}{a^2} + \dfrac{y^2}{b^2} = 1 \ (a, b > 0)$ to the right by x_0 units and upward by y_0 units, the equation of the new ellipse is

$$\frac{(x - x_0)^2}{a^2} + \frac{(y - y_0)^2}{b^2} = 1.$$

The new ellipse has center (x_0, y_0); the vertices are $(x_0 \pm a, y_0)$, $(x_0, y_0 \pm b)$; the axes of symmetry are $x = x_0$ and $y = y_0$. Its major axis is on $y = y_0$ if $a > b$ and on $x = x_0$ if $a < b$. Any equation of the form $Ax^2 + Cy^2 + Dx + Ey + F = 0$ with $A, C > 0$ can be transformed into an equation of the above form by completing squares for x and y separately. As a result, it is the equation of an ellipse and we can find the vertices, axes of symmetry and foci accordingly.

Examples. Find the center, vertices, foci and axes of symmetry for the ellipses. Sketch the graph.

(1)$4x^2 + y^2 = 4$

(2)$4x^2 + 8x + 9y^2 - 36y + 4 = 0$

Solutions.

(1) To transform the equation into the standard form, divide by 4 on both sides to make the right side of the equation the constant 1. The equation becomes $\dfrac{x^2}{1} + \dfrac{y^2}{4} = 1$. Immediately we get:

- the axes of symmetry are $y = 0$ and $x = 0$ (x and y-axis) and the center is the origin;

- the length of the semi-major axis is $a = \sqrt{4} = 2$, the length of semi-minor axis is $b = \sqrt{1} = 1$ and the half focal length is $c = \sqrt{4 - 1} = \sqrt{3}$;

- the major axis is on the y-axis.

As a result, the four vertices are $(0, \pm 2)$, $(\pm 1, 0)$ and the foci are $(0, \pm\sqrt{3})$. We sketch the graph by connecting the four vertices using a smooth curve.

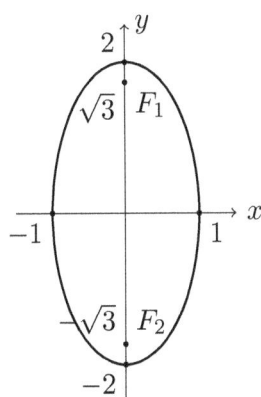

Problem (1)

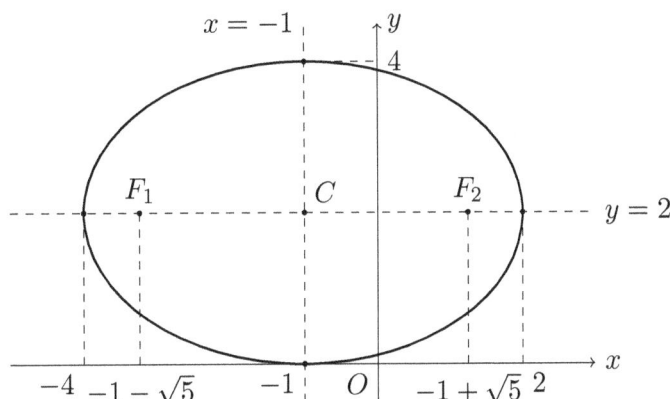

Problem (2)

(2) Completing the squares for x and y separately, we get $4(x + 1)^2 + 9(y - 2)^2 = 36$. Dividing by 36 on both sides to make the right side 1, the equation becomes $\dfrac{(x + 1)^2}{9} + \dfrac{(y - 2)^2}{4} = 1$, so the ellipse can be obtained by shifting $\dfrac{x^2}{9} + \dfrac{y^2}{4} = 1$ to the left by 1 unit and upward by 2 units. From the equation,

- the major axis is on $y = 2$ and the minor axis is on $x = -1$;

- the length of the semi-major axis is $a = \sqrt{9} = 3$, the length of semi-minor axis is $\sqrt{4} = 2$ and the half focal length is $c = \sqrt{9 - 4} = \sqrt{5}$.

As a result, the center C is $(-1, 2)$; the vertices are $(-1 \pm 3, 2)$ and $(-1, 2 \pm 2)$, i.e. $(-4, 2), (2, 2), (-1, 0), (-1, 4)$; the foci are $(-1 \pm \sqrt{5}, 2)$; the axes of symmetry are $x = -1$ and $y = 2$. Then we can sketch the graph (see above).

The equation in the form $Ax^2 + Cy^2 + Dx + Ey + F = 0$ with $A, C > 0$ is called the general equation of an ellipse. Similar to the general equations of circles, the graphs of such equations may degenerate to a point or not exist at all.

Exercises.

1. Find the focal length, the length of the semi-major and semi-minor axes, and the center, foci, vertices and axes of symmetry for the ellipses.

(1) $\dfrac{x^2}{36} + \dfrac{y^2}{25} = 1$

(2) $\dfrac{x^2}{9} + \dfrac{y^2}{16} = 1$

(3) $3x^2 + 5y^2 + 3x - 10y = 24.25$

(4) $4x^2 + 6y^2 - 24x + 34 = 0$

2. Sketch the ellipses.

(1) $\dfrac{x^2}{4} + \dfrac{y^2}{1} = 1$

(2) $\dfrac{(x+3)^2}{16} + \dfrac{(y-4)^2}{25} = 1$

(3) $2x^2 + 3y^2 + 8x + 6y + 5 = 0$

(4) $2x^2 + y^2 + 4x + 2y + 2 = 0$

11.4   Hyperbolas

A hyperbola is the trajectory of points in the plane such that the absolute value of the difference of distances to two fixed points is a constant. The two fixed points are called foci (focus points) of the hyperbola.

If we denote the two foci by F_1, F_2 and assume the constant is $2a$ $(a > 0)$, then a point P is on the hyperbola if and only if

$$||PF_1| - |PF_2|| = 2a.$$

Assume the focal length $|F_1F_2|$ is $2c$ $(c > a)$, and build a coordinate system such that the coordinates of F_1 and F_2 are $(-c, 0)$ and $(c, 0)$ separately. By the distance formula, $P(x, y)$ is a point on the hyperbola if and only if $|\sqrt{(x+c)^2 + y^2} - \sqrt{(x-c)^2 + y^2}| = 2a$. If we let $b = \sqrt{c^2 - a^2}$ (i.e. $b > 0, c^2 = a^2 + b^2$), the equation simplifies into the standard form of the hyperbola

$$\frac{x^2}{a^2} - \frac{y^2}{b^2} = 1 \; (a, b > 0).$$

The graph of the hyperbola is given below.

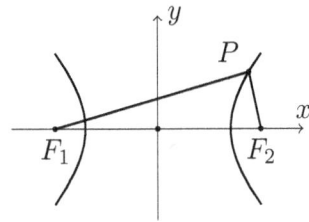

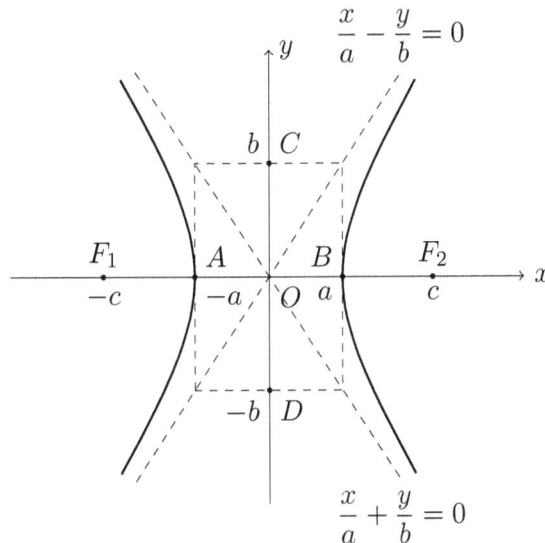

A hyperbola has two branches. The points A, B are called vertices; the line segment AB is called the transverse or real axis, and the line segment CD is called the conjugate or imaginary axis. Similar to an ellipse, $|OA| = |OB| = a$ is the length of the semi-real axis and $|OC| = |OD| = b$ the length of the semi-imaginary axis.

The two lines $\frac{x}{a} \pm \frac{y}{b} = 0$ are solutions to the equation $\frac{x^2}{a^2} - \frac{y^2}{b^2} = 0$ (setting the right side of the standard equation to be 0). They are called the asymptotes of the hyperbola as the difference of the y-coordinates of the points on the hyperbola and asymptotes approaches 0 when the x-coordinate approaches infinity. The asymptotes do not intersect with the hyperbola.

$-\frac{x^2}{b^2} + \frac{y^2}{a^2} = 1 \; (a, b > 0)$ also determines a hyperbola whose real axis is on the y-axis and whose imaginary axis is on the x-axis. The standard equations of hyperbolas can be classified by the following.

$\dfrac{x^2}{a^2} - \dfrac{y^2}{b^2} = 1 \ (a, b > 0)$	$-\dfrac{x^2}{b^2} + \dfrac{y^2}{a^2} = 1 \ (a, b > 0)$

$$c = \sqrt{a^2 + b^2}$$

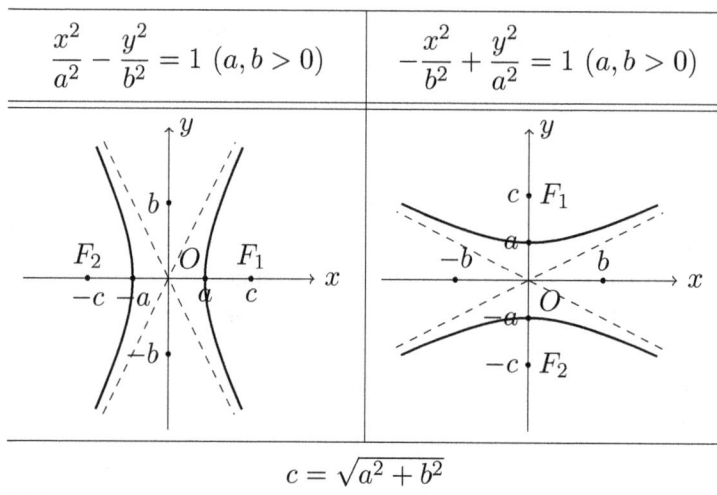

Like ellipses, any hyperbola in the standard equation is symmetric about the origin, x-axis and y-axis. $O(0,0)$, the midpoint of the real axis, is called the center of the hyperbola; the x and y-axes, the lines where real and imaginary axes live, are called the axes of symmetry.

If $a = b \neq 0$, the hyperbola $\dfrac{x^2}{a^2} - \dfrac{y^2}{a^2} = 1$ (or $x^2 - y^2 = a^2$) is called equilateral and the asymptotes are always $y = \pm x$, the angle bisector of all four quadrants.

In general, any equation of the form $Ax^2 + Cy^2 + De + Ey + F = 0$ with $AC < 0$ determines a hyperbola. Such equations are called the general equations of hyperbolas. We can make any equation in the general form into the standard form

$$\pm \left(\dfrac{(x - x_0)^2}{a^2} - \dfrac{(y - y_0)^2}{b^2} \right) = 1$$

by completing squares for x and y. It is easy to find the vertices, foci, center and axes of symmetry from the equation. For asymptotes, the equations are always $\dfrac{x - x_0}{a} \pm \dfrac{y - y_0}{b} = 0$.

Example. Find the center, vertices, foci and axes of symmetry for the hyperbola $-x^2 + 4y^2 + 2x + 4y - 2 = 0$. Sketch the graph.

Solutions. By completing the squares, we get $-(x - 1)^2 + 4(y + \dfrac{1}{2})^2 = 4$. Divide by 4 on both sides, the equation becomes $-\dfrac{(x - 1)^2}{4} + \dfrac{(y + \dfrac{1}{2})^2}{1} = 1$. The hyperbola can be obtained by shifting $-\dfrac{x^2}{4} + \dfrac{y^2}{1} = 1$ to the right by 1 unit and downward by 0.5 units. From the equation,

- the center is $(1, -\dfrac{1}{2})$;

- the real axis lies on $x = 1$, and the imaginary axis lies on $y = -\dfrac{1}{2}$;

- the length of the semi-real axis is $a = \sqrt{1} = 1$, the length of semi-imaginary axis is $b = \sqrt{4} = 2$, and the focal length is $2c = 2\sqrt{1 + 4} = 2\sqrt{5}$. As a result, the vertices are $(1, -\dfrac{1}{2} \pm 1)$, i.e. $(1, \dfrac{1}{2})$ and $(1, -\dfrac{3}{2})$; the foci are $(1, -\dfrac{1}{2} \pm \sqrt{5})$;

- the asymptotes are $\dfrac{x-1}{2} \pm \dfrac{y \mp \frac{-}{2}}{1} = 0$, i.e. lines with slope $\pm\dfrac{1}{2}$ through the center $(1, -\dfrac{1}{2})$. After simplifications we get $y = \dfrac{1}{2}x - 1$ and $y = -\dfrac{1}{2}x$.

With the center, vertices and asymptotes, we can sketch the graph as follows.

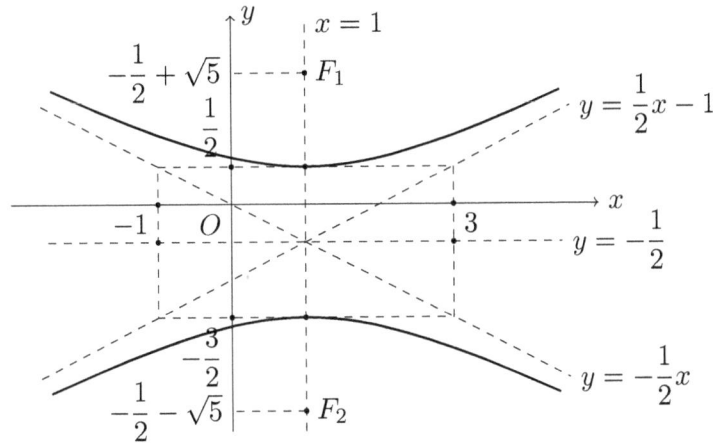

A hyperbola may degenerate to two intersecting lines. The reason is that from the general equation $Ax^2 + Cy^2 + Dx + Ey + F = 0$ $(AC < 0)$, it is possible to get $\dfrac{(x-x_0)^2}{a^2} - \dfrac{(y-y_0)^2}{b^2} = 0$ after completing squares, but this is just two intersection lines $\dfrac{x-x_0}{a} \pm \dfrac{y-y_0}{b} = 0$.

Exercises. Find the focal length, the length of the semi-real and semi-imaginary axes, and the center, foci, vertices, axes of symmetry and asymptotes for the hyperbolas. Sketch the graphs.

(1) $\dfrac{x^2}{4} - \dfrac{y^2}{9} = 1$

(2) $-\dfrac{(x+1)^2}{12} + \dfrac{(y-2)^2}{4} = 1$

(3) $2x^2 - 6y^2 + 4x - 12y = 16$

(4) $-x^2 + y^2 + 4x + 6y + 2 = 0$

11.5 Parabolas From A New Viewpoint

We know the graph of $y = ax^2 + bx + c$ $(a \neq 0)$ is a parabola, but as a conic section, a parabola is the trajectory of points in the plane equidistant to a fixed point and a fixed line. The fixed point is called the focus of the parabola and the fixed line is called the directrix. Suppose the distance between the focus, denoted by F, and the directrix, denoted by l, is $p > 0$; then build a coordinate system such that F has coordinate $(\frac{p}{2}, 0)$ and l has equation $x = -\frac{p}{2}$. A point $P(x, y)$ is on the parabola if and only if

$$\text{dist}(P, l) = |PF|,$$

$|PQ| = |PF|$

i.e. $|x - (-\frac{p}{2})| = \sqrt{(x - \frac{p}{2})^2 + y^2}$. Simplifying the equation we get the standard equation of a parabola,

$$y^2 = 2px \ (p > 0).$$

In general, the axis of symmetry of a parabola is the line passing through the focus and perpendicular to the directrix, while the vertex is the intersection of the parabola with the axis of symmetry. For the current case $y^2 = 2px$, if we solve x to get $x = \frac{y^2}{2p}$, immediately it is of the form $y = ax^2$ (here $a = \frac{1}{2p} > 0$) by exchanging x and y. As a result, the graph of $x^2 = 2py$ is the mirror image of $y = \frac{x^2}{2p}$ about the line $y = x$. Then it is easy to see that the parabola $y^2 = 2px$ opens to the right with vertex at the origin and the axis of symmetry is the x-axis.

There are three other possible standard equations of parabolas. All of them together with their graphs are listed as follows. Assume $p > 0$.

Equation	$x^2 = 2py$	$x^2 = -2py$	$y^2 = 2px$	$y^2 = -2px$
Graph				
Focus	$F(\frac{p}{2}, 0)$	$F(-\frac{p}{2}, 0)$	$F(0, \frac{p}{2})$	$F(0, -\frac{p}{2})$
Directrix	$l : x = -\frac{p}{2}$	$l : x = \frac{p}{2}$	$l : y = -\frac{p}{2}$	$l : y = \frac{p}{2}$
Vertex	$O(0, 0)$			
Axis of Symmetry	$y = 0$		$x = 0$	

For example, $y^2 = -2x$ is the equation of a parabola which is centered at the origin

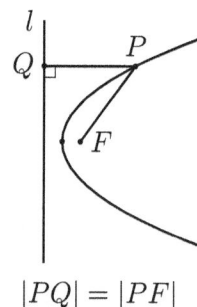

opening to the left. Here the equation is of the form $y^2 = -2px$ with $p = 1$, so the focus is $(-\frac{1}{2}, 0)$ and the directrix is $x = \frac{1}{2}$.

In general, the equations $Ax^2 + Dx + Ey + F = 0$ $(A \neq 0)$ and $Cy^2 + Dx + Ey + F = 0$ $(C \neq 0)$ also define parabolas. Such parabolas may not be centered at the origin, but we can treat them as the shifting graphs of parabolas centered at the origin by the method learned in Section 11.2. Besides, when $A \neq 0$ and $C = D = E = 0$, the equation become $Ax^2 + F = 0$. Depending on the sign of A and F, the equation may define two parallel lines (including coincidence) or have no graph at all. Thus a parabola may degenerate to two parallel lines.

Exercises.

1. Find the focus, directrix, vertex and axis of symmetry of the parabolas. Sketch the graphs.

(1) $y^2 = -6x$

(2) $-3y^2 = 6x - 4$

(3) $-y^2 - 2y + 5x - 11 = 0$

(4) $4x^2 - 4x + 9y - 17 = 0$

2. Given the focus F and directrix l, find an equation of the parabola.

(1) $F(1, 0)$, $l : x = -1$

(2) $F(3, 0)$, $l : x = -2$

(3) $F(-2, 3)$, $l : x = 1$

(4) $F(-5, -7)$, $l : y = 3$

(5) $F(1, 1)$, $l : x + y = -1$.

11.6   Unified Definition of Conic Sections

The conic sections - ellipses, hyperbolas and parabolas - are defined by their own features in the previous three sections. The common feature of these curves is that their equations are all quadratic equations in two variables. The most general quadratic equation in two variables x and y is $Ax^2 + Bxy + Cy^2 + Dx + Ey + F = 0$. The graph of a quadratic equation is called a quadratic curve. In fact, it can be proved that ellipses, hyperbolas and parabolas with their degenerates are all possible quadratic curves. The quadratic curves have a unified definition.

Fix a line l, called the directrix, and a point F, called the focus, on a plane π. Let $e > 0$ be a fixed number, called the eccentricity. The trajectory of the points P on π such that

$$\frac{|PF|}{\text{dist}(P,l)} = e$$

is a quadratic curve C. In fact,

- when $0 < e < 1$, C is an ellipse;

- when $e = 1$, C is a parabola;

- when $e > 1$, C is a hyperbola.

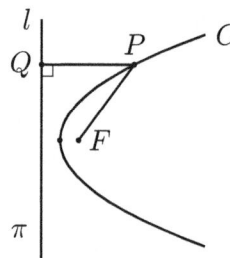

From the previous section, it should be clear that the case $e = 1$ corresponds to parabolas. If placing the focus F and the directrix l nicely in a coordinate system, by the distance formula, it is not hard to derive the equation of the curve and verify the above correspondence. In fact, for ellipses and hyperbolas in their standard equations $\frac{x^2}{a^2} \pm \frac{y^2}{b^2} = 1$, the eccentricity is always $e = \frac{c}{a}$, and the focus and directrix are always $(-c, 0)$, $x = -\frac{a^2}{c}$ or $(c, 0)$, $x = \frac{a^2}{c}$.

	Ellipse	Parabola	Hyperbola
Equation	$\frac{x^2}{a^2} + \frac{y^2}{b^2} = 1 \ (a > b > 0)$	$y^2 = 2px$	$\frac{x^2}{a^2} - \frac{y^2}{b^2} = 1 \ (a, b > 0)$
	$c = \sqrt{a^2 - b^2}$		$c = \sqrt{a^2 + b^2}$
Graph $\frac{|PQ|}{|PF|} = e$			
Eccentricity	$e = \frac{c}{a} \in (0, 1)$	$e = 1$	$e = \frac{c}{a} \in (1, +\infty)$
Focus	$(-c, 0)$ (or $(c, 0)$)	$(\frac{p}{2}, 0)$	$(-c, 0)$ (or $(c, 0)$)
Directrix	$x = -\frac{a^2}{c}$ (or $x = \frac{a^2}{c}$)	$x = -\frac{p}{2}$	$x = -\frac{a^2}{c}$ (or $x = \frac{a^2}{c}$)

Example. Tell the name of the quadratic curve whose equation is $x^2 - 4y^2 + 2x - 16y + 1 = 0$. Find the eccentricity and directrixes.

Solutions. Completing squares for x and y separately, the original equation becomes $(x+1)^2 - 4(y-2)^2 = 16$, or equivalently $\dfrac{(x+1)^2}{16} - \dfrac{(y-2)^2}{4} = 1$. The curve is a hyperbola whose lengths of semi-real and semi-imaginary axes are $a = 4$ and $b = 2$, so $c = \sqrt{a^2 + b^2} = 2\sqrt{5}$ and the eccentricity e is $\dfrac{c}{a} = \dfrac{2\sqrt{5}}{4} = \dfrac{\sqrt{5}}{2}$. The hyperbola can be obtained by shifting $\dfrac{x^2}{16} - \dfrac{y^2}{4} = 1$ to the left by 1 unit and upward by 2 units, so the directrixes are $x = -1 \pm \dfrac{a^2}{c} = -1 \pm \dfrac{4^2}{2\sqrt{5}} = -1 \pm \dfrac{8\sqrt{5}}{5}$.

Exercises. Name the following quadratic curves. Find the eccentricity, foci and directrixes when possible.

(1) $\dfrac{x^2}{6} - \dfrac{y^2}{10} = -1$

(2) $\dfrac{x^2}{4} + \dfrac{y}{2} = 1$

(3) $3x^2 + 4y^2 = 24$

(4) $-y^2 - 4x - 6y + 11 = 0$

(5) $2x^2 + 2y^2 - 3x + 5y + 2 = 0$

(6) $2x^2 - y^2 + 4y = 7$

(7) $4x^2 + y^2 + 4y - 4x + 5 = 0$ (8) $3y^2 + 2x^2 - 6y + 9x + 7.125 = 0$

Functions

Back to elementary algebra, we have learned algebraic expressions and algebraic equations, both of which are built on variables. However, the capability of variables is not limited to these aspects. We can also construct functions. The idea of functions is natural. Given an expression, if we plug different numbers into variables, we get different results. Such kind of mechanisms, are essentially functions.

Corresponding to algebraic expressions, we have algebraic functions — polynomial functions, rational functions and radical functions, which serve as basic examples. With the help of coordinate geometry, functions be may visualized by graphs. Other basics, such as domains, operations of functions, inverse functions, graph shifting, are all presented.

Power functions are functions constructed from exponentiations. As special algebraic functions, they are of great importance in the function theory. Power functions consist of one important class of elementary functions. The other two terms missed from the dictionary of elementary functions are exponential and logarithmic functions, which will be introduced in the last chapter.

12.1 Basic Concepts

Suppose A, B are two nonempty sets. A map from A to B is that for each $a \in A$, there is a unique element $b \in B$ corresponding to a. By convention, we denote a map from A to B by $f : A \to B$ where A is called the domain, denoted by $\text{dom}(f)$, B is called the range, denoted by $\text{Ran}(f)$, and f is the name of the map. For any $a \in A$, the unique element $b \in B$ that a corresponds to is called the image of a, denoted by $f(a)$. In order to specify the rule of correspondence, the map above is also denoted by

$$f : A \to B$$
$$a \mapsto f(a).$$

$\text{Im} f = \{f(a) | a \in A\}$ is called the image of f. From the definition, we must have $\text{Im} f \subset \text{Ran}(f)$. Given $A = \{1, 2, 3\}$, $B = \{a, b, c\}$, the correspondence

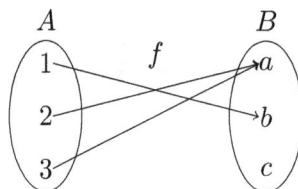

is a map for which the image of 1 is b, the images of both 2 and 3 are a, and $\mathrm{Im} f = \{a, b\}$. The correspondences

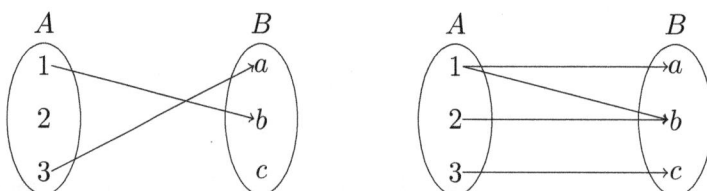

are not maps as 2 has no image in the first example, and 1 has more than one images in the second example.

A map is completely determined by its domain, range and the rule of correspondence. As a result, two maps are equal if they have the same domain, the same range and the same rule of correspondence. Maps with different domains or ranges are not considered equal even if they have the same rule of correspondence. For example $f : \mathbb{R} \to \mathbb{R} \; x \mapsto x^2$ and $g : (-\infty, +\infty) \to [0, +\infty) \; x \mapsto x^2$ have the same domain and rule of correspondence, but they are still not equal since their ranges are different.

Given any set $A \neq \emptyset$, the map $A \to A$ defined by $a \mapsto a$ is called the identity map of A, denoted Id_A. In particular, $Id_\mathbb{R}$ is the identity map of $\mathbb{R}$: $Id_\mathbb{R}(x) = x$ for any real number $x \in \mathbb{R}$.

A map is called a function if its range is a subset of $\mathbb{R}$. For example, $f : \mathbb{R} \to \mathbb{R} \; x \mapsto x^2$ and $g : [1, +\infty) \to (0, +\infty) \; x \mapsto \sqrt{x + 2}$ are both functions. We are only interested in functions whose domains are also number sets. In the following contents, the word 'function' refers to functions whose domains are also subsets of $\mathbb{R}$.

If a function is defined by giving the rule of correspondence only, the domain is implicitly involved in the rule. We need to find the natural (maximal possible) domain explicitly.

Examples. Find the natural domains of the functions.

(1) $f(x) = 2x^3 - 3x^2 + 5$

(2) $g(x) = \sqrt{x^2 - 1}$

(3) $h(x) = \dfrac{x^2 - 49}{x - 7}$

Solutions.

(1) For any real number a, $f(a) = 2a^3 - 3a^2 + 5$ is well-defined, so the maximal domain of f is the set $\mathbb{R}$ of all real numbers.

(2) $g(x)$ is a square root which requires the radicand nonnegative. As a result, $g(x)$ is well-defined only when $x^2 - 1 \geq 0$. Solving the inequality, we get $x \geq 1$ or $x \leq -1$, so the

natural domain is $(-\infty, -1) \cup (1, +\infty)$.

(3) As a fraction, the denominator cannot be 0, so $x - 7 \neq 0$ or $x \neq 7$. The domain is the set of all real numbers except 7, i.e. $\mathbb{R} - \{7\} = (-\infty, 7) \cup (7, +\infty)$. Notice that $\dfrac{x^2 - 49}{x - 7}$ can be simplified to $x + 7$, but as functions $h(x) = \dfrac{x^2 - 49}{x - 7}$ and $\tilde{h}(x) = x + 7$ are not equal, since their domains are different: the domain of $h(x) = \dfrac{x^2 - 49}{x - 7}$ is $\mathbb{R} - \{7\}$, but the domain of $\tilde{h}(x) = x + 7$ is $\mathbb{R}$.

The above three examples represent three classes of elementary functions — polynomial functions, rational functions and radical (or irrational) functions.

- A polynomial function is of the form $x \mapsto P(x)$, where $P(x)$ is a polynomial in one variable x. The domain of a polynomial function is always $\mathbb{R}$.

- A rational function is of the form $x \mapsto \dfrac{P(x)}{Q(x)}$, where $P(x)$ and $Q(x)$ are two polynomials in x and $Q(x) \neq 0$. The domain of a rational function is the set of all real numbers such that the denominator is nonzero, i.e $\{x \in \mathbb{R} | Q(x) \neq 0\}$.

- An irrational function is a function involving radicals with the variable appearing in radicands. To find the domain, we must make sure the radicands are nonnegative for roots of even order.

In the previous example, f is a polynomial function, h is a rational function, and g is an irrational function.

We also use the notation $y = f(x)$ to indicate y is a function of x. In this case the domain of f is understood to be its natural domain and the range is just $\text{Im} f$ (i.e. the natural range), if not specified otherwise.

Given $A \subseteq \mathbb{R}$ and $kin\mathbb{R}$ a fixed number, the function defined by $A \to \mathbb{R} \quad x \mapsto a$ is called a constant function over A.

Let f be a function whose domain $A \subseteq \mathbb{R}$ is symmetric about the origin, i.e. $x \in A$ if and only if $-x \in A$. f is an odd function if $f(-x) = -f(x)$ for any $x \in A$, and an even function if $f(x) = f(-x)$ for any $x \in A$. For example, $f(x) = x^3$ defined over $\mathbb{R}$ is an odd function as $f(-x) = (-x)^3 = -x^3 = -f(x)$; $g(x) = x^2$ defined over $\mathbb{R}$ is an even function as $g(-x) = (-x)^2 = x^2 = g(x)$; $h(x) = x^2$ defined over $[-3, 3]$ is also an even function.

In fact, any polynomial in one variable containing odd powers only defines an odd function over any subset of $\mathbb{R}$ symmetric about the origin. Similarly, any polynomial containing even powers only defines an even function. Not all functions are even or odd even if they have domains symmetric about the origin. For example, $f(x) = x + 1$ is neither even nor odd.

Exercises.

1. True or false.

(1) The functions $f(x) = x$ and $g(x) = \dfrac{x^2}{x}$ are equal.

(2) $f(x) = 1$ is not a function.

(3) Since $\dfrac{x^2-1}{x+1}$ can be simplified to $x-1$, they define the same function.

(4) $f(x)=x$ is the same function as $g(x)=\sqrt{x^2}$.

(5) The domain of a polynomial function is $(-\infty,+\infty)$.

(6) The three elements to define a function are domain, range and the rule of correspondence.

(7) If $f(x)=\sqrt{x^2-1}$, the natural domain of f is $[1,+\infty)$ and $\mathrm{Im}f=[0,+\infty)$.

(8) $f(x)=\sqrt{x^2+1}$ is an even function over $\mathbb{R}$.

(9) If $g(x)=\dfrac{x^2}{x}$, $\mathrm{Im}g=\mathbb{R}$.

(10) $(-1,1]$ can be the domain of even functions.

(11) The domain of odd functions must be $\mathbb{R}$.

(12) If both f and g are odd functions over $\mathbb{R}$, then $\dfrac{f(x)}{g(x)}$ is an even function over $\{x\in\mathbb{R}|g(x)\neq 0\}$.

2. Multiple choice.

(1) Which of the following is not a map?

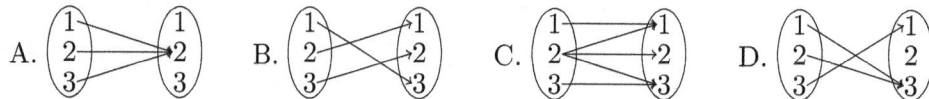

A.   B.   C.   D.

(2) Which of the following is the same as the function $f(x)=x$?

A. $g_1(x)=(\sqrt{x})^2$ B. $g_2(x)=\sqrt{x^2}$ C. $g_3(x)=\sqrt[3]{x^3}$ D. $g_4(x)=\dfrac{x^3}{x^2}$

(3) If the domain of a function $y=f(x)$ is $[0,1]$, then the domain of $y=f(x^2)$ is

A. $[0,1]$ B. $[-1,1]$

C. $[-1,0]$ D. cannot be determined

(4) The domain of $f(x)=\sqrt{x+2}-\dfrac{1}{x-1}$ is

A. $[-2,1)\cup(1,+\infty)$ B. $[-2,+\infty)$

C. $[-2,1)$ D. $(-2,1)\cup(1,+\infty)$

(5) If $f(x)=\sqrt{x+1}$, then $f(8)$ is

A. ±3 B. 3 C. -3 D. $2\sqrt{2}$

(6) If $f(x)$ is an odd function defined over $\mathbb{R}$, which of the following may not be correct?

A. $f(-x)+f(x)=0$ B. $f(-x)-f(x)=-2f(x)$

C. $f(-x)f(x)\leq 0$ D. $\dfrac{f(x)}{f(-x)}=-1$

(7) Given four groups of functions, which group has two functions equal?

A. $f(x) = \dfrac{x}{x}$ and $f(x) = 1$

B. $f(x) = x^2 - 3x + 4$ and $g(t) = t^2 - 3t + 4$

C. $f(x) = \sqrt{x^2}$ and $f(x) = (\sqrt{x})^2$

D. $f(x) = \sqrt{x-1}\sqrt{x+1}$ and $f(x) = \sqrt{x^2-1}$

(8) If $y = x^2 + 4x + c$, which of the following is correct?

A. $f(1) < c < f(-2)$ B. $f(1) > c > f(-2)$

C. $c > f(1) > f(-2)$ D. $c < f(-2) < f(1)$

(9) Given four statements:

(i) $f(x) = x^2 - 1$ is an even function over $[-1, 1]$;

(ii) $f(x) = x + \dfrac{1}{x}$ is an odd function;

(iii) If f is an even function and g is an odd function, then $y = f(x)g(x)$ is an odd function;

(iv) $f(x) = x - x^3 + 1$ is an odd function.

How many of them are correct?

A. 1   B. 2   C. 3   D. 4

(10) The image of $f(x) = x^2 + 2x - 3$ is

A. $[-1, +\infty)$ B. $[-3, +\infty)$ C. $[4, +\infty)$ D. $[-4, +\infty)$

(11) Which of the following is not correct?

A. The domain of an odd function is symmetric about the origin.

B. The image of an odd function is symmetric about the origin.

C. The domain of an even function is symmetric about the origin.

D. The image of an even function is symmetric about the origin.

3. Find the domains of the functions.

(1) $f(x) = \sqrt{x^2 - 9}$ (2) $f(x) = \dfrac{\sqrt{x-2}}{x^2 - 2x - 3}$

(3) $f(x) = \sqrt{2x+1} + \sqrt{3x-4}$ (4) $f(x) = \sqrt{1 - \left(\dfrac{x-1}{x+1}\right)^2}$

(5) $f(x) = \dfrac{\sqrt{x^2 - 2x - 15}}{|x+3| - 3}$ (6) $f(x) = \dfrac{\sqrt{-x^2 - 3x + 4}}{x}$

(7) $f(x) = \dfrac{3x^2}{\sqrt{1-x}}$

(8) $f(x) = -x^3 + 2x^2 + \dfrac{x-2}{2x^2 - 3x - 2}$

(9) $f(x) = \sqrt{3-x}\sqrt{3+x}$

(10) $f(x) = \sqrt{25 - 4x^2}$

(11) $f(x) = \sqrt{x^2 - 4x + 5}$

(12) $f(x) = \dfrac{\sqrt{2x+1}}{\sqrt{1-x}}$

(13) $f(x) = \sqrt{x^2 + 5x + 6}$

(14) $f(x) = \sqrt[3]{x^2 - x - 6}$

4. Determine if the functions are even, odd or neither.

(1) $f(x) = x^2$, $x \in [-1, 2]$

(2) $f(x) = \dfrac{1}{x^2 + 1}$

(3) $f(x) = \sqrt{3 - x^4}$

(4) $f(x) = x + \sqrt[3]{x}$

(5) $f(x) = (x+1)(x-1)$

(6) $f(x) = \sqrt{2x-1} + \sqrt{1-2x}$

(7) $f(x) = x^2 - 2|x| + 1$

(8) $f(x) = \dfrac{x}{2 + x^6}$

(9) $f(x) = \dfrac{|x-3| - 3}{\sqrt{20 - x^4}}$

(10) $f(x) = \dfrac{(2x - 1)^0}{\sqrt{x^2 - 2x + 1}}$

5. Find the natural ranges of the functions.

(1) $f(x) = 3x + 2$, $x \in (-\infty, +\infty)$

(2) $g(x) = 7 - 5x$, $x \in (-2, 4]$

(3) $f(x) = -3x^2 + 6x$

(4) $g(x) = 2x^2 + 5x - 2$, $x \in (-2, 1]$

(5) $f(x) = \sqrt{x^2 + |x| + 2}$

(6) $g(x) = \sqrt[3]{27 - x^4}$

(7) $f(x) = \sqrt{8 - 3x^2}$

(8) $g(x) = \sqrt{3 + x}\sqrt{5 - x}$

(9) $f(x) = \dfrac{1}{x^2 + 1}$

(10) $g(x) = \dfrac{x + 1}{x - 1}$

6. If f is a function defined on $(-\infty, +\infty)$ which satisfies $f(x) = f(x + 7)$ for all x and $f(x) = \sqrt[3]{x}$ when $x \in [1, 8)$, find $f(8)$ and $f(123)$.

7. Find the maximum value of $y = \dfrac{1}{1 - x(1 - x)}$.

8. If f is a function with domain $[-3, 5)$, find the domain of $y = f(2x + 1)$ and $y = f(\sqrt{x^2 + 1})$.

9. $f(x) = a_n x^n + a_{n-1} x^{n-1} + \cdots + a_1 x + a_0$ is a polynomial function. Prove (i) f is odd if and only if $a_{2k} = 0$ for all $k \leq \dfrac{n}{2}$; (ii) f is even if and only if $a_{2k-1} = 0$ for all $k \leq \dfrac{n+1}{2}$.

10. Assume g is an arbitrary function defined over a subset of $(-\infty, +\infty)$ symmetric about the origin. Determine if the following functions are even, odd or neither.

(i) $y = g(x) + g(-x)$

(ii) $y = g(x) - g(-x)$

11. Show (i) any polynomial function $f(x) = a_n x^n + a_{n-1} x^{n-1} + \cdots + a_1 x + a_0$ is the sum of an odd function and an even function; (ii) any function g on a subset of $(-\infty, +\infty)$ symmetric about the origin is the sum of an odd function and an even function.

12. Find a such that the domain of $y = \dfrac{x + 7a}{ax^2 + 4ax + 3}$ is $(-\infty, +\infty)$.

12.2   Graphs

Given a map $f : A \to B$, the subset $\{(a, f(a)) | a \in A\} \subseteq A \times B$ is called the graph of f, denoted by $\mathrm{gr}(f)$. If f is a function, i.e. $A, B \subseteq \mathbb{R}$, then $(a, f(a))$ is an ordered pair of real numbers which can be identified as a point in the coordinate plane. In this case the graph $\mathrm{gr}(f)$ of f can be visualized by plotting all points in a coordinate plane. A point (x, y) is on the graph of f if and only if x and y satisfy the relation $y = f(x)$, so the graph of f is the same as the graph of the equation $y = f(x)$ $(x \in A)$.

Examples. Sketch the graphs of the following functions.

(1) $f(x) = 2x + 1$

(2) $g(x) = x^2 - 2x$

(3) $h(x) = |x|$

(4) $r(x) = \dfrac{1}{x}$

Solutions. The domains of these functions are not specified, so they are understood to be the natural domains. Let us find them first: $\mathrm{dom}(f) = \mathbb{R}$, $\mathrm{dom}(g) = \mathbb{R}$, $\mathrm{dom}(h) = \mathbb{R}$,

$\text{dom}(r) = \{x \in \mathbb{R} | x \neq 0\}$.

(1) The graph of f is just the graph of $y = 2x + 1$, which is a line with slope 2 and y-intercept 1. The graph is sketched as follows.

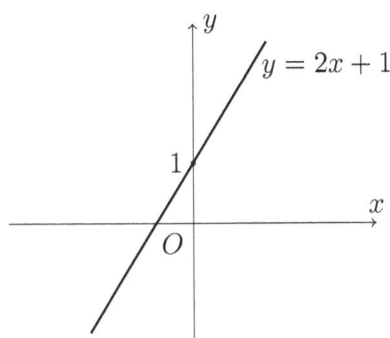

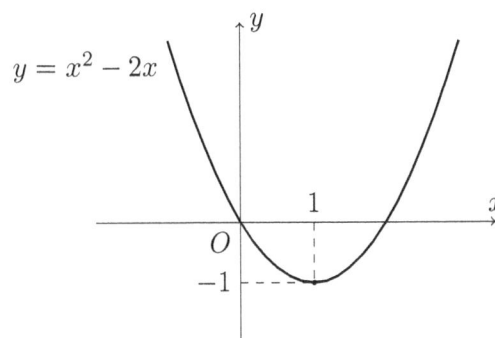

(2) The graph of g is the same as the graph of the equation $y = x^2 - 2x$, which is a parabola. It is easy to see that the parabola opens upward with vertex $(1, -1)$. The graph can then be sketched as above by finding several points on the graph symmetric about $x = 1$.

(3) $h(x) = |x|$ can be simplified to $h(x) = \begin{cases} x & \text{when } x \geq 0 \\ -x & \text{when } x < 0 \end{cases}$, so the graph of h is the line $y = x$ when $x \geq 0$ and $y = -x$ when $x < 0$.

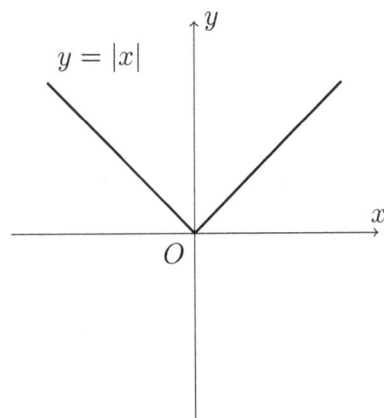

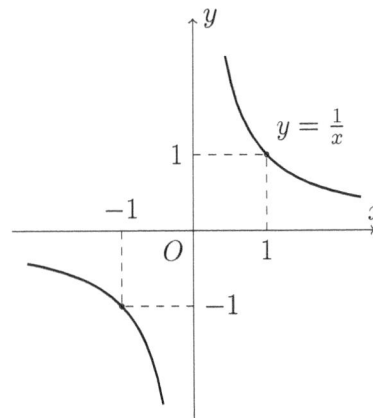

(4) The graph of $r(x)$ is an equilateral hyperbola with vertices $(1, 1)$ and $(-1, -1)$. The asymptotes of the hyperbola are $x = 0$ and $y = 0$. It is sketched as above.

The graph of an odd function is symmetric about the origin, while the graph of an even function is symmetric about the y-axis. Suppose $y = f(x)$ is an odd function and $(x, f(x))$ is a point on $\text{gr}(f)$. Since $f(-x) = -f(x)$, $(-x, -f(x))$ is also a point on $\text{gr}(f)$, but $(x, f(x))$ and $(-x, -f(x))$ are symmetric about the origin, $\text{gr}(f)$ is symmetric about the origin. Similarly, if f is even, i.e. $f(-x) = f(x)$ for any x in the domain, $(x, f(x))$ and $(-x, f(x))$ are both on $\text{gr}(f)$. Since they are symmetric about the y-axis, so is $\text{gr}(f)$. In the previous example, $h(x) = |x|$ is an even function and $r(x) = \dfrac{1}{x}$ is an odd function. From the graphs we can see that $\text{gr}(h)$ is symmetric about the y-axis and $\text{gr}(r)$ is symmetric about the origin. The following are two more examples of graphs of odd and even functions.

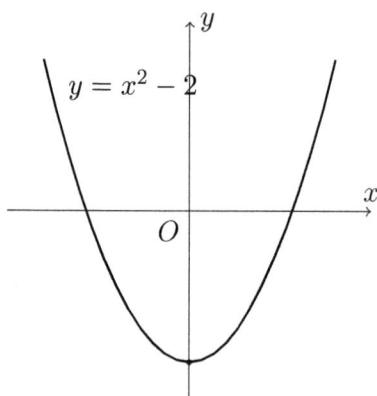

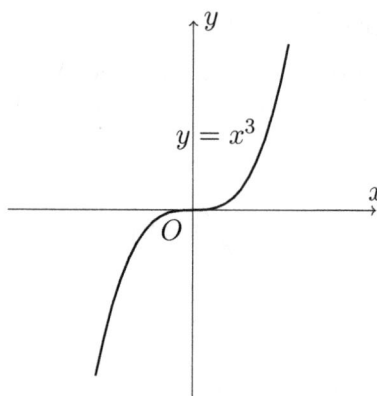

Consider the function $f : A \to B$ $(A, B \subset \mathbb{R})$. If we shift the graph S of $y = f(x)$ to the right by a units and upward by b units, by the result in Section 11.2, the equation defining the new graph S' should be $y - b = f(x - a)$, i.e. S' is the graph of the function $y = \tilde{f}(x) = f(x - a) + b$. Notice the domain and range of $\tilde{f}$ are also shifted: $\mathrm{dom}(\tilde{f}) = \{x + a | x \in A\}$, i.e. shifting A to the right by a units; $\mathrm{ran}(\tilde{f}) = \{y + b | y \in B\}$, i.e. shifting B upward by b units. For example, if we shift the graph of $g(x) = \sqrt{x}$ to the right by 2 units and downward by 1 unit, the new graph is defined by the function $\tilde{g}(x) = \sqrt{x - 2} - 1$. The domain of g is $[0, +\infty)$; the domain of $\tilde{g}$ is $[2, +\infty)$.

Given a function f, the graphs of $y = f(x)$, $y = f(-x)$, $y = -f(x)$ and $y = -f(-x)$ have the following relations.

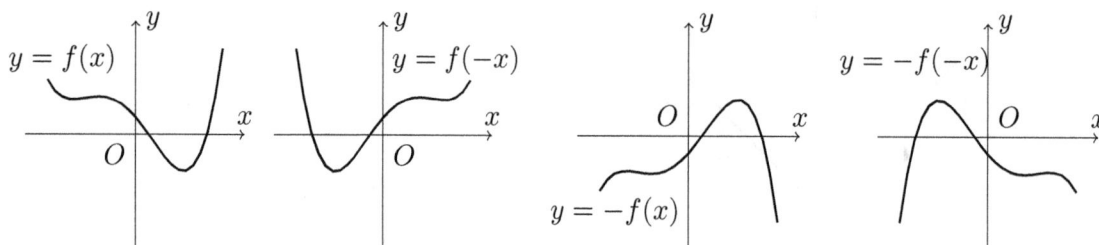

• The graphs of $y = f(x)$ and $y = f(-x)$ are symmetric about the y-axis, so we can get the graph of $y = f(-x)$ by flipping the graph of $y = f(x)$ about the y-axis.

• The graphs of $y = f(x)$ and $y = -f(x)$ are symmetric about the x-axis, so we can get the graph of $y = -f(x)$ by flipping the graph of $y = f(x)$ about the x-axis.

• The graphs of $y = f(x)$ and $y = -f(-x)$ are symmetric about the origin.

Now we can draw graphs of more functions from the ones we learned. For example, to find the graph of $y = 1 - \dfrac{1}{x + 1}$, first write it as $(y - 1) = -\dfrac{1}{x + 1}$. Immediately we see that the graph can be obtained by shifting the graph $y = -\dfrac{1}{x}$ to the left by 1 unit and upward by 1 unit. But the graph of $y = -\dfrac{1}{x}$ is the mirror image of the hyperbola $y = \dfrac{1}{x}$ about the x-axis (or y-axis), so the graph of $y = 1 - \dfrac{1}{x + 1}$ is a hyperbola and we can find it by flipping and shifting the equilateral hyperbola $y = \dfrac{1}{x}$.

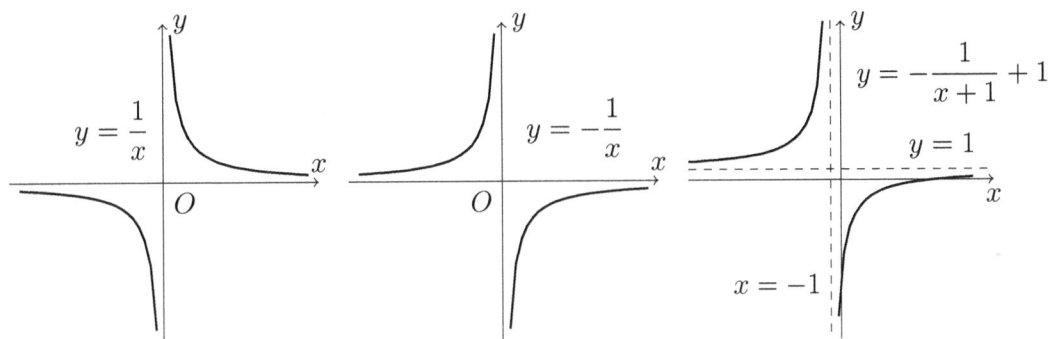

Consider a function f defined over some interval $D \subseteq \mathbb{R}$. We define f to be increasing or decreasing as follows.

- f is increasing if for any $x_1, x_2 \in D$ and $x_1 < x_2$, we have $f(x_1) \leq f(x_2)$;
- f is strictly increasing if for any $x_1, x_2 \in D$ and $x_1 < x_2$, we have $f(x_1) < f(x_2)$;
- f is decreasing if for any $x_1, x_2 \in D$ and $x_1 < x_2$, we have $f(x_1) \geq f(x_2)$;
- f is strictly decreasing if for any $x_1, x_2 \in D$ and $x_1 < x_2$, we have $f(x_1) > f(x_2)$.

For example, given the graphs of three functions f_1, f_2, f_3 below, it is easy to see that f_1 is increasing but not strictly increasing, f_2 is neither increasing nor decreasing, and f_3 is strictly decreasing.

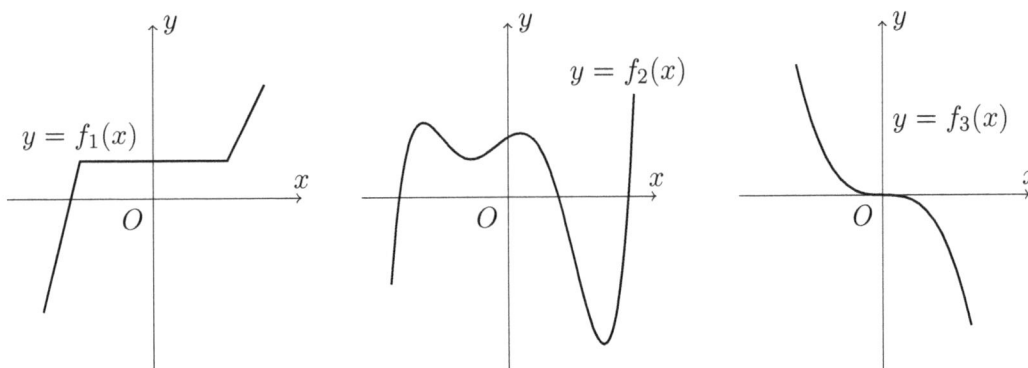

To prove a function is increasing or decreasing we have to check the definition. For example, to show $f(x) = \dfrac{1}{x}$ is decreasing over $(0, +\infty)$, first pick any $a, b \in (0, +\infty)$ and assume $a < b$. Then check the sign of $f(b) - f(a)$: since $a, b > 0$ and $a < b$, $f(b) - f(a) = \dfrac{1}{b} - \dfrac{1}{a} = \dfrac{a-b}{ab} < 0$. As a result, $f(a) > f(b)$ when $a < b$, so f is strictly decreasing over $(0, +\infty)$. This is also clear from the graph of the equilateral hyperbola $y = \dfrac{1}{x}$ (see above).

Exercises.

1. True or false.

(1) To get the graph of $y = 3(x-1)^2 + 4$, we can shift the graph of $y = 3x^2$ to the left by 1 unit and upward by 4 units.

(2) If f is an odd function having definition at 0, then $f(0)$ must be 0.

(3) If f is an even function, $\text{gr}(f)$ must have an even number of intersections with the x-axis.

(4) If $y = f(x)$ and $y = g(x)$ are both odd functions over $(-\infty, 0) \cup (0, +\infty)$ and $g(x) \neq 0$, then the graph of $y = \dfrac{f(x)}{g(x)}$ must be symmetric about the y-axis.

(5) The graph of $y = \dfrac{1}{x-1}$ is symmetric about the origin.

(6) The graph of $y = \dfrac{1}{1-x^2}$ is symmetric about the y-axis.

(7) There does not exist a function whose graph is symmetric about the x-axis.

(8) To get the graph of $y = -2x^2 - 8x - 11$, we can shift the graph of $y = 2x^2$ to the left by 2 units, then flip the graph about the x-axis and move the graph downward by 3 units.

(9) $y = \dfrac{1}{x}$ is an increasing function over $(-\infty, 0)$ and a decreasing function over $(0, +\infty)$.

(10) Given $f(x) = ax^2 + bx + c$ where $a > 0$, the graph of f is decreasing over the interval $(-\dfrac{b}{2a}, +\infty)$.

(11) If flipping the graph of $y = 3x^2 + 2$ about the x-axis, we get the graph of $y = -3x^2 + 2$.

(12) If the point (m, n) is on the graph of $f(x) = ax^2 + a$, then $(-m, n)$ is also on the graph of f.

(13) The graph of $y = \dfrac{2x}{\sqrt{1 + \dfrac{1}{x^2}}}$ is symmetric about the y-axis.

(14) If f is decreasing and $a < b$, then $\dfrac{1}{(f(a))^2} \leq \dfrac{1}{(f(b))^2}$ whenever $f(a), f(b) \neq 0$.

2. Multiple choice.

(1) If $y = f(x)$ is an even function over $\mathbb{R}$, which of the following points must be on the graph of $y = f(x)$?

    A. $(-a, -f(a))$ B. $(a, -f(a))$ C. $(-a, f(a))$ D. $(-a, -f(-a))$

(2) To get the graph of $y = |x + 3| - 2$, we need to shift the graph of $y = |x|$

    A. to the right by 3 units and downward by 2 units

    B. to the left by 2 units and upward by 3 units

    C. to the left by 3 units and downward by 2 units

    D. to the right by 2 units and downward by 2 units

(3) The graph of $y = x + \dfrac{1}{x}$ is symmetric about

    A. x-axis   B. y-axis   C. the origin   D. $y = 1$

(4) To get the graph of $y = 3x^2 - 12x + 14$, we need to shift the graph of $y = 3x^2$

    A. to the right by 2 units and upward by 14 units

    B. to the left by 2 units and upward by 2 units

C. to the left by 2 units and upward by 14 units

D. to the right by 2 units and upward by 2 units

(5) The increasing interval of $y = \dfrac{1}{x-3} + 1$ is

    A. $(-3, +\infty)$ B. $(3, +\infty)$ C. $(-\infty, 3)$ D. none

(6) $y = \sqrt{x^2 + x - 6}$ is increasing over the interval

    A. $(-\infty, 3]$ B. $[-2, +\infty)$ C. $[2, +\infty)$ D. $(-\infty, -3]$

(7) The graph of $y = x^2 + \dfrac{a}{x}$ is symmetric about the y-axis when

    A. $a > 0$ B. $a < 0$ C. $a = 0$ D. $a \neq 0$

(8) Which of the following functions is increasing over $(0, +\infty)$?

    A. $y = \dfrac{1}{x} + 1$ B. $y = -2x + 2$ C. $y = |x + 2| - 1$ D. $y = -\sqrt{x}$

(9) Which of the following graphs is symmetric about the origin?

    A. $y = \dfrac{\sqrt{1 - x^2}}{|x + 2| - 2}$ B. $y = \dfrac{2x^2 + 2x}{x + 1}$

    C. $y = \sqrt{4 - x^2} + (x - 2)^0$ D. $y = x - \dfrac{1}{x}$

(10) If $f(x) = kx + b(k \neq 0)$ is decreasing over $\mathbb{R}$, then the point (k, b) is located

    A. above the x-axis   B. below the x-axis

    C. to the left of the y-axis   D. to the right of the y-axis

(11) How many of the following statements are true?

    (i) The graph of an even function must intersect with the y-axis;

    (ii) The graph of an odd function must go through the origin;

    (iii) The graph of an even function is symmetric about the y-axis;

    (iv) If a function is both even and odd, then it must satisfy $f(x) = 0$.

    A. 4   B. 3   C. 2   D. 1

(12) Suppose the even function $y = f(x)$ is defined over $\mathbb{R}$ and when $x \in [0, +\infty)$, $f(x)$ is increasing. Then $f(-2)$, $f(\pi)$, $f(-3)$ should be ordered as

    A. $f(\pi) > f(-3) > f(-2)$ B. $f(\pi) > f(-2) > f(-3)$

    C. $f(\pi) < f(-3) < f(-2)$ D. $f(\pi) < f(-2) < f(-3)$

(13) If the increasing interval of f is $(-4, 7)$, then the decreasing interval of $y = -f(x+3)$ is

    A. $(-7, 4)$ B. $(-1, 10)$ C. $(-1, 7)$ D. $(-4, 10)$

(14) If an odd function $y = f(x)$ has a maximum over $[a, b](0 < a < b)$, then over $[-b, -a]$ the function has

    A. a minimum   B. a maximum   C. no minimum   D. no maximum

(15) If f is an odd function increasing over $(-\infty, -1)$ but decreasing over $(-1, 0)$ with

$f(-1) < 0$, then the maximum value of $y = -\dfrac{1}{f(x)}$ over $\mathbb{R} - \{0\}$ is

   A. $f(1)$ B. $-f(1)$ C. $\dfrac{1}{f(1)}$ D. $-\dfrac{1}{f(1)}$

3. Sketch the graphs of the following functions.

(1) $y = -3x + 5$ (2) $y = -|x - 3| + 1$

(3) $y = \dfrac{2x}{x - 2}$ (4) $y = x^2 - 4x + 7$, $x \in [-1, 2]$

(5) $y = \dfrac{x^2 + 2x - 8}{x - 2}$ (6) $y = -\dfrac{1}{|x + 1|}$, $x \neq -1$

4. Tell if the function is increasing or decreasing in the given interval.

(1) $y = 2x - 7$, $x \in (-5, 7)$ (2) $y = x^2 - 3$, $x \in (0, +\infty)$

(3) $y = \sqrt{x}$, $x \in [0, +\infty)$ (4) $y = \sqrt{4 - 2x}$, $x \in (-\infty, 2]$

(5) $y = \sqrt{x^2 + 1}$, $x \in [-10, -1]$

(6) $y = x^2 - 2x - 3$, $x \in (-\infty, 1]$

(7) $y = \dfrac{1}{x}$, $x \in (-\infty, 0)$

(8) $y = \dfrac{1}{(x+4)(x-2)}$, $x \in (-4, -1]$

5. Find the intervals in which the function is increasing or decreasing.

(1) $y = -2x - 4$

(2) $y = \dfrac{1}{x^2}$

(3) $y = -x^2 - 10x - 26$

(4) $y = |x^2 - 2x - 3|$

(5) $y = \sqrt{x^2 + x - 2}$

(6) $y = -\dfrac{1}{x+3} + 2$

6. Suppose f and g have the same domain. Tell if the functions defined by f and g as follows are increasing or decreasing.

(1) $y = -f(x)$ where f is increasing

(2) $y = f(-x)$ where f is decreasing

(3) $y = f(x) - g(x)$ where f is decreasing, g is increasing

(4) $y = -f(x) - g(x)$ where f, g are both decreasing

(5) $y = \dfrac{1}{f(x)}$ where f is positive and increasing

(6) $y = \dfrac{1}{g(x)}$ where g is negative and increasing

7. Fill in the blanks using 'increasing' or 'decreasing'.

(1) If $y = ax$ and $y = -\dfrac{b}{x}$ are strictly decreasing over $(0, +\infty)$, then $y = ax^2 + bx$ is _____ over $(0, +\infty)$.

(2) If $y = f(x)$ is defined over $\mathbb{R}$ satisfying $f(a+x) = f(a-x)$ and f is decreasing over $(a, +\infty)$, then f is _____ over $(-\infty, a)$.

8. If $f(x) = x^2 + 2(a-1)x + 2$ is decreasing on $(-\infty, 4)$, find a.

9. Find the interval in which $f(x) = -\sqrt{5 - 4x - x^2}$ is decreasing.

10. Prove $f(x) = -x^3 - x - 1$ is decreasing over $\mathbb{R}$.

12.3 Operations on Functions

Given two function f and g, we can form the sum, difference, product and quotient functions from them.

- Addition: $f + g : x \mapsto f(x) + g(x)$;
- Subtraction: $f - g : x \mapsto f(x) - g(x)$;
- Multiplication: $f \cdot g : x \mapsto f(x) \cdot g(x)$;
- Division: $f/g : x \mapsto \dfrac{f(x)}{g(x)}$ whenever $g(x) \neq 0$.

If the domains of f and g are A and B separately, the domains of $f \pm g$ and $f \cdot g$ are all $A \cap B$, where both functions are defined; the domain of f/g is $A \cap B - \{b \in B | g(b) \neq 0\}$ where both f and g are defined and the denominator is not zero.

Given the functions $f : A \to B$ and $g : B \to C$, the composition of g and f is a function $g \circ f : A \to C$ defined by $x \mapsto g(f(x))$. Notice that $f(x) \in B$ and $\mathrm{dom}(g) = B$, so $g(f(x))$ is well-defined.

Examples. Given $f(x) = x^2 + 1$, $g(x) = x - 1$, find $f - g$, $f \cdot g$, f/g, $g \circ f$ and $f \circ g$.

Solutions. From the definition, immdediately we get:

- $(f - g)(x) = f(x) - g(x) = (x^2 + 1) - (x - 1) = x^2 - x + 2$;
- $f \cdot g(x) = f(x)g(x) = (x^2 + 1)(x - 1) = x^2 - x^2 + x - 1$;
- $f/g(x) = \dfrac{f(x)}{g(x)} = \dfrac{x^2 + 1}{x - 1}$;
- $g \circ f(x) = g(f(x)) = g(x^2 + 1) = (x^2 + 1) - 1 = x^2$;
- $f \circ g(x) = f(g(x)) = f(x - 1) = (x - 1)^2 + 1 = x^2 - 2x + 2$.

As $\mathrm{dom}(f) = \mathrm{dom}(g) = \mathbb{R}$, it should be clear that all the functions above except f/g have the domain $\mathbb{R}$, but the domain of f/g is $\{x \in \mathbb{R} | x \neq 1\}$, where the denominator $g(x)$ is not 0.

Exercises.

1. Multiple choice.

(1) If $f(x) = 3x^2 - 2x - 1$, $g(x) = -x - 1$, then $g \circ f(-1)$ is

    A. 2   B. -3 C. 1   D. -5

(2) Given $f(x) = \sqrt{x}$, $g(x) = \dfrac{1}{x^2 + 1}$, then $f \circ g(0) + g \circ f(0)$ is

    A. -2 B. -1 C. 0   D. 2

(3) Given $f(x) = 2x - 1$, $g(x) = 9 - x^2$, then $\dfrac{f + g}{f - g}(-1)$ is

    A. $\dfrac{7}{9}$ B. $-\dfrac{7}{9}$ C. $-\dfrac{5}{11}$ D. $\dfrac{5}{11}$

(4) If $f(x) = x + \dfrac{1}{x}$, $g(x) = \dfrac{1}{x}$, then $g \circ f$ is

    A. $\dfrac{x^2}{x + 1}$ B. $\dfrac{x}{x^2 + 1}$ C. $\dfrac{x}{x + 1}$ D. $\dfrac{x + 1}{x}$

(5) If $f(2x + 1) = x^2 - 2x$, then $f(2\sqrt{2} + 3)$ is

    A. 1   B. $\sqrt{2}$ C. $2\sqrt{2} + 2$ D. $\sqrt{2} - 1$

(6) If $y = f(x)$ is an even function and $y = g(x)$ is an odd function, then $f \circ g$ must be

    A. an even function   B. an odd function

    C. neither even nor odd   D. cannot be determined

(7) $f(x) = \sqrt{x}$, $g(x) = x^2 - x - 2$; find the domain of $f \circ g$

    A. $[-1, 2]$ B. $(-\infty, -2] \cup [1, +\infty)$

    C. $(-\infty, -1] \cup [2, +\infty)$ D. $[2, +\infty)$

(8) The image of $y = |x - 2| + |x + 8|$ is

    A. $(-\infty, 10]$ B. $[10, +\infty)$ C. $[6, 10]$ D. $[6, +\infty)$

2. Find $f - g$, $f \cdot g$, $f \div g$, $f \circ g$ and $g \circ f$.

(1) $f(x) = 3x - 2$, $g(x) = x^2 - x + 1$

(2) $f(x) = \sqrt{x+1}$, $g(x) = \dfrac{1}{x}$

(3) $f(x) = \dfrac{2}{3-x}$, $g(x) = -2x - 1$

(4) $f(x) = x + 1$, $g(x) = \dfrac{1}{x^2}$

(5) $f(x) = x^2 - 3x + 1$, $g(x) = \dfrac{1}{\sqrt{1+x}}$

3. Assume all functions are well-defined. Complete the table by telling if the function is even, odd or neither under given conditions.

f	g	$f \pm g$	$f \cdot g$	f/g	$f \circ g$
odd	odd				
odd	even				
even	odd				
even	even				

4. Assume all functions are well-defined. Complete the table by telling if the function is increasing($\nearrow$), decreasing($\searrow$) or neither($-$) under given conditions.

f	g	$f+g$	$f-g$	$fg(f,g>0)$	$fg(f,g<0)$	$f/g(f,g>0)$	$f/g(f,g<0)$	$f \circ g$
$\nearrow$	$\nearrow$							
$\nearrow$	$\searrow$							
$\searrow$	$\nearrow$							
$\searrow$	$\searrow$							

5. $f(x) = \dfrac{ax}{x^2 - 1}\,(a \neq 0)$ is a function defined over $(-1, 1)$. Find the intervals in which f is increasing and decreasing.

6. If the function $f(x) = ax^2 - (3a-1)x + a^2$ is increasing on $[-1, +\infty]$, find a.

7. Sketch the graph of $f(x) = |x^2 + 2x - 3|$ and $g(x) = |x|^2 + 2|x| - 3$.

12.4   Power Functions

Power functions form another important class of elementary functions. A power function is of the form

$$x \mapsto x^a,$$

where a is a fixed real number. For example, $y = x^0$, $y = x^{\frac{1}{3}}$, $y = x$, $y = x^2$, $y = x^{\sqrt{2}}$, $y = x^3$, $y = x^{-\frac{1}{2}} = \dfrac{1}{\sqrt{x}}$, $y = x^{-1} = \dfrac{1}{x}$, $y = x^{-2} = \dfrac{1}{x^2}$ are all power functions.

The domain of the power function $f(x) = x^a$ depends on a. For a general $a \in \mathbb{R}$,

$$\mathrm{dom}(f) = \begin{cases} [0, +\infty) & \text{if } a > 0, \\ (0, +\infty) & \text{if } a \leq 0. \end{cases}$$

There is an exception: when $a \in \mathbb{Q}$ and the denominator of a in lowest terms is odd, (i) $\mathrm{dom}(f) = \mathbb{R}$ if $a > 0$; (ii) $\mathrm{dom}(f) = \mathbb{R} - \{0\}$ if $a \leq 0$.

For example, $y = x$, $y = x^2$, $y = x^3$, $y = x^{\frac{1}{3}} = \sqrt[3]{x}$, $y = x^{\frac{2}{5}} = \sqrt[5]{x^2}$ can all be defined over $\mathbb{R}$, which verifies conclusion (i). For conclusion (ii), if writing $a = -b$ with $b > 0$, then $x^a = x^{-b} = \dfrac{1}{x^b}$. x^b is defined everywhere (case (i)) and equals 0 only when $x = 0$. Immediately $y = x^a$ is only undefined at $x = 0$. $y = x^{-1} = \dfrac{1}{x}$, $y = x^{-2} = \dfrac{1}{x^2}$, $y = x^{-\frac{2}{3}} = \dfrac{1}{\sqrt[3]{x^2}}$ can serve as examples. 0^0 is not defined, so $y = x^0$ is also only undefined at $x = 0$ (when $x \neq 0$, $x^0 = 1$).

When $a = \dfrac{p}{q}$ $(p, q \in \mathbb{Z}, q \neq 0)$ is in lowest terms and q is even, x^a is not defined for negative x, since negative numbers have no roots of even order. Thus the domain of these power functions is $(0, +\infty)$. $y = x^{\frac{3}{4}} = \sqrt[4]{x^3}$, $y = x^{-\frac{1}{2}} = \dfrac{1}{\sqrt{x}}$ demonstrate as examples of such functions.

Irrational exponents are also allowed in power functions. If $0 < a \in \mathbb{R} - \mathbb{Q}$, x^a is defined only for nonnegative numbers, so the domain of $y = x^a$ is $[0, +\infty)$; if $0 > a \in \mathbb{R} - \mathbb{Q}$, the domain of $y = x^a$ is $(0, +\infty)$.

The graphs of power functions are sketched as follows.

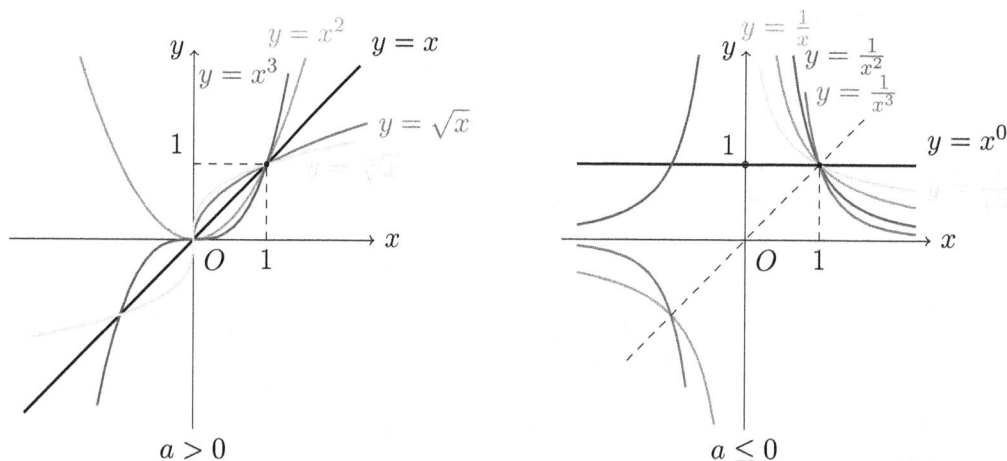

From the graph, the properties of the power function $f(x) = x^a$ can be summarized as follows:

• the graph of f always passes the point $(1, 1)$ (since $1^a = a$ for any $a \in \mathbb{R}$); it also passes $(0, 0)$ if $a > 0$ (since $0^a = 0$ for any $a > 0$);

• on the interval $(0, +\infty)$, $f(x)$ is increasing if $a > 0$ and decreasing if $a < 0$;

• when $a = \dfrac{p}{q}$ is a fraction in lowest terms with q an odd number, f is an odd function if p is odd, and an even function if p is even; (by this property, we can complete the graph of a power function by knowing its graph in the first quadrant.)

• when $a \neq 0$, the graphs of $y = x^a$ and $y = x^{\frac{1}{a}}$ are symmetric about $y = x$;

• if $a > b > 0$, $x^a < x^b$ when $x \in (0, 1)$ and $x^a > x^b$ when $x \in (1, +\infty)$; if $a < b < 0$, $x^a > x^b$ when $x \in (0, 1)$ and $x^a < x^b$ when $x \in (1, +\infty)$.

The monotonic and the last property can be used to compare numbers. For example, (i) $\pi^{-\frac{3}{2}} > 4^{-\frac{3}{2}}$ since $y = x^{-\frac{3}{2}}$ is strictly decreasing and $\pi < 4$; (ii) since $-\dfrac{3}{2} < -\sqrt{2} < 0$ and $\pi > 1$, we have $\pi^{-\sqrt{2}} > \pi^{-\frac{3}{2}}$.

Exercises.

1. True or false.

(1) $y = x^2$ is both a power function and a quadratic function.

(2) The graph of $y = x^0$ is a horizontal line that passes through $(0, 1)$.

(3) $y = x^{-3} - x^{-2} + x^{-1} + x^0$ is a power function.

(4) The domain of a power function depends on its exponent.

(5) The domain of $y = x^{\frac{2}{3}}$ is $\mathbb{R}$ while the domain of $y = x^{\frac{1}{2}}$ is $(0, +\infty)$.

(6) The graph of any power function passes through the points $(0,0)$ and $(1,1)$.

(7) The power function $y = x^a$ is decreasing on $(0, +\infty)$ when $a < 0$.

(8) The function $y = -\dfrac{1}{x^3}$ has maximum value of -8 over $[\frac{1}{2}, 2]$.

(9) The graphs of $y = x^2$ and $y = x^{-2}$ are symmetric about the x-axis.

(10) $y = x|x|$ is an odd function which is increasing over $\mathbb{R}$.

2. Multiple choice.

(1) How many of the following are power functions?

(i) $y = \dfrac{1}{\sqrt{x+1}}$; (ii) $y = \dfrac{x^2+1}{x}$; (iii) $y = x^2 - x^3$; (iv) $y = x^0$; (v) $y = \dfrac{1}{x^3}$

   A. 0   B. 1   C. 2   D. 3

(2) The graph of a power function must pass through the point

   A. $(1,1)$ B. $(0,1)$ C. $(0,0)$ D. $(1,0)$

(3) The domain of $y = x^{-\frac{2}{3}}$ is

   A. $(-\infty, 0)$ B. $(0, +\infty)$

   C. $(-\infty, +\infty)$ D. $(-\infty, 0) \cup (0, +\infty)$

(4) If the power function $y = x^a$ is strictly decreasing over $(0, +\infty)$, which of the following conditions holds?

   A. $a < 0$ B. $a \geq 0$ C. $a > 0$ D. $a \leq 0$

(5) If $a = 1.1^{-\frac{1}{2}}$, $b = 0.9^{-\frac{1}{2}}$, which of the following is correct?

   A. $a < 1 < b$ B. $b < 1 < a$ C. $b < a < 1$ D. $a < b < 1$

(6) Which of the following function in the given domain is increasing?

   A. $y = x^{-4}$ $(x > 0)$ B. $y = \sqrt{-x}$ $(x \leq 0)$

   C. $y = -x + \dfrac{1}{x}$ $(x \neq 0)$ D. $y = x^2 - 16x + 9$ $(x \geq 10)$

(7) If $y = x^{m - \frac{1}{2}}$ is increasing over $(0, +\infty)$, then m is

   A. $m > \dfrac{1}{2}$ B. $m < \dfrac{1}{2}$ C. $m \geq \dfrac{1}{2}$ D. $m \leq \dfrac{1}{2}$

(8) If $a \in (-1, 0)$, which of the following is correct?

   A. $3^a > 3^{-a} > 0.3^a$ B. $0.3^a > 3^{-a} > 3^a$

   C. $3^{-a} > 0.3^a > 3^a$ D. $3^a > 0.3^a > 3^{-a}$

(9) Which of the following functions is even as well as increasing over $(-\infty, 0)$?

   A. x^{-2} B. $x^{-\frac{1}{2}}$ C. x^2 D. $x^{\frac{1}{2}}$

(10) If a function of the form $y = ax^b$ passes through $(1,2)$ and $(8, \dfrac{1}{\sqrt{2}})$, then $f(\dfrac{1}{2})$ is

   A. $2\sqrt{2}$ B. $\dfrac{\sqrt{2}}{4}$ C. $4\sqrt{2}$ D. $\dfrac{\sqrt{2}}{2}$

(11) If the graph of a power function $y = x^a$ passes through the third quadrant, then

    A. $a > 0$

    B. $a < 0$

    C. $a = \dfrac{p}{q}$, $p, q \in \mathbb{Z}$ and both p, q are odd

    D. $a = \dfrac{p}{q}$, $p, q \in \mathbb{Z}$ and q is odd

3. Order the numbers.

(1) $2.5^{\frac{2}{3}}, (-1.4)^{\frac{2}{3}}, (-3)^{\frac{2}{3}}$

(2) $0.16^{-\frac{3}{4}}, 0.5^{-\frac{3}{2}}, 6.25^{\frac{3}{8}}$

(3) $4^{-\frac{1}{2}}, \sqrt[3]{\dfrac{1}{4}}, (\dfrac{1}{4})^{\frac{2}{5}}, 4^{-\frac{1}{4}}$

(4) $\left(\dfrac{2}{3}\right)^{\frac{1}{3}}, \left(\dfrac{2}{5}\right)^{\frac{1}{2}}, \left(\dfrac{3}{5}\right)^{\frac{1}{3}}, 3^{-\frac{1}{3}}, \left(\dfrac{3}{2}\right)^{-\frac{2}{3}}$

4. Sketch the graphs.

(1) $y = \sqrt[3]{x + 2} - 1$

(2) $y = (x - 2)^{\frac{3}{2}}$

(3) $y = \dfrac{x^2 + 6x + 10}{x^2 + 6x + 9}$

(4) $y = (x - 4)^{-\frac{3}{2}} - 2$

5. Solve the inequality $\sqrt[3]{x} > x$ using the graph of power functions.

6. If $f(x) = x^4 + ax^3 + bx - 2$ and $f(-1) = 6$, find $f(1)$.

7. Find the domain of $y = (x-1)^0 + (2-x)^{\frac{1}{2}}$.

8. Find the biggest natural number n such that $n^{200} < 5^{300}$.

9. Find the intervals in which $y = \dfrac{-3x^2 - 12x - 11}{x^2 + 4x + 4}$ is increasing and decreasing.

12.5 Inverse Functions

Consider a map $f : A \to B$. f is injective (or 'one-to-one') if different elements in A have different images, i.e. for any $a_1, a_2 \in A$, if $f(a_1) = f(a_2)$, then $a_1 = a_2$. f is surjective (or 'onto') if every element of B is the image of some element of A under f, i.e. $\mathrm{Im} f = B$, or equivalently, for any $b \in B$, there exists some $a \in A$ such that $f(a) = b$. If f is both injective and surjective, we say f is bijective or f is a 1-1 correspondence. For example, among the maps:

$g_1 : (-\infty, +\infty) \to (-\infty, +\infty) \quad x \mapsto x^2$

$g_2 : (-\infty, +\infty) \to [0, +\infty) \quad x \mapsto x^2$

$g_3 : [0, +\infty) \to (-\infty, +\infty) \quad x \mapsto x^2$

$g_4 : [0, +\infty) \to [0, +\infty) \quad x \mapsto x^2$

g_1 is neither injective — a and $-a$ always have the same image, nor surjective — negative numbers are not images at all; g_2 is surjective but not injective; g_3 is injective but not surjective; g_4 is both injective and surjective and hence bijective.

Given the graph S of an equation in x and y, we can use the vertical line test to see if it is the graph of a function $y = f(x)$. If each vertical line intersects with S at no more than one point, then it is the graph of a function; otherwise it is not. For example, the unit circle $x^2 + y^2 = 1$ is not the graph of any function; as in the following picture the vertical line l intersects with the unit circle at two points, which means the corresponding x has more than one image and this violates the definition of a function.

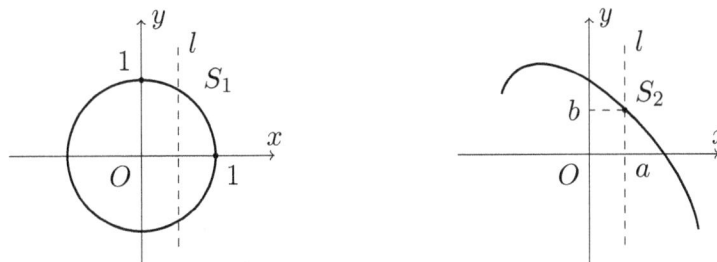

The second graph S_2 is the graph of some function f. The function f should be defined in the following way: for any $a \in \mathbb{R}$, if the line $l : x = a$ intersects with S_2 at some point whose y-coordinate is b, define $f(a) = b$. The domain of f should be the set of numbers a such that the intersection of $x = a$ and S_2 is nonempty.

Moreover, we use the horizontal line test to see if a function $y = f(x)$ is injective. If each horizontal line intersects with the graph of f at no more than one point, then it is injective; otherwise it is not injective. For example, the function f defining S_2 in the previous example is not injective. As in the following picture, the horizontal line k intersects S_2 at 2 points, which means different numbers a_1 and a_2 have the same value under f (i.e. $f(a_1) = f(a_2)$).

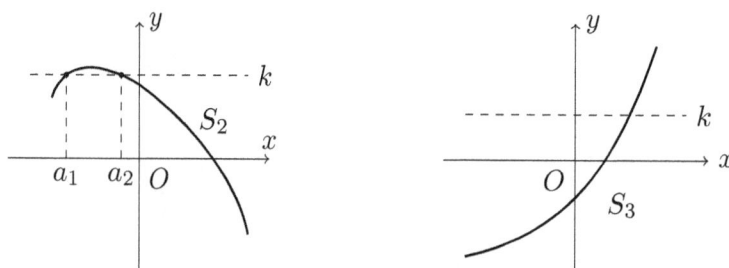

According to the picture above, S_3 is the graph of some function g by the vertical line test. g is also injective by the horizontal line test as each horizontal line k intersects with S_3 at no more than 1 point.

Given a map $f : A \to B$, a map $g : B \to A$ is called the inverse of f if $g \circ f = Id_A : A \to A$ and $f \circ g = Id_B : B \to B$. If g is the inverse of f, f is also the inverse of g, and we say f and g are inverse maps. For example, the maps f and g in the following are inverse maps.

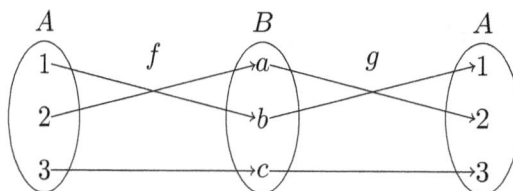

From the example, the inverse map g just reverses the arrows of f. In general, the inverse map exchanges the domain and range of the original map and reverses the rule of correspondence. Not every map has an inverse map. For example, the map h in the following has no inverse as if we reverse the arrows, what we get is not a map; (i) c has no image (resulted from h is not surjective) and (ii) a has more than one images (resulted from h is not injective).

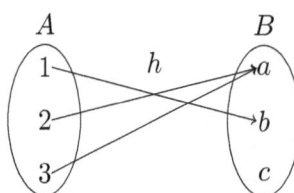

A map $f : A \to B$ has an inverse if and only if it is both injective and surjective, i.e. bijective. The inverse of a map f is unique if existing and it is denoted by f^{-1}. If a function has an inverse, we also say it is invertible. The following are more examples of inverse functions.

$$f_1 : (-\infty, +\infty) \to (-\infty, +\infty) \quad x \mapsto 2x$$
$$g_1 : (-\infty, +\infty) \to (-\infty, +\infty) \quad x \mapsto \frac{x}{2}$$

are inverse functions since $g_1 \circ f_1(x) = g_1(2x) = \frac{2x}{2} = x$ and $f_1 \circ g_1(x) = f_1(\frac{x}{2}) = 2 \cdot \frac{x}{2} = x$. Similarly,

$$f_2 : [0, +\infty) \to [0, +\infty) \quad x \mapsto \sqrt{x}$$
$$g_2 : [0, +\infty) \to [0, +\infty) \quad x \mapsto x^2$$

are inverse functions as well. In fact, any power function $p : (0, +\infty) \to (0, +\infty) \quad x \mapsto x^a$ ($a \neq 0$, and 0 can be included in both domain and range if $a > 0$) is invertible and its inverse $p^{-1}(x) = x^{\frac{1}{a}}$ is also a power function.

From the definition of inverse functions, given $f : A \to B$ invertible,

$$f^{-1} \circ f(a) = a \quad \forall a \in A$$
$$f \circ f^{-1}(b) = b \quad \forall b \in B.$$

If a function f is bijective, to find f^{-1}, we solve for x from the equation $y = f(x)$. The solution must be of the form $x = g(y)$ for some function g. Then f^{-1} is just $y = g(x)$ by exchanging x and y in $x = g(y)$.

Example. Tell if f has an inverse. If so, find its inverse.

$$f : (-\infty, 1) \to (-1, +\infty)$$
$$x \mapsto x^2 - 2x$$

Solutions. $f(x) = (x - 1)^2 - 1$ by completing squares.

(i) f is injective in its domain $(-\infty, 1)$ as $x - 1$ is always negative and different negative numbers have different squares.

(ii) Since $x < 1$, we have $(x - 1)^2 > 0$, and hence $f(x) = (x - 1)^2 - 1 > -1$. Obviously $\text{Im} f \subseteq \text{gr}(f)$. To check if f is surjective, pick an arbitrary $y > -1$ in the range and set $y = f(x)$ to see if it has solutions for x. First we get $y = (x - 1)^2 - 1$, or equivalently $(x - 1)^2 = y + 1$. As $y > -1$, $y + 1 > 0$, we can find square roots on both sides to get $\sqrt{y + 1} = \sqrt{(x - 1)^2} = |x - 1|$. Since $x < 1$, $x - 1$ is negative, $|x - 1| = -(1 - x) = 1 - x$. Finally we get $\sqrt{y + 1} = 1 - x$ and $x = 1 - \sqrt{y + 1}$. Since $y > -1$, we have $\sqrt{1 + y} > 0$ and $x = 1 - \sqrt{1 + y} < 1$; thus x is in the domain of f. For any y in the range, there is an $x = 1 - \sqrt{y + 1} \in \text{dom}(f)$ such that $y = f(x)$, so f is surjective.

(iii) f is both injective and surjective, so f is invertible. To find f^{-1}, we need to solve $y = f(x)$ for x. In fact, we have already found that the solution is $x = 1 - \sqrt{y + 1}$ in step (ii). Exchanging x and y we get $y = 1 - \sqrt{x + 1}$, so $f^{-1}(x) = 1 - \sqrt{x + 1}$. Notice that f^{-1} must exchange the domain and range of f, so $\text{dom}(f^{-1}) = (-1, +\infty)$ and $\text{ran}(f^{-1}) = (-\infty, 1)$.

The graphs of any inverse functions f and f^{-1} are symmetric about the line $y = x$. The reason is as follows: if (x_0, y_0) is a point on $\text{gr}(f)$, then $y_0 = f(x_0)$. By the definition of inverse function, $f^{-1}(y_0) = x_0$ and hence (y_0, x_0) is on $\text{gr}(f^{-1})$. As the points (x_0, y_0) and (y_0, x_0) are symmetric about $y = x$, $\text{gr}(f)$ and $\text{gr}(f^{-1})$ must be symmetric about $y = x$ as well. For example, the graphs of $y = x^2$ ($x \geq 0$) and $y = \sqrt{x}$ ($x \geq 0$) are symmetric about the line $y = x$ since they are graphs of inverse functions (see below).

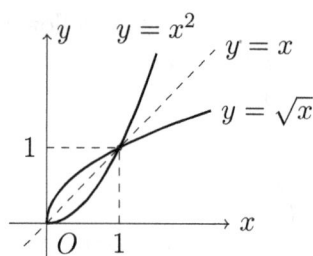

According to this property, we should be able to find the graph of f^{-1} from the graph of f.

Exercises.

1. True or false.

(1) Only bijective functions have inverse functions.

(2) $x^2 = 2py$, where $p \neq 0$, is not invertible.

(3) $y = \sqrt{25 - 16x^2}$ is invertible.

(4) $f : [0, +\infty) \to (-\infty, +\infty) \quad x \mapsto x^2 - 4$ is invertible.

(5) If $f(x) = 2x + 6$, then $f^{-1}(x) = \dfrac{1}{2}x - 3$.

(6) The natural range of $f(x) = \dfrac{3x + 2}{5x - 7}$ is $\mathbb{R}$.

(7) If $f(x)$ is invertible, then $f(f^{-1}(x)) = x$.

(8) The graphs of $y = x^3$ and $y = \dfrac{1}{\sqrt[3]{x}}$ are symmetric about $y = x$.

(9) The inverse function of $y = \dfrac{1}{x}$ is itself.

(10) When f is invertible, $x = f(y)$ and $y = f^{-1}(x)$ define the same function.

2. Multiple Choice.

(1) The inverse of $f(x) = -x^2$ over $[0, +\infty)$ is

    A. $f^{-1}(x) = -\sqrt{x}$, $x \geq 0$ B. $f^{-1}(x) = \sqrt{x}$, $x \geq 0$

    C. $f^{-1}(x) = -\sqrt{-x}$, $x \leq 0$ D. $f^{-1}(x) = \sqrt{-x}$, $x \leq 0$

(2) If $f(x) = -x(x - 3)$ is defined over $(-\infty, \dfrac{3}{2}]$, the domain of f^{-1} is

    A. $(-\infty, +\infty)$ B. $[\dfrac{9}{4}, +\infty)$ C. $(-\infty, \dfrac{9}{4}]$ D. $(-\infty, -\dfrac{9}{4}]$

(3) If $f(x) = \sqrt{x + 5} - 2$, then f^{-1} and its domain are

    A. $f^{-1} = x^2 + 4x - 1$, $x \in [-5, +\infty)$

    B. $f^{-1}(x) = x^2 - 4x + 1$, $x \in [-5, +\infty)$

    C. $f^{-1}(x) = -x^2 - 4x + 1$, $x \in (-\infty, -2]$

    D. $f^{-1}(x) = x^2 + 4x - 1$, $x \in [-2, +\infty)$

(4) Which of the following could be inverse functions?

A. $f(x) = x$ and $g(x) = \dfrac{1}{x}$ B. $f(x) = x^{\frac{1}{2}}$ and $g(x) = x^{-\frac{1}{2}}$

C. $f(x) = \dfrac{2x+1}{x-1}$ and $g(x) = \dfrac{x+1}{2x-1}$ D. $y = x^2 (x > 0)$ and $y = \sqrt{x}(x > 0)$

(5) If (m, n) is on the graph of f, which point is on the graph of f^{-1}?

     A. $(m, f^{-1}(m))$ B. $(n, f^{-1}(n))$ C. $(f^{-1}(m), m)$ D. $(f^{-1}(n), n)$

(6) If the inverse function of $y = f(x)$ is $y = g(x)$, then the inverse function of $y = f(-x)$ is

     A. $y = -g(x)$ B. $y = g(-x)$ C. $y = -g(-x)$ D. $y = g(x)$

(7) Denote the inverse function of $f(x) = \sqrt{\dfrac{1}{x}}$ by g. Which of the following is correct?

     A. $g(2) > g(\dfrac{1}{2}) > g(1)$ B. $g(\dfrac{1}{2}) > g(1) > g(2)$

     C. $g(1) > g(2) > g(\dfrac{1}{2})$ D. $g(1) > g(\dfrac{1}{2}) > g(2)$

(8) If f is invertible, which of the following is correct?

     A. If f is increasing over $[0, 1]$, then f^{-1} is increasing over $[0, 1]$.

     B. If f is an odd function, then f^{-1} is also an odd function.

     C. If f is an even function, then f^{-1} is also an even function.

     D. If $\mathrm{gr}(f)$ intersects with the y-axis, then $\mathrm{gr}(f^{-1})$ also intersects with the y-axis.

(9) If $f(x) = x^2 - 1(x \le -1)$, then $f^{-1}(7)$ is

     A. $-\sqrt{2}$ B. $-\sqrt{6}$ C. $-2\sqrt{2}$ D. -2

(10) Suppose f is invertible. If $f^{-1}(x) = 3x + 2$, then $f(3)$ is

     A. $\dfrac{1}{3}$ B. 11 C. -3 D. $-\dfrac{1}{3}$

3. Given the function and its domain, assume its range is just its image (i.e. natural range). Tell if the function is invertible and find the inverse when possible.

(1) $f(x) = \dfrac{2x+5}{3x-1}, x \ne \dfrac{1}{3}$ (2) $g(x) = x^2 - 3x - 4, x \in (-\infty, 1]$

(3) $h(x) = \dfrac{1}{x^2-1}, x \in [0, 1) \cup (1, +\infty)$ (4) $p(x) = \sqrt{x-2} - 1, x \in [2, +\infty)$

(5) $q(x) = \dfrac{1}{x^3 + 4}$, $x \neq \sqrt[3]{-4}$

(6) $r(x) = \dfrac{4}{x^4 + 9}$, $x \in (-\infty, +\infty)$

(7) $l(x) = \begin{cases} \sqrt{x+1} & x \in [-1, 0) \\ -\sqrt{x} & x \in (0, 1] \end{cases}$

4. Given that f is invertible, find f^{-1} and sketch $\mathrm{gr}(f)$, $\mathrm{gr}(f^{-1})$ in the same coordinate system.

(1) $f(x) = \sqrt{x - 1} - 1$, $x \in [1, +\infty)$

(2) $f(x) = -3x^2 - 2$, $x \in (-\infty, 0]$

5. If f is an odd function which is invertible, show that f^{-1} is also an odd function.

6. Show that $f(x) = \dfrac{x-1}{x+2}$ with its natural domain and natural range is invertible. Find $f^{-1}(\sqrt{2})$.

7. Assume f is invertible. (1) Show that if f is increasing or decreasing, so is f^{-1}. (2) If f is decreasing with $\text{dom}(f) = (-1, 9)$, find the domain and range of f^{-1}.

8. (1) Given $f(x) = \dfrac{ax+b}{cx+d}$ with $a, b, c, d \neq 0$, find the condition that a, b, c, d should satisfy such that f is invertible in its natural domain and $f^{-1} = f$.

(2) Find a such that $f(x) = \dfrac{3x+1}{x+a}$ such that f is invertible and $f = f^{-1}$.

(3) Show the graph of $y = \dfrac{x-1}{ax-1}$ where $a \neq 0, 1$ is symmetric about $y = x$.

12.6 Asymptotes

We have met the asymptotes when learning hyperbolas. From the graph, the hyperbola and its asymptotes get closer and closer when x approaches $\pm\infty$. In general, an asymptote of a graph S is a line l such that the distance between l and S approaches zero as they approach to infinity (x or y coordinate approaches $\pm\infty$). A line l is called an asymptote of a function $y = f(x)$ if it is the asymptote of $\text{gr}(f)$. For example, (1) $y = \dfrac{1}{x}$ has two asymptotes, $x = 0$

and $y = 0$; (2) $y = x + \dfrac{1}{x}$ has two asymptotes, $x = 0$ and $y = x$; (3) $y = \sqrt{4x^2 - 4}(x \geq 1)$ has one asymptote, $y = 2x$; (4) $y = x^2 - 1$ has no asymptotes at all.

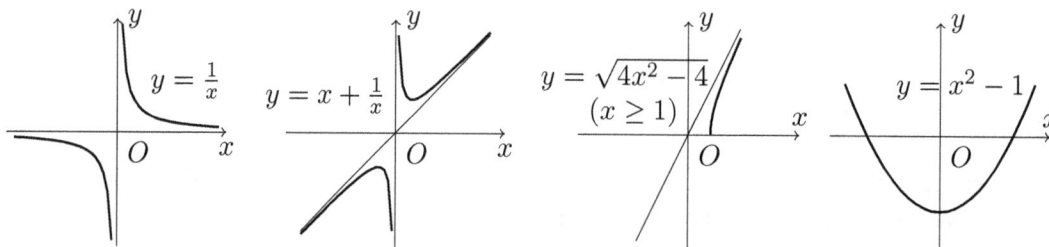

From the above examples, there are three kinds of asymptotes: horizontal asymptotes, vertical asymptotes and slant asymptotes. Given a rational function $f(x) = \dfrac{P(x)}{Q(x)}$ where $P(x), Q(x)$ are polynomials in x and $Q(x) \neq 0$, the asymptotes of f can be classified as follows. Assume the quotient of the long division $P(x) \div Q(x)$ is $R(x)$ (the remainder does not matter here).

- If $\deg P < \deg Q$, $R(x)$ must be 0 and f has a horizontal asymptote $y = 0$.

- If $\deg P = \deg Q$, then $R(x) = r$ is a constant number and f has a horizontal asymptote $y = r$.

- If $\deg P = \deg Q + 1$, then $R(x)$ is a liner polynomial $ax + b$. In this case, f has a slant asymptote $y = ax + b$.

- If $\deg P > \deg Q + 1$, $R(x)$ is of degree greater than 1. In this case, f has no horizontal or slant asymptote.

- If $\dfrac{P(x)}{Q(x)}$ is in its lowest terms, for any root x_0 of $Q(x)$, f has a vertical asymptote $x = x_0$.

In brief, if the quotient $R(x)$ of the long division $P(x) \div Q(x)$ is of degree less than or equal to 1, $f(x) = \dfrac{P(x)}{Q(x)}$ has a horizontal or slant asymptote $y = R(x)$. The reasoning of these results uses the concept of limits, which is out of the scope of our book. The readers can simply accept our results at the moment and wait for the proof during the learning of calculus. We only focus on how to apply these results.

Examples. Find all asymptotes of the following rational functions.

(1) $f(x) = \dfrac{5x^3 - 4x^2 + 3x - 2}{16 - 4x^2}$

(2) $g(x) = \dfrac{-3x^4 + 3x^3 + 18x^2}{x^2 - 2x - 3}$

(3) $h(x) = \dfrac{2x - 1}{x^2 + 1}$

Solutions.

(1) The degree of the numerator is 1 bigger than the degree of the denominator, so the function has a slant asymptote. The quotient of the long division $(5x^3 - 4x^2 + 3x - 2) \div$

$(16 - 4x^2)$ is $-\dfrac{5}{4}x + 1$, so the slant asymptote is $y = -\dfrac{5}{4}x + 1$.

The denominator can be factored as $16 - 4x^2 = -4(x+2)(x-2)$. The denominator and the numerator have no common factors, since $x = \pm 2$ are roots of the denominator but not the roots of the numerator $5x^3 - 4x^2 + 3x - 2$ (plugging in to check). As a result, the fraction is in its lowest terms and $x = \pm 2$ are two vertical asymptotes.

(2) The degree of the numerator is 2 bigger than the degree of the denominator, so the function has no slant or horizontal asymptotes. $-3x^4 + 3x^3 + 18x^2 = -3x^2(x^2 - x - 6) = -3x^2(x-3)(x+2)$, $x^2 - 2x - 3 = (x-3)(x+1)$, so the fraction can be simplified:

$$\frac{-3x^4 + 3x^3 + 18x^2}{x^2 - 2x - 3} = \frac{-3x^2(x-3)(x+2)}{(x-3)(x+1)}$$
$$= \frac{-3x^2(x+2)}{x+1}.$$

-1 is the only root of the denominator, so the function has a vertical asymptote $x = -1$. Notice that although $x = 3$ is a root of the denominator $x^2 - 2x - 3$, it does not give an asymptote.

(3) The degree of the numerator is less than the degree of the denominator, so the function has a horizontal asymptote $y = 0$. The function has no vertical asymptotes as the denominator $x^2 + 1 > 0$ has no real roots.

From the second example, it is crucial to simplify the rational expressions before finding vertical asymptotes.

Exercises.

1. Find vertical asymptotes.

(1) $y = \dfrac{x-2}{2x^2 - 3x - 2}$

(2) $y = \dfrac{x^3}{x^2 + 2x - 15}$

(3) $y = \dfrac{2x - 1}{4x^2 - 1}$

(4) $y = \dfrac{6 - 10x}{-10x^2 + 41x - 21}$

(5) $y = -\dfrac{2x^4 + 2x^3 - 2x^2 + 2x}{7x^2 + 13x - 2}$

(6) $\dfrac{(2x^2 - 5x + 3)(5x^2 - x - 6)}{(3x^2 - x - 2)(8x^2 - 4x - 12)}$

2. Find horizontal asymptotes.

(1) $y = \dfrac{x+2}{3x^2(x+3)}$

(2) $y = \dfrac{1 + 4x - \sqrt{2}x^2}{3x(x+5)}$

(3) $y = \dfrac{x^3 - 1}{x + 1}$

(4) $y = \dfrac{-3x(x-7)}{(x-1)(2x+3)(x-6)}$

(5) $y = \dfrac{x(1-x)(x+3)}{(x+4)(5-x)}$

(6) $y = \dfrac{(3x+1)(2x-1)(2-3x)}{(5x^2 - 3x + 2)(3x - 2)}$

3. Find slant asymptotes.

(1) $y = \dfrac{x^3}{x^2 + 2x - 3}$

(2) $y = -\dfrac{2x^3}{4x^2 + 1}$

(3) $y = -3x + \dfrac{x}{x^2 - 1}$

(4) $y = \dfrac{x^3}{3(x+1)^2}$

(5) $y = \dfrac{x^2 - 1}{x}$

(6) $y = \dfrac{2x^3 + x - 2}{x - 3}$

4. Find all asymptotes.

(1) $y = \dfrac{x^4 - 1}{2(x-1)^3}$

(2) $y = \dfrac{(6x^2 - 3x - 18)(x^3 - 1)}{(x^4 - 25)(x^2 - 1)(6x^2 + 5x - 6)}$

Exponents and Logarithms

In this last chapter, we shall introduce exponential and logarithmic functions. Exponential functions are also constructed from exponentiations, by allowing variables in exponents. Logarithmic functions are inverse functions of exponential functions. Both of these functions are not algebraic. Instead, they are classified as transcendental functions.

Power functions, exponential functions and logarithmic functions are unified as elementary functions. By definition, elementary functions are functions built from a finite number of exponentials, logarithms, constants, and n-th roots through composition and combinations using the four elementary operations $(+, -, \times, \div)$. Elementary functions are widely used in mathematics. Other functions, which you may have high chances to encounter, are trigonometric functions. They are taught in the course of trigonometry.

We will also learn to solve exponential and logarithmic equations and inequalities in this chapter.

13.1   Exponential Functions

Exponents are not new to us. For any positive real number $a > 0$, we have defined the power a^b for any $b \in \mathbb{R}$:

(1) if $b > 0$

    (i) when b is a positive integer $n \in \mathbb{N}$, $a^n = \overbrace{a \cdots a}^{n}$;

    (ii) when b is a positive rational number $\dfrac{p}{q} \in \mathbb{Q}$ where $p, q \in \mathbb{N}$, $a^{\frac{p}{q}} = \sqrt[q]{a^p} = (\sqrt[q]{a})^p$;

    (iii) when b is a positive irrational number, we can approximate b by positive rational numbers $\tilde{b}$ and approximate a^b by $a^{\tilde{b}}$;

(2) if $b = 0$, $a^b = 0$;

(3) if $b < 0$, $a^b = \dfrac{1}{a^{-b}}$ ($-b > 0$, so a^{-b} is defined in (1)).

In the power a^b, a is called the base and b is called the exponent. The powers satisfy the following propositions. Fix $a, b > 0$ and $\eta, \delta \in \mathbb{R}$.

- $a^\eta b^\eta = (ab)^\eta$, $\dfrac{a^\eta}{b^\eta} = \left(\dfrac{a}{b}\right)^\eta$;

- $(a^\eta)^\delta = a^{\eta\delta}$;

- $a^\eta a^\delta = a^{\eta+\delta}$, $\dfrac{a^\eta}{a^\delta} = a^{\eta-\delta}$.

Besides, we know $1^\eta = 1$ for all η.

Exponential functions are closely related with powers. In the power function $f(x) = x^a$, the exponent is fixed and the base is the variable. In an exponential function, the base is fixed and the exponent becomes the variable. Fix $a > 0, a \neq 1$, the exponential function with base a is defined by

$$x \mapsto a^x \quad (a > 0, a \neq 1).$$

Here we exclude the case $a = 1$, as $1^x = 1$ is a constant function which is not injective. The graph of exponential functions are sketched as follows.

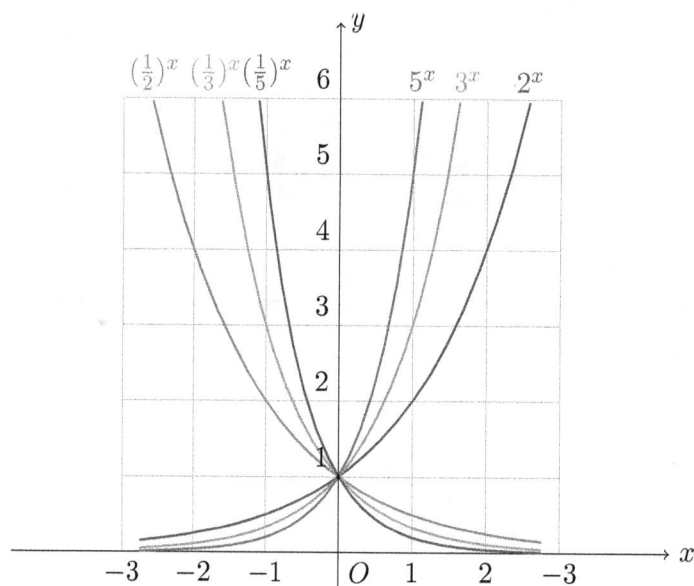

The exponential functions $f(x) = a^x$ $(a > 0, a \neq 1)$ have the following propositions:

- $\mathrm{dom}(f) = (-\infty, +\infty)$ and $\mathrm{Im} f = (0, +\infty)$; as a corollary, a^x is always greater than 0;

- f is injective;

- when $a \in (0, 1)$, f is decreasing; when $a \in (1, +\infty)$, f is increasing;

- the graph of f always passes the point $(0, 1)$, which represents the proposition $a^0 = 1$ for all $a > 0, a \neq 1$;

- f has one horizontal asymptote $y = 0$ (x-axis);

- the graphs of $y = a^x$ and $y = \left(\dfrac{1}{a}\right)^x$ are symmetric about the y-axis. (This should be clear as $y = (\dfrac{1}{a})^x = a^{-x}$, but the graphs of $y = f(x)$ and $y = f(-x)$ are always symmetric about the y-axis.)

Since the exponential function is injective, for any $a > 0, a \neq 1$, $a^b = a^c$ if and only if $b = c$, i.e. the exponents are equal. We can solve some exponential equations by this proposition. Exponential equations are equations with variables in exponents.

Examples. Solve the equations.

(1) $5^x = 125$

(2) $4^{2x-1} = 64$

Solutions.

(1) Since $125 = 5^3$, $5^x = 125$ is the same as $5^x = 5^3$, but $5^x = 5^3$ only when $x = 3$, so the solution is $x = 3$.

(2) As $64 = 4^3$, $4^{2x-1} = 64 = 4^3$ if and only if $2x - 1 = 3$. Solving this linear equation we get $x = 2$.

Exercises.

1. True or false.

(1) $y = a^x$ is an increasing function over its domain when $0 < a < 1$.

(2) The graph of any exponential function passes through the point $(0, 1)$.

(3) The domain of $y = 2^x + 1$ is $(1, +\infty)$.

(4) If $y = (a - 1)^x$ decreases over $\mathbb{R}$, then $a \in (1, 2)$

(5) $\dfrac{a^m}{a^n} = a^{\frac{m}{n}}$, for all $a \in \mathbb{R}$, $m, n \in \mathbb{Z}$ and $n \neq 0$.

(6) $y = x^{-2}$ decreases over $\mathbb{R}$.

(7) $(-27)^{\frac{2}{3}} = 9$.

(8) None of $y = (-3)^x$, $y = -4^x$, $y = 4^{x+1}$ and $y = (-\dfrac{1}{3})^x$ is an exponential function.

(9) Since the graphs of $f(x) = 3^x$ and $g(x) = 3^{-x}$ are symmetric about the y-axis, $f(x)$ is an even function.

(10) If $0.2^m < 0.2^n$, then $m > n$.

2. Multiple choice.

(1) Among the functions (i)$y = 2^x + 1$, (ii)$y = 3^{x-1}$, (iii)$y = -3^x$ (iv)$y = 2^x$, how many of them are exponential functions?

    A. 0   B. 1   C. 2   D. 3

(2) Given $f(x) = a^x (a > 0, a \neq 1)$, which of the following is correct for all $x, y \in \mathbb{R}$?

    A. $f(xy) = f(x)f(y)$ B. $f(xy) = f(x) + f(y)$

    C. $f(x + y) = f(x)f(y)$ D. $f(x + y) = f(x) + f(y)$

(3) The natural range of $f(x) = \dfrac{1}{3^x - 2}$ is

    A. $(-\infty, -\dfrac{1}{2})$ B. $(-\dfrac{1}{2}, +\infty)$

    C. $(-\infty, -\dfrac{1}{2}) \cup (0, +\infty)$ D. $(0, +\infty)$

(4) If the sum of the maximum and minimum values of $y = a^x$ over $[0, 1]$ is 4, then a is

A. $\dfrac{1}{4}$　　　　　B. 3　　　　　C. $\dfrac{1}{2}$　　　　　D. 4

(5) Which of the following satisfies $f(x+1) = \dfrac{1}{4}f(x)$?

A. $f(x) = \dfrac{1}{4}(x+1)$　　　　　B. $f(x) = x + \dfrac{1}{4}$

C. $f(x) = 4^x$　　　　　D. $f(x) = 4^{-x}$

(6) Given $a > 0$ and $a \neq 1$, $f(x) = \dfrac{1}{1 - a^x}$ is

A. an even function　　　　　B. an odd function

C. neither even nor odd　　　　　D. cannot be determined

(7) If $y = (2a^2 - 1)^x$ decreases over $\mathbb{R}$, then

A. $|a| > \dfrac{\sqrt{2}}{2}$　　　B. $|a| < a$　　　C. $\dfrac{\sqrt{2}}{2} < a < 1$　　　D. $\dfrac{\sqrt{2}}{2} < |a| < 1$

(8) Given $a > b$ and $ab \neq 0$, how many of the following are correct?

(i) $a^2 > b^2$; (ii) $2^a > 2^b$; (iii) $\dfrac{1}{a} < \dfrac{1}{b}$; (iv) $(\dfrac{1}{2})^a < (\dfrac{1}{2})^b$; (v) $a^{\frac{1}{3}} > b^{\frac{1}{3}}$.

A. 4　　　　　B. 3　　　　　C. 2　　　　　D. 1

(9) $f(x) = (\dfrac{1}{2})^{\sqrt{-x^2+x+2}}$ decreases over

A. $[-1, \dfrac{1}{2}]$　　　B. $(-\infty, -1]$　　　C. $[2, +\infty)$　　　D. $[\dfrac{1}{2}, 2]$

(10) Given $4 < (\dfrac{1}{4})^x < 64$, then

A. $-1 < x < 3$　　　　　B. $x < -1$ or $x > 3$

C. $-3 < x < -1$　　　　　D. $1 < x < 3$

(11) If $a = 2^{\frac{1}{2}}$, $b = (\dfrac{2}{3})^{-1}$, $c = 3^{\frac{1}{3}}$, then

A. $b > a > c$　　　B. $c > a > b$　　　C. $b > c > a$　　　D. $a > c > b$

(12) The graph of $y = \dfrac{4^x + 1}{2^x}$ is symmetric about

A. the origin　　　B. $y = x$　　　C. the x-axis　　　D. the y-axis

(13) The graph of $y = (\dfrac{\pi - 1}{2})^{x-1}$ is

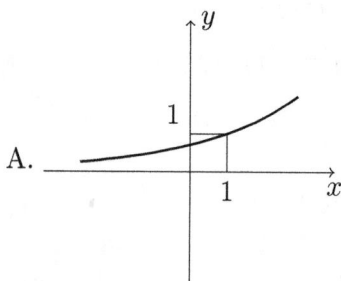

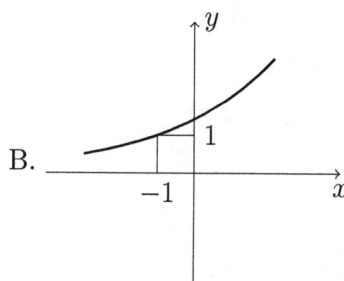

C.

D.

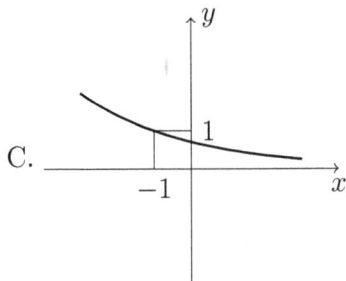

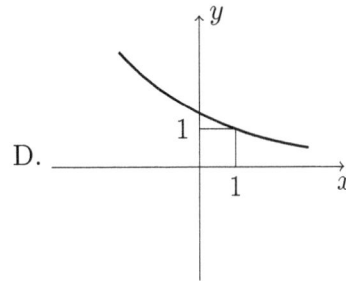

(14) Which of the following function has $(0, +\infty)$ as its image?

A. $y = 3^{\frac{1}{2-x}} + 1$

B. $y = (\frac{1}{5})^{1-x}$

C. $y = \sqrt{(\frac{1}{3})^x - 1}$

D. $y = \sqrt{1 - x^2}$

(15) Give the graphs of exponential functions, which of the following is correct?

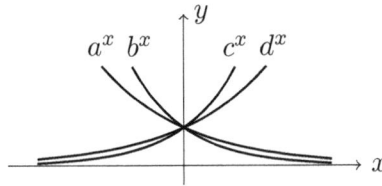

$$a^x \quad b^x \qquad c^x \quad d^x$$

A. $a < b < 1 < c < d$

B. $a < b < 1 < d < c$

C. $b < a < 1 < c < d$

D. $b < a < 1 < d < c$

3. Solve the equations.

(1) $(\frac{1}{3})^{2x} = 81$

(2) $(5^{3x-2}) \cdot (5^{x+1}) = 125$

(3) $(\frac{4}{3})^{\frac{1}{2}x-2} = \frac{64}{27}$

(4) $6^{3x-1} - 6 = 30$

(5) $4^x - 3 \cdot 2^x + 2 = 0$

(6) $3^{2x+1} - 5 \cdot 3^x - 12 = 0$

4. Find the natural domain and natural range of the given function.

(1) $y = 3^{\frac{1}{2-x}}$

(2) $y = \sqrt{2^{x+2} + 1}$

(3) $y = \sqrt{3 - 3^{x-1}}$

(4) $y = \dfrac{2}{5^x - 1}$

5. Compare and order the numbers.

(1) $0.6^{-\frac{4}{5}}, \left(\dfrac{5}{3}\right)^{\frac{1}{2}}$

(2) $4.5^{4.1}, 3.7^{3.6}$

(3) $\sqrt{2}, \sqrt[3]{2}, \sqrt[5]{4}, \sqrt[8]{8}, \sqrt[9]{16}$

(4) $\sqrt[n-1]{a^n}, \sqrt[n]{a^{n+1}}$ $(a > 0, a \neq 1, 1 < n \in \mathbb{N})$

6. Solve the following problems.

(1) If $a^{-0.5} < a^{0.2}$, find a.

(2) If $a^{7.5} < a^{5.9}$, find a.

(3) Given $a > 0$, if $a^{\frac{2}{3}} > a^{\frac{3}{5}}$, find a.

(4) Given $x^{\frac{1}{2}} + a^{-\frac{1}{2}} = 3$, find $\dfrac{x^{\frac{3}{2}} + x^{-\frac{3}{2}} + 2}{x^2 + x^{-2} + 3}$.

(5) If $y = \dfrac{1}{2^x - 1} + a$ is an odd function, find a.

(6) If $y = f(x)$ is an even function over $[-2, 4^a - 2^a]$, find a.

7. Calculate.

(1) $[(0.027^{\frac{1}{3}})^{-2.5}]^{\frac{2}{5}} - [256^{0.125} + (-32)^{\frac{3}{5}} + 0.1^{-1}]^{\frac{1}{2}}$

(2) $\dfrac{1}{\sqrt{5}+2} - (\sqrt{3}-1)^0 - \sqrt{9-4\sqrt{5}}$

8. Simplify.

(1) $(a^4 - b^4)\sqrt{\dfrac{a+b}{a-b}} - \dfrac{b^2}{a-b}\sqrt{a^2b^4 - b^6} - (a^2 + b^2)\sqrt{(a+b)^3(a-b)}$ $(a > b, a + b > 0)$

(2) $\dfrac{a^{\frac{5}{3}} - 8a^{\frac{2}{3}}b}{a^{\frac{2}{3}} + 2\sqrt[3]{ab} + 4b^{\frac{2}{3}}} \cdot \dfrac{a^{\frac{1}{3}}}{a^{\frac{1}{3}} - 2b^{\frac{1}{3}}}$ $(a \neq 8b, a^2 + b^2 \neq 0)$

9. Sketch the graphs.

(1) $y = (\dfrac{1}{2})^{x+1}$ (2) $y = 2^x - 2$

10. Given $f(x) = (\dfrac{3}{4})^{x^2 - 5x + 6}$, find the natural domain and range of f as well as the intervals over which f is increasing or decreasing.

11. Consider the function $f(x) = \dfrac{a^x - 1}{a^x + 1}(a > 1)$.

(1) Tell if f is an odd function, even function, or neither;

(2) Find $\mathrm{Im}f$;

(3) Prove f is increasing over $\mathbb{R}$.

12. Consider the function $f(x) = x\left(\dfrac{1}{2^x - 1} + \dfrac{1}{2}\right)$.

(1) Tell if f is an odd function, even function, or neither;

(2) Prove $f(x) > 0$ when $x \neq 0$.

13.2 Logarithms

Fix a number $a > 0, a \neq 1$. For any $b > 0$, the equation $a^x = b$ has one unique solution x_0 as shown in the following picture.

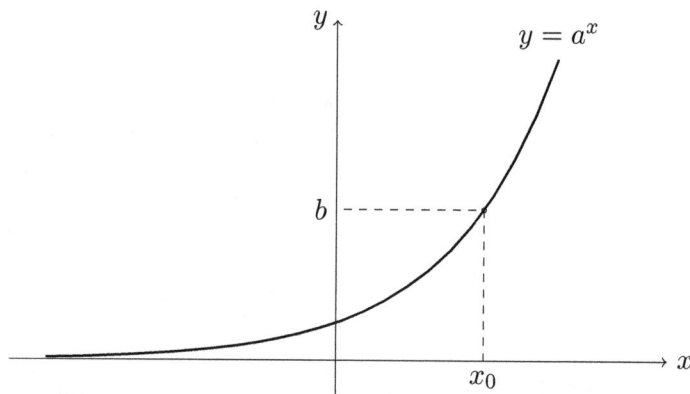

The solution x_0 is called the logarithm of b with base a, denoted by $\log_a b$. When $b \leq 0$, $a^x = b$ has no solutions, so $\log_a b$ makes sense only when $a > 0, a \neq 1$ and $b > 0$. From the definition, immediately we have, for any $a > 0, a \neq 1$ and $b > 0$,

(i) $a^{\log_a b} = b$ ($x_0 = \log_a b$ is the solution to $a^x = b$, thus $a^{\log_a b} = a^{x_0} = b$);

(ii) $\log_a a^b = b$ (the solution to $a^x = a^b$ is b, which is also denoted by $\log_a a^b$).

As a corollary of proposition (ii), $\log_a a = 1$ and $\log_a 1 = 0$.

From the definition or the above properties, sometimes we can simplify the symbol $\log_a b$. For example, $\log_4 64$ stands for the solution to $4^x = 64$, but the solution to this equation is $x = 3$, so $\log_4 64 = 3$. Similarly, $\log_2 \dfrac{1}{4} = -2$ as $\dfrac{1}{4} = 2^{-2}$ and $\log_2 \dfrac{1}{4} = \log_2 2^{-\frac{1}{2}} = -\dfrac{1}{2}$ by proposition (ii). Not all logarithms can be simplified to rational numbers. In fact, most of them are irrational numbers. For example, $\log_2 3$, $\log_5 4$ and $\log_9 100$ are all irrational.

With logarithms we can solve more exponential equations.

Examples. Solve the equations.

(1) $5^{2x-1} = 3$

(2) $3^{2x} - 5 \times 3^x + 6 = 0$

Solutions.

(1) The idea is substitution. Let $y = 2x - 1$; then the equation becomes $5^y = 3$, so $2x - 1 = y = \log_5 3$ (we can find logarithms of the two sides with base 5 to get $2x - 1 = \log_5 3$ as well); then, solving for x we get $x = \dfrac{1}{2}(\log_5 3 + 1)$.

(2) The equation can also be solved by substitution. Let $y = 3^x$; then the equation transforms into $y^2 - 5y + 6 = 0$. It is easy to get $y = 2$ or 3, so the original equation is equivalent to $3^x = 3$ or $3^x = 2$. Immediately, we get the solutions $x = 1$ and $x = \log_3 2$.

As logarithms are opposite operation of exponentiations, we can derive more propositions of logarithms from the propositions of powers or exponents. Given $a > 0, a \neq 1$, $b, c > 0$ and $r \in \mathbb{R}$, the logarithms also satisfy

(iii) $\log_a(bc) = \log_a b + \log_a c$, $\log_a \dfrac{b}{c} = \log_a b - \log_a c$;

(iv) $\log_a b^r = r \log_a b$.

The formulas can be proved by taking the exponentiation over a. As $a^{\log_a b + \log_a c} = a^{\log_a b} \cdot a^{\log_a c} = b \cdot c = a^{\log_a bc}$, so the exponents must be equal, i.e. $\log_a bc = \log_a b + \log_a c$. Similarly, $a^{r \log_a b} = (a^{\log_a b})^r = b^r = a^{\log_a b^r}$, so the exponents $\log_a b^r$ and $r \log_a b$ are equal as well. With all four propositions we can do more simplifications.

Example. Simplify.

(1) $\log_3 72$

(2) $\log_a \sqrt{\dfrac{x^2 y^3}{z^4}}$ $(x, y, z > 0, a > 0, a \neq 1)$

Solutions.

(1) 72 can be factored as $2^3 \cdot 3^2$, so

$$
\begin{aligned}
\log_3 72 &= \log_3(3^2 \cdot 2^3) \\
&= \log_3 3^2 + \log_3 2^3 \quad \text{(proposition (iii))} \\
&= \log_3 3^2 + 3 \log_3 2 \quad \text{(proposition (iv))} \\
&= 2 + 3 \log_3 2 \quad \text{(proposition (ii))}.
\end{aligned}
$$

(2) By the propositions above,

$$
\begin{aligned}
\log_a \sqrt{\dfrac{x^2 y^3}{z^4}} &= \log_a \left(\dfrac{x^2 y^3}{z^4}\right)^{\frac{1}{2}} \\
&= \dfrac{1}{2} \log_a \dfrac{x^2 y^3}{z^4} \\
&= \dfrac{1}{2}\left(\log_a x^2 + \log_a y^3 - \log_a z^4\right) \\
&= \dfrac{1}{2}\left(2 \log_a x + 3 \log_a y - 4 \log_a z\right) \\
&= \log_a x + \dfrac{3}{2} \log_a y - 2 \log_a z.
\end{aligned}
$$

If we fix the base to be 10, the logarithms $\log_{10} b$ are called common logarithms, denoted by $\log b$. For example, $\log 100 = \log_{10} 10^2 = 2$, $\log 1000 = \log 10^3 = 3$. From these examples, the common logarithm $\log b$ can be used to test how many digits there are in the number $b > 0$.

If we fix the base to be e, an irrational number which is approximately 1.718281828, the logarithms $\log_e b$ are called natural logarithms, denoted by $\ln b$. The reason to introduce natural logarithms is not clear at the moment. After learning calculus, the readers may get a better understanding of the number e as well as the importance of the natural logarithms.

Given a logarithm $\log_a c$ ($a > 0, a \neq 1, c > 0$), we can change the base to any number b ($b > 0, b \neq 1$) by the formula

$$\log_a c = \frac{\log_b c}{\log_b a}.$$

The formula can be proved as follows: first, since $a, b > 0, b \neq 1$, $\log_b a$ is well-defined and it is not 0, i.e. the fraction $\dfrac{\log_b c}{\log_b a}$ exists. Secondly, the formula is equivalent to $\log_b a \log_a c = \log_b c$, but this is true since if taking exponentiations with base b, we get an identity $b^{\log_b a \log_a c} = (b^{\log_b a})^{\log_a c} = a^{\log_a c} = c = b^{\log_b c}$. As an application, if setting $b = c$ in the formula, we get $\log_a b = \dfrac{1}{\log_b a}$ whenever $a, b > 0$ and $a, b \neq 1$.

A logarithmic equation is an equation with variables appearing in logarithms. The only thing to pay attention to is to make sure all expressions inside logarithms are positive. The following examples show how to solve logarithmic equations.

Example. Solve the equations.

(1) $\ln(x - 3) + \ln(x - 1) = \ln(7 - x)$

(2) $\log_2(x + 3) - \log_2(x - 2) = -1$

Solutions.

(1) First, $x - 3$, $x - 1$ and $7 - x$ must be all positive, i.e. $x - 3 > 0$, $x - 1 > 0$, $7 - x > 0$, so $x \in (3, 7)$. Secondly, by the proposition of logarithms, the equation is equivalent to $\ln(x - 3)(x - 1) = \ln(7 - x)$, so we must have $(x - 3)(x - 1) = 7 - x$ (taking exponentiation with base e). The equation simplifies to $x^2 - 4x + 3 = 7 - x$, or $x^2 - 3x - 4 = 0$. Solving the equation we get $x = 4$ or -1. Together with the requirement $x \in (3, 7)$, the only solution is $x = 4$.

(2) As $\log_2(x + 3) - \log_2(x - 2) = \log_2 \dfrac{x + 3}{x - 2}$ and $\log_2 \dfrac{1}{2} = -1$, the equation is equivalent to

$$\begin{cases} x + 3 > 0 \\ x - 2 > 0 \\ \dfrac{x + 3}{x - 2} = \dfrac{1}{2}. \end{cases}$$

Solving the last equation we get $x = -8$, but $x = -8$ does not satisfy the first two inequalities, so the equation has no solutions.

Exercises.

1. True or false.

(1) $\log_2 \dfrac{1}{8} = -3$.

(2) $\log_{-3} 27 = 3$.

(3) $\log 4 + \log \dfrac{1}{4} = 0.$

(4) $\dfrac{\log_3 36}{\log_3 9} = \log_3 4.$

(5) $\ln e = 1$ and $\ln 1 = 0.$

(6) $\log_2 48 = \log_2 3 + 4.$

(7) $\log_3 6 = \dfrac{\log_2 6}{\log_2 3} = \log_2 2 = 1.$

(8) $(\dfrac{1}{2})^a$ and $\log_a \dfrac{1}{2}$ have the same value if $a > 0$ and $a \neq 1.$

(9) $\log(x-1) + \log(x+1) = \log 2x$ has two solutions.

(10) $2^{5x-1} = 6$ has an irrational solution.

2. Multiple choice.

(1) Given $f(x) = 3^x + 5$, the domain of f^{-1} is

    A. $(0, +\infty)$ B. $(5, +\infty)$ C. $(6, +\infty)$ D. $(-\infty, +\infty)$

(2) If $a = \log_3[26 - (2 - \sqrt{3})(7 - 4\sqrt{3})]$, then

    A. $a \in \mathbb{N}$ B. $a \in \mathbb{Z} - \mathbb{N}$ C. $a \in \mathbb{Q} - \mathbb{N}$ D. $a \in \mathbb{R} - \mathbb{Q}$

(3) If $\log_2 x = \log_3 y = \log_5 z > 0$, then

    A. $x^{\frac{1}{2}} > y^{\frac{1}{3}} > z^{\frac{1}{5}}$ B. $y^{\frac{1}{3}} > x^{\frac{1}{2}} > z^{\frac{1}{5}}$ C. $y^{\frac{1}{3}} > z^{\frac{1}{5}} > x^{\frac{1}{2}}$ D. $z^{\frac{1}{5}} > x^{\frac{1}{2}} > y^{\frac{1}{3}}$

(4) If $ab = m$ where $a, b > 0, m \neq 1$ and $\log_m b = x$, then $\log_m a$ is

    A. $1 + x$ B. $1 - x$ C. $\dfrac{1}{x}$ D. $x - 1$

(5) If $f(\log x) = x$, then $f(3)$ is

    A. $\log 3$ B. 3 C. 1000 D. 3000

3. Fill in the blanks.

(1) $25^{\log_5 3} = \underline{\qquad}.$

(2) $(\sqrt{2})^{\log_2 3} = \underline{\qquad}.$

(3) If $\log_{\sqrt{3}} 2 = \dfrac{1-a}{a}$, then $\log_{\sqrt{3}} 12 = \underline{\qquad}.$

(4) $\log_{\frac{1}{2}} 16 = \underline{\qquad};$ $\log 0.01 = \underline{\qquad}.$

(5) If $f(x) = a^{x - \frac{1}{2}}$ and $f(\log a) = \sqrt{10}$, then $a = \underline{\qquad}.$

4. Rewrite the following identities into logarithms.

(1) $5^3 = 125$ (2) $2^{-6} = \dfrac{1}{64}$

(3) $e^0 = 1$ (4) $(\dfrac{3}{2})^3 = \dfrac{27}{8}$

(5) $(0.1)^{-2} = 100$

(6) $(\frac{1}{8})^3 = \frac{1}{512}$

5. Evaluate.

(1) $\log_4 \frac{1}{64}$

(2) $\log \frac{1}{100}$

(3) $\log_2 \sqrt{2} + \log_3 \sqrt[3]{3} + \log_4 \sqrt[4]{4}$

(4) $\frac{\log_8 9}{\log_2 3}$

(5) $6^{\log_6 10}$

(6) $27^{\frac{2}{3} - \log_3 2}$

(7) $\log_3 18 - \log_3 2$

(8) $\log_9 16 \cdot \log_8 81$

(9) $2^{3 + \log_2 3} + 3^{5 - \log_3 9}$

(10) $\log_{24} 2 + \log_{24} 3 + \log_{24} 4$

(11) $\log_{25} 625 + e^{\ln 2}$

(12) $\log 5 \cdot \log 20 + (\log 2)^2$

(13) $3 \cdot \log_7 2 - \log_7 9 + 2 \cdot \log_7 \frac{3}{2\sqrt{2}}$

(14) $(\log 2)^2 + \log 4 \cdot \log 50 + \log^2 50$

(15) $\dfrac{\log_2 \sqrt{7} \cdot \log_5 27}{\log_5 \frac{1}{9} \cdot \log_2 \sqrt[3]{49}}$

(16) $(\ln \sqrt{27} + \ln 8 - \ln \sqrt{1000}) \div \ln 1.2$

6. Simplify.

(1) $\log_2 54$

(2) $\log_9 27$

(3) $\log 20 + \log_{100} 25$

(4) $\log_5 175 - \log_5 14$

(5) $\sqrt{\log^2 99 - 2 \log 99^2 + 4}$

(6) $\ln[4x^3(2x - 1)^2] \ (x > \frac{1}{2})$

(7) $\log_3 \dfrac{x^2yz^3}{\sqrt{x^3yz^7}}$ $(x, y, z > 0)$

(8) $\log_3 \sqrt{3x^2 - 4x - 15}$ $(x > 3)$

(9) $\ln \left(\dfrac{x^2 - 1}{x^2} \right)^5$ $(|x| > 1)$

(10) $\log_2(3x^2 + 5x - 2) - \log_2(x + 2)$ $(x > \dfrac{1}{3})$

(11) $\dfrac{1}{2}\log(2x + 2\sqrt{x^2 - 1}) - \log 2 + \log(\sqrt{x + 1} - \sqrt{x - 1})$ $(x > 1)$

7. Solve the equations.

(1) $2^{6x-2} = 5$

(2) $\log_3 x = -\dfrac{3}{4}$

(3) $\log_x 2 = \dfrac{7}{8}$

(4) $(\dfrac{1}{3})^{6x-3} = 24$

(5) $\log_{2x^2-1}(3x^2 + 2x - 1) = 1$

(6) $\log x^2 - \log x = 0$

(7) $2 \cdot \log_6 x = 1 - \log_6 3$

(8) $\log_2(x - 1) + \log_2 x = 1$

(9) $\ln(x - 1) + \ln(x - 2) = \ln(3 - x)$

(10) $\log_2(x - 1) = \log_2(2x + 1)$

(11) $\dfrac{1 + 3^{-x}}{1 + 3^x} = 3$

(12) $\ln^2 x - (\ln 5 + \ln 7)\ln x + \ln 5 \cdot \ln 7 = 0$

(13) $\log_x(2x^2 - 3) = 2$

(14) $2^{3\ln x}5^{\ln x} = 1600$

(15) $\log_2(\log_3(\log_5 x)) = 0$

(16) $2\log_x 4 + \log_4 x^2 = 5$

(17) $\log_x x^x = 2$

(18) $\dfrac{1}{2}(\ln x - \ln 5) = \ln 2 - \dfrac{1}{2}\ln(9 - x)$

(19) $2^{2x+1} - 11 \cdot 2^x + 14 = 0$ \hspace{2cm} (20) $6e^{2x} - 7e^x - 20 = 0$

8. If $10^x = 2, 10^y = 3$, find 100^{2x-y}.

9. If $a, b > 0$ satisfy $a^2 - 2ab - 8b^2 = 0$, find $\log(a^2 + ab - 6b^2) - \log(a^2 + 4ab - 15b^2)$.

10. If $\log_8 9 = a, \log_2 5 = b$, denote $\log 3$ by a, b.

11. If $8^x = 9^y = 6^z$, prove $\dfrac{x}{2} + \dfrac{y}{3} = \dfrac{z}{6}$.

12. If $M = \{0, 1\}$, $N = \{11 - a, \log a, 2^b, b\}$, are there $a, b \in \mathbb{R}$ such that $M = N$?

13.3   Logarithmic Functions

Fixing $a > 0, a \neq 1$, the function

$$x \mapsto \log_a x$$

is called the logarithmic function with base a. Notice $\log_a x$ is defined only when $x > 0$, so the domain of the logarithmic function is $(0, +\infty)$.

The exponential function and the logarithmic function with the same base are inverse functions. Let

$$f(x) : (-\infty, +\infty) \to (0, +\infty)$$
$$x \mapsto a^x$$

be the exponential function with base a. We already showed in the last section that f is injective; f is also surjective since $\mathrm{Im} f = (0, +\infty)$ coinciding with its range. As a result, f

is invertible. To find the inverse, solving x from the equation $y = a^x$ we get $x = \log_a y$, so the inverse function f^{-1} is the logarithmic function $g(x) = \log_a x$.

As f^{-1} exchanges the domain and range of f, the domain of a logarithmic function must be $(0, +\infty)$, which is already found at the beginning of this section; the range of a logarithmic function is always $(-\infty, +\infty)$.

As inverse functions, the graphs of exponential functions and logarithmic functions with the sames bases are symmetric about the line $y = x$. The graphs of logarithmic functions are sketched as follows.

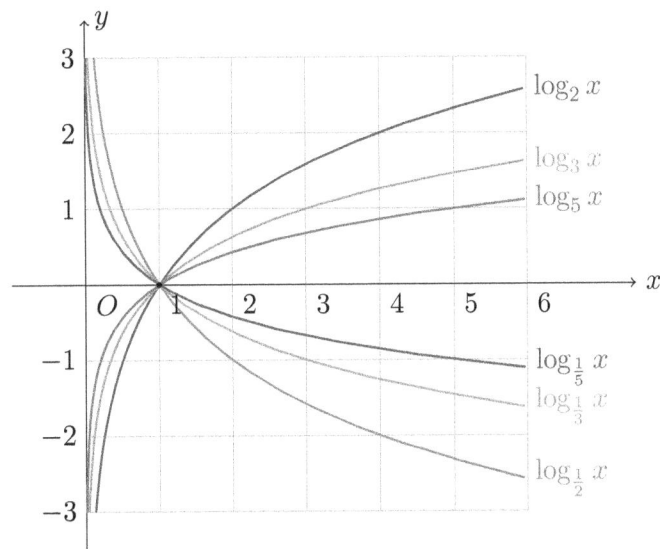

The logarithmic functions $g(x) = \log_a x$ $(a > 0, a \neq 1)$ have the following propositions:

- $\text{dom}(g) = (0, +\infty)$ and $\text{Im}g = (-\infty, +\infty)$;

- g is injective;

- when $a \in (0, 1)$, g is decreasing; when $a \in (1, +\infty)$, g is increasing;

- the graph of g always passes through the point $(0, 1)$, which represents the proposition $\log_a 1 = 0$ for all $a > 0, a \neq 1$;

- g has one vertical asymptote $x = 0$ (y-axis);

- the graphs of $y = \log_a x$ and $y = \log_{\frac{1}{a}} x$ are symmetric about the x-axis.

The graphs and the propositions of logarithmic functions give us more tools to handle problems with logarithms.

Examples.

(1) Compare and order the numbers $\log_2 a$, $\ln a$ and $\log_3 a$ when $a > 1$;

(2) Solve the equation $\log_x 2x + 3 = 2$;

(3) Solve the inequality $\log_{a+1} 3 - a > 1$.

Solutions.

(1) Since $2 < e < 3$, the relative position of the graphs of $y = \log_2 x$, $y = \ln x$, $y = \log_3 x$

is similar to the relative position of the graphs of $y = \log_2 x$, $y = \log_3 x$ and $y = \log_5 x$. When $a < 1$, we should have $\log_2 a < \ln a < \log_3 a$.

(2) The equation $\log_x 2x + 3 = 2$ is equivalent to

$$\begin{cases} x > 0 \\ x \neq 1 \\ 2x + 3 > 0 \\ 2x + 3 = x^2. \end{cases}$$

Here x should satisfy $x > 0$ and $x \neq 1$ since it is the base of a logarithm. $2x + 3$ should be positive since logarithms are defined only for positive numbers. The last equation is obtained by taking exponentiation with base x.

Solving the last equation $x^2 - 2x - 3 = 0$, we get $x = -1$ or 3. Since $x = -1$ does not satisfy the first inequality, it is not a solution. However, $x = 3$ satisfies all the inequalities, so it is the only solution to the equation.

(3) As the monotonic property of logarithmic functions depends on the base, we should make discussions on $a + 1$.

• When $a + 1 > 1$, i.e. $a > 0$, $y = \log_{a+1} x$ is increasing. As a result, the inequality $\log_{a+1} 3 - a > 1 = \log_{a+1} a + 1$ is equivalent to

$$\begin{cases} a > 0 \\ 3 - a > 0 \\ 3 - a > a + 1. \end{cases}$$

The solution set for this case is $(0, 1)$.

• When $0 < a + 1 < 1$, i.e. $-1 < a < 0$, $y = \log_{a+1} x$ is decreasing. As a result, the inequality $\log_{a+1} 3 - a > 1 = \log_{a+1} a + 1$ is equivalent to

$$\begin{cases} -1 < a < 0 \\ 3 - a > 0 \\ 3 - a < a + 1. \end{cases}$$

There is no solution for this case.

Combining the two cases, the solution set to the inequality is $(0, 1)$.

Exercises.

1. True or false.

(1) The graphs of $y = \log_a x$ and $y = \log_{\frac{1}{a}} x$ are symmetric about the x-axis as long as $a > 0, a \neq 1$.

(2) $\log_2 3.4 < \log_2 8.5$ and $\log_{0.3} 1.8 < \log_{0.3} 2.7$.

(3) The graph of $y = \log_a(x + 2) - 1$ passes through the point $(-1, -2)$.

(4) $y = \log_{0.2}(x^2 - x - 2)$ increases over $(2, +\infty)$.

(5) The domain of $y = \log_2(x^2 + x + 1)$ is $\mathbb{R}$.

(6) $y = \log_{\frac{1}{2}} \dfrac{x+1}{x-1}$ is an even function over $(-\infty, -1) \cup (1, +\infty)$.

(7) If $f(x) = 2^x - 1$, then $f^{-1}(x) = \log_2(x+1)$.

(8) $f(x) = \log_{0.3}(x+2) - 3$ has a vertical asymptote $x = 3$.

(9) The graphs of $y = 2^x$ and $y = \log_{\frac{1}{2}} x$ are symmetric about $y = x$.

(10) The graph of $y = \log_3(x-1) + 5$ does not intersect with the y-axis.

2. Multiple choice.

(1) Which of the following is a logarithmic function?

    A. $y = \log_2(x-3)$ B. $y = \log_{3x-4}(1-x)$

    C. $y = \log x^2$ D. $y = \ln x$

(2) The domain of $y = \dfrac{1}{\log_2(-x^2 + 2x)}$ is

    A. $(0, 2)$ B. $(0, 1) \cup (1, 2)$ C. $(-1, 0) \cup (0, 2)$ D. $(-1, 2)$

(3) The range of $y = \log_2(x^2 + 2x + 5)$ is

    A. $(-\infty, 2)$ B. $(4, +\infty)$ C. $[2, +\infty)$ D. $(-\infty, 4]$

(4) If $\log_a \dfrac{2}{3} < 1$, then a

    A. $(0, \dfrac{2}{3}) \cup (1, +\infty)$ B. $(\dfrac{2}{3}, +\infty)$

    C. $(\dfrac{2}{3}, 1)$ D. $(0, \dfrac{2}{3}) \cup (\dfrac{2}{3}, +\infty)$

(5) Given $f(x) = 0.2^{-x} + 1$, then $f^{-1}(x)$ is

    A. $\log_5 x + 1$ B. $\log_5(x-1)$ C. $\log_5 x - 1$ D. $\log_5(x+1)$

(6) Simplify $(\sqrt{3})^{\log_3(-a)^2}$, $a \neq 0$

    A. $-a$ B. a^2 C. $|a|$ D. a

(7) Given that $m = \log_{2n-6}(4-n)$ exists, then

    A. $n \in (-\infty, 3) \cup (4, +\infty)$ B. $n \in (-3, 4)$

    C. $n \in (3, 4)$ D. $n \in (3, 3.5) \cup (3.5, 4)$

(8) Given $a > 1$, if the difference of the maximum and minimum value of the function $f(x) = \log_a x$ over $[a, 2a]$ is $\dfrac{1}{2}$, then a is

    A. $\dfrac{1}{4}$ B. 4 C. $-\dfrac{1}{4}$ D. -4

(9) $y = \log_{\frac{1}{3}}\left(\dfrac{x+1}{x-1}\right)$ increases over

    A. $(1, +\infty)$ B. $(-\infty, 1)$

    C. $(-\infty, -1) \cup (1, +\infty)$ D. $(-1, 1)$

(10) If the domain of $y = \log(ax^2 + 2x + 1)$ is $\mathbb{R}$, then

    A. $0 \leq a \leq 1$ B. $a > 1$ C. $0 < a \leq 1$ D. $a < 0$

(11) The graph of $y = \log(\dfrac{2}{1+x} - 1)$ is symmetric about

    A. the x-axis   B. the y-axis   C. the origin   D. the line $y = x$

(12) If the natural domains of $y = \log_a(x+1), y = \log_a x, y = \log_a(-x), y = \log_a(1 - |x|)$ are M, N, P, Q separately, then $(M \cup N) \cap (P \cup Q)$ is

    A. $(-1, +\infty)$ B. $(-\infty, 0)$ C. $(-1, 1)$ D. $(-1, 0)$

(13) The domain of $y = \log_{2x-1} \sqrt{3x - 2}$ is

    A. $(\dfrac{2}{3}, 1) \cup (1, +\infty)$ B. $(\dfrac{1}{2}, 1) \cup (1, +\infty)$

    C. $(\dfrac{2}{3}, +\infty)$ D. $(\dfrac{1}{2}, +\infty)$

(14) The image of $y = \log_{\frac{1}{2}}(x^2 - 6x + 17)$ is

    A. $(-\infty, +\infty)$ B. $(-\infty, -3]$ C. $[-3, +\infty)$ D. $[3, +\infty)$

(15) If $f(x) = \log_a |x + 1|$ is positive when $x \in (-1, 0)$, then

    A. $f(x)$ is increasing over $(-\infty, 0)$ B. $f(x)$ is decreasing over $(-\infty, 0)$

    C. $f(x)$ is increasing over $(-\infty, -1)$ D. $f(x)$ is decreasing over $(-\infty, -1)$

(16) If the graphs of f and $g(x) = (\dfrac{1}{3})^x$ are symmetric about $y = x$, then the interval over which $F(x) = f(2x - x^2)$ is increasing is

    A. $[1, +\infty)$ B. $(-\infty, 1]$ C. $(0, 1]$ D. $[1, 2)$

(17) If $\log_a(\sqrt{2} + 1) + \log_b(\sqrt{2} + 1) = 0$ where $a, b > 0$ and $a, b \neq 1$, then

    A. $a = b$ B. $b < a$ C. $a < b$ D. $ab = 1$

3. Fill in the blanks.

(1) The inverse function of $f(x) = 5^x + 1$ is _____.

(2) If $\log_{12} 27 = a$, then $\log_6 16 =$_____.

(3) If f is an odd function and $f(x) = x^2 + \ln(x + 1)$ when $x > 0$, then when $x < 0$, $f(x) =$_____.

(4) If $\log_3 4 \cdot \log_4 8 \cdot \log_8 m = \log_4 16$, them $m =$_____.

(5) If $\log_a c + \log_b c = 0$, then $ab + c - abc =$_____.

(6) If f is a function with domain $(-3, 3)$ satisfying $f(x^2 - 3) = \log_a \dfrac{x^2}{6 - x^2}$, then f is an _____(even, odd or neither) function.

4. Find domains of the following functions.

(1) $y = \sqrt{\log_{\frac{1}{2}} \dfrac{3x - 2}{2x - 1}}$ (2) $\dfrac{1}{\sqrt{1 - \log_a(x + a)}}$

(3) $y = f[\log_{\frac{1}{3}}(3 - x)]$ where f is a function with domain $[0, 1]$.

(4) $y = f(\ln(x^2 - 1))$ where f is a function whose domain is $[0, 1]$.

5. Find the inverse of the function $f(x) = \dfrac{e^x}{1 + e^x}$ as well as the domain and range of f^{-1}.

6. Solve the equations.

(1) $\log_2(x^2 - 5x - 2) = 2$

(2) $2^{\log_3 x} = \dfrac{1}{4}$

(3) $\log(2x - 1)^2 - \log(x - 3)^2 = 2$

(4) $(\ln x)^2 - 3 \cdot \ln x - 4 = 0$

(5) $\log(4^x + 2) = \log 2^x + \log 3$

(6) $\log(y - 1) - \log y = \log(2y - 2) - \log(y + 2)$

(7) $\log(x + 10)^2 - \log(x + 10)^3 - 1 = 0$ (8) $\log_9 x + \log_{x^2} 3 = 1$

(9) $\log_3(1 - 2 \cdot 3^x) = 2x + 1$

(10) $3^{x+2} - 3^{2-x} = 80$

(11) $10^{\log(x+10)} = 100$

(12) $2^{x^2 - 6x - \frac{5}{2}} = 16\sqrt{2}$

(13) $x^{1+\log x} = 0.1^{-2}$

(14) $\log_{x-1}(x^2 - 5x - 10) = 2$

7. Sketch the graphs.

(1) $y = \log_{\frac{1}{2}}(-x)$

(2) $y = \log_2(x - 3) + 1$

(3) $y = |\log_{\frac{1}{2}}(x - 1)|$

(4) $y = \ln|x + 1|$

8. Compare the numbers.

(1) $\log_a \dfrac{a}{b}, \log_b \dfrac{b}{a}, \log_b a, \log_a b$ if $1 < a < b < a^2$

(2) $|\log_a(1 - x)|, |\log_a(1 + x)|$ when $a > 1, 0 < x < 1$

(3) $(\log_2 x)^2, \log_2 x^2, \log_2(\log_2 x)$ when $1 < x < 2$

(4) $a, b, 1$ if $\log_a 3 > \log_b 3$

9. Tell if $y = \log_a(x + \sqrt{1 + x^2})(a > 0, a \neq 1)$ is an even function, an odd function, or neither.

10. Given f below, find the natural domain and range of f as well as the intervals over which f is increasing or decreasing.

(1) $f(x) = \log_2 \dfrac{x}{1-x}$

(2) $f(x) = -\log_{\frac{1}{2}}^2 x - 3\log_{\frac{1}{2}} x - 2$

11. If x, y satisfy $\log_a x + 3\log_x a - \log_x y = 3$ where $0 < a < 1$ and the maximum value of y is $\dfrac{\sqrt{2}}{4}$, find a and x when y attains its maximum value.

12. If $x^2 \log_2 \dfrac{4(a+1)}{a} + 2x \log_2 \dfrac{2a}{a+1} + \log_2 \dfrac{(a+1)^2}{4a^2} > 0$ for all $x \in \mathbb{R}$, find a.

13. Suppose $f(x) = \ln(a^x - b^x)$ where $0 < b < 1 < a$.
(1) Find the domain of f;

(2) Prove that f is an strictly increasing function;

(3) If $f(x) > 0$ exactly when $x \in (1, +\infty)$ and $f(2) = \ln 2$, find a, b.

14. If $f(x) = \log_2[(3 - 2k)x^2 - 2kx - k + 1]$ is decreasing over $(-\infty, 0)$ and increasing over $(1, +\infty)$, find k.

15. If A, B, C are three points on the graph of $y = \log_a x (0 < a < 1)$ with x-coordinates $m, m + 2, m + 4 (m \geq 1)$ separately,

(1) find the area $S(m)$ of $\triangle ABC$;

(2) determine where $S(m)$ is increasing or decreasing;

(3) find the maximum value of $S(m)$.

Keys

Chapter 1

Section 1.1

1. (1) F (2) T (3) F (4) T (5) T (6) F (7) T (8) F (9) T (10) F

2. (1) $\{a|a$ is a nonnegative integer, $a < 6\}$

    (2) $\{x|x$ is an odd number$\}$

    (3) $\{x|x$ is a rational number$\}$

3. (1) $A = \{0, 1, 2, 3, 4, 5, 6, 7, 8, 9\}$

    (2) $B = \{2, 5\}$

    (3) $C = \{4, 5, 6, 7, 8, 9, 10, 11, 12\}$

    (4) $D = \emptyset$

    (5) $E = \{(-1, 0), (0, 0), (1, 0), (-1, 1), (0, 1), (1, 1)\}$

4. $\emptyset, \{1\}, \{2\}, \{3\}, \{1, 2\}, \{2, 3\}, \{1, 3\}, \{1, 2, 3\}$

Section 1.2

1. (1) $\{3, 4, 6\}$ (2) $\{4\}$ (3) $\{1, 3, 5, 6\}$ (4) $\{2, 3, 4, 5, 6\}$ (5) $\{1, 2, 4\}$ (6) $\{5\}$

2. (1) $\{0, 1, 2, 4, 6, 7, 8, 9, 10, 13, 16, 19, 25, 32, 36, 49\}$ (2) $\{1\}$

    (3) $\{0, 9, 25, 36, 49\}$ (4) $\{1, 2, 4, 6, 7, 8, 10, 13, 16, 19, 32\}$

3. (1) (2)

    (3) (4)

4. (1) B (2) A (3) $\emptyset$ (4) $\emptyset$

Chapter 2

Section 2.1

1. (1) F (2) T (3) T (4) F (5) F (6) F (7) F (8) T
(9) T (10) F (11) F (12) F (13) T (14) T (15) F (16) F
(17) T (18) T (19) F (20) T

2. (1) D (2) C (3) B (4) C (5) B (6) C (7) A (8) C
(9) B (10) D

3. (1) 56 (2) 42 (3) 63 (4) 132 (5) 12 (6) 92
(7) 104 (8) 78 (9) 60 (10) 51 (11) 17 (12) 75
(13) 35 (14) 204 (15) 14 (16) 4 (17) 92 (18) 195
(19) 7 (20) 288 (21) 6 (22) 55 (23) 224 (24) 132
(25) 672 (26) 273 (27) 209 (28) 18 (29) 179 (30) 26

4. (1) 49 (2) 16 (3) 27 (4) 125 (5) 81 (6) 1000
(7) 64 (8) 1024 (9) 81 (10) 512 (11) 216 (12) 625
(13) 121 (14) 144 (15) 169 (16) 196 (17) 225 (18) 256
(19) 289 (20) 324 (21) 361 (22) 441 (23) 625 (24) 1369
(25) 1764 (26) 4225 (27) 2809 (28) 10201 (29) 5929 (30) 7921

5. (1) 16 (2) 4 (3) 233 (4) 325 (5) 388 (6) 904
(7) 93 (8) 282 (9) 264 (10) 249 (11) 518 (12) 4
(13) 100 (14) 1740 (15) 15 (16) 1247 (17) 295 (18) 185
(19) 5726 (20) 668 (21) 17 (22) 124 (23) 255 (24) 54
(25) 425 (26) 42 (27) 10 (28) 17 (29) 0 (30) 13

6. (1) m can be any natural number.

(2) The ones digit of m is even.

(3) The sum of all digits of m can be divided by 3.

(4) The number formed by the last two digits of m can be divided by 4.

(5) The ones digit of m is 5 or 0.

(6) m can be both divided by 2 and 3.

(7) The number formed by the last three digits of m can be divided by 8.

(8) The sum of all digits of m can be divided by 9.

(9) The ones digit of m is 0.

(10) The alternating sum of the digits of m can be divided by 11.

Section 2.2

1. (1) T   (2) F   (3) F   (4) F   (5) T   (6) F   (7) T   (8) F
   (9) F   (10) F   (11) F   (12) F   (13) F   (14) T   (15) F   (16) F
   (17) F   (18) F   (19) T   (20) F

2. (1) D   (2) C   (3) D   (4) A   (5) B   (6) C   (7) C   (8) D
   (9) A   (10) D   (11) C   (12) C   (13) C   (14) B   (15) C   (16) C

3. (1) 1   (2) negative; both positive or both negative   (3) 1; 2
   (4) $4, 6, 8, 9, 10$ (5) 10; 210   (6) 1; mn (7) $b; a$ (8) 1; $n; n; 0$
   (9) 1   (10) $-45 < -21 < -11 < 0 < 17 < 63$

4. (1) -2 (2) 3   (3) -5 (4) 3   (5) 8   (6) -16 (7) -19 (8) -61

5. Primes: $-2, -13, 19, 47, -61, 97$
   Composites: $12, 26, -33, 54, -57, 75, -91, 99, -143, -159, -221$

6. (1) 1   (2) -24 (3) 9   (4) -49 (5) 48   (6) -34
   (7) 1   (8) -11 (9) -180 (10) -24 (11) 730   (12) 5
   (13) 189   (14) -350 (15) 32   (16) -8 (17) 143   (18) -253

7. (1) $-5 < 0 < 4$ (2) $-101 < -100 < -99$ (3) $-73 < -53 < -22 < 69 < 121$
   (4) $2^3 - 3^3 < 3^2 - 5^2 < 1^3 - 2^3 < 2^2 - 3^2 < 1^2 - 2^2$
   (5) $-3^2 < (-2)^3 < (-2)^1 < -(-3)^0 < 2^0 < -(-3)^1 < (-2)^2$

8. Format: GCD; LCM
   (1) 2, 12   (2) 1; 45   (3) 13; 78   (4) 1; 78   (5) 15; 120   (6) 12; 180
   (7) 9; 216   (8) 29; 174   (9) 1; 0   (10) 1; 30   (11) 6; 180   (12) 12; 144

9. (1) $\pm 1, \pm 2, \pm 3, \pm 4, \pm 6, \pm 8, \pm 12, \pm 24$ (2) $\pm 1, \pm 3, \pm 7, \pm 21$

10. 3

11. 20km/h

Section 2.3

1. (1) F   (2) T   (3) F   (4) F   (5) F   (6) T   (7) F   (8) T
   (9) F   (10) F   (11) F   (12) F   (13) F   (14) T   (15) T   (16) T
   (17) F   (18) T   (19) T   (20) F   (21) F   (22) F   (23) T   (24) T

2. (1) C   (2) B   (3) C   (4) D   (5) D   (6) A   (7) C   (8) C
   (9) D   (10) D   (11) D   (12) C

3. Opposites: $1, -7, \dfrac{1}{11}, -\dfrac{2}{3}, \dfrac{9}{4}, \dfrac{1}{2}, -\dfrac{9}{7}, \dfrac{50}{13}, 0, \dfrac{123}{44}, \dfrac{168}{53}$

   Reciprocals: $-1, \dfrac{1}{7}, -11, \dfrac{3}{2}, -\dfrac{4}{9}, -2, \dfrac{7}{9}, -\dfrac{13}{50},$ D.N.E, $-\dfrac{44}{123}, -\dfrac{53}{168}$

4. (1) $\dfrac{3}{4}$ (2) $-\dfrac{4}{5}$ (3) $\dfrac{5}{12}$ (4) $-\dfrac{11}{7}$ (5) $-\dfrac{4}{9}$ (6) $-\dfrac{3}{7}$

   (7) $\dfrac{1}{2}$ (8) $\dfrac{12}{5}$ (9) $-\dfrac{8}{7}$ (10) $-\dfrac{6}{25}$ (11) $\dfrac{11}{17}$ (12) $\dfrac{5}{7}$

5. (1) 4.5 (2) -1.25 (3) -0.875 (4) 0.6 (5) 0.65 (6) -0.24

   (7) -1.5625 (8) 3.025 (9) 0.6 (10) -1.625 (11) $-1.\overline{4}$ (12) $1.\overline{45}$

6. (1) $\dfrac{3}{10}$ (2) $-\dfrac{17}{10}$ (3) $-\dfrac{28}{5}$ (4) $\dfrac{16}{25}$ (5) $\dfrac{71}{50}$ (6) $-\dfrac{17}{8}$

   (7) $\dfrac{9}{4}$ (8) $-\dfrac{35}{8}$ (9) $-\dfrac{31}{20}$ (10) $\dfrac{79}{25}$

7. (1) $\dfrac{1}{6}$ (2) $-\dfrac{26}{5}$ (3) $-\dfrac{3}{11}$ (4) $\dfrac{20}{3}$ (5) $\dfrac{3}{11}$ (6) $-\dfrac{3}{14}$

   (7) $\dfrac{5}{12}$ (8) $-\dfrac{15}{13}$ (9) $\dfrac{9}{4}$ (10) $-\dfrac{19}{21}$ (11) $\dfrac{4}{5}$ (12) $-\dfrac{44}{45}$

   (13) $-\dfrac{10}{7}$ (14) $-\dfrac{10}{33}$ (15) $\dfrac{36}{49}$ (16) $-\dfrac{25}{12}$ (17) $-\dfrac{9}{11}$ (18) $\dfrac{2}{21}$

   (19) $-\dfrac{4}{5}$ (20) $-\dfrac{10}{9}$ (21) $-\dfrac{35}{12}$ (22) $-\dfrac{65}{51}$

8. (1) $\dfrac{9}{6}, -\dfrac{5}{6}$ (2) $\dfrac{9}{42}, \dfrac{4}{42}$ (3) $-\dfrac{9}{24}, -\dfrac{28}{24}, \dfrac{10}{24}$ (4) $\dfrac{51}{36}, -\dfrac{63}{36}, -\dfrac{20}{36}, -\dfrac{24}{36}$

9. (1) -7.2 (2) -1.2 (3) $-\dfrac{29}{12}$ (4) $-\dfrac{8}{3}$ (5) $\dfrac{5}{6}$ (6) -1.3

   (7) $\dfrac{77}{6}$ (8) $\dfrac{19}{12}$ (9) $-\dfrac{8}{15}$ (10) -5.3 (11) $-\dfrac{11}{12}$ (12) $-\dfrac{1}{3}$

   (13) 40 (14) -6 (15) -49.2 (16) $-\dfrac{2}{5}$ (17) -50 (18) -250

   (19) $\dfrac{5}{6}$ (20) $-\dfrac{44}{15}$ (21) $-\dfrac{16}{77}$ (22) $-\dfrac{25}{4}$ (23) $-\dfrac{5}{2}$ (24) -1

10. (1) $\dfrac{9}{4}$ (2) $\dfrac{1}{49}$ (3) $-\dfrac{27}{64}$ (4) $\dfrac{25}{9}$

   (5) $\dfrac{8}{125}$ (6) $\dfrac{256}{121}$ (7) $-\dfrac{144}{25}$ (8) $\dfrac{1000}{27}$

11. (1) $-\dfrac{90}{7}$ (2) 6 (3) -125 (4) $-\dfrac{25}{4}$ (5) $\dfrac{1}{4}$

   (6) -26 (7) $\dfrac{3}{5}$ (8) $\dfrac{1}{4}$ (9) $-\dfrac{5}{4}$ (10) $-\dfrac{41}{5}$

12. (1) $-1 < -0.01 < 0 < 0.5$ (2) $-\dfrac{1}{2} < -\dfrac{1}{3} < \dfrac{1}{3} < \dfrac{1}{2}$ (3) $-\dfrac{6}{7} < -1 < -\dfrac{7}{6}$

   (4) $-\dfrac{1}{3} < -0.3 < -\dfrac{1}{4}$ (5) $\dfrac{4}{15} < \dfrac{3}{11}$ (6) $-\dfrac{2}{3} < -\dfrac{5}{8}$

   (7) $-\dfrac{18}{17} < -\dfrac{21}{20} < -\dfrac{24}{23}$ (8) $\dfrac{17}{69} > \dfrac{15}{67}$ (9) $-\dfrac{13}{15} < -\dfrac{11}{13} < -\dfrac{7}{9}$

   (10) $-\dfrac{46}{94} < -\dfrac{32}{66} < -\dfrac{36}{75}$

13. $-4, -3, -2, -1, 0, 1, 2$

14. $-\dfrac{1}{3}$

15. $-\dfrac{125}{28}$

16. $\dfrac{625}{81}$

Section 2.4

1. (1) F (2) T (3) F (4) T (5) F (6) T (7) F (8) F

(9) F (10) F (11) F (12) F (13) T (14) T (15) F (16) F

(17) F (18) F (19) T (20) T (21) F (22) T (23) T (24) F

(25) T (26) T (27) F (28) T (29) F (30) F

2. (1) D (2) C (3) B (4) D (5) C (6) D (7) C (8) A

(9) B (10) B (11) C (12) B (13) A (14) D (15) D (16) C

(17) C (18) B (19) C (20) D (21) C (22) A (23) D (24) C

(25) D (26) C (27) C (28) C (29) B (30) D (31) D (32) B

3. negative numbers: $\{-1, -\sqrt{3}, -\dfrac{1}{3}, -1.24, -0.6, -41, -\pi\}$

fractions: $\{-1, \dfrac{1}{2}, -\dfrac{1}{3}, 0, 2, 1\%, -1.24, \dfrac{9}{7}, 0.2\overline{43}, -0.6, -41, 127\%, 4.13\overline{5}\}$

integers: $\{-1, 0, 2, -41\}$

nonnegative numbers: $\{\sqrt{2}, \dfrac{1}{\pi}, \dfrac{1}{2}, 0, 2, 1\%, \dfrac{9}{7}, 0.2\overline{43}, 127\%, 4.13\overline{5}\}$

natural numbers: $\{2\}$

whole numbers: $\{0, 2\}$

positive rational numbers: $\{\dfrac{1}{2}, 2, 1\%, \dfrac{9}{7}, 0.2\overline{43}, 127\%, 4.13\overline{5}\}$

nonpositive irrational numbers: $\{-\sqrt{3}, -\pi\}$

4. A real number line is a line provided with the origin, positive direction and unit length.

5. (1) $<$ (2) $<$ (3) $<, >$ (4) $<, >$ (5) $>$ (6) $<$

(7) $>, <$ (8) $<$ (9) $>, <$ (10) $>$ (11) $\leq$ (12) $\geq$

6. (1) $>$ (2) $=$ (3) $=$ (4) $<$ (5) $>$ (6) $=$ (7) $\leq$ (8) $\geq$

7. (1) $-\dfrac{49}{12}$ (2) ± 4 (3) 3 (4) D.N.E (5) 1 (6) 3

(7) 5 (8) 4 (9) 3 (10) 2.3, 1.7 (11) $\pm 1, \pm 7$ (12) ± 5

(13) 8 (14) 3, 13 (15) $-\dfrac{23}{6}$ (16) ± 1

8. (1) 0 (2) 1.8 (3) $-4\dfrac{7}{8}$ (4) 0.4

9. (1) commutative property of addition (2) commutative property of multiplication

(3) inverse property of addition

(4) inverse property of multiplication

(5) identity property of addition

(6) identity property of multiplication

(7) associative property of addition

(8) associative property of multiplication

(9) distributive property

(10) multiplication property of -1

(11) multiplication property of 0

(12) multiplication property of opposites

10. (1) $-a+b$ (2) $a-b$ (3) $a+b$ (4) $a-b+c+d$

(5) $a+b+c$ (6) $-a+bc$ (7) $a-b+c$ (8) $-a+b-c-d$

(9) $a-b+c$ (10) $-a-b+c$

11. (1) $-ab+ac$ (2) $ab-ac$ (3) $ab+ac$ (4) $-ab-ac+ad$

12. (1) 0   (2) -4 (3) 2   (4) 7   (5) -11.3 (6) -4

(7) -2 (8) $\dfrac{7}{5}$ (9) 0   (10) 37   (11) 5   (12) 10

(13) -279.2 (14) $\dfrac{29}{7}$ (15) $-\dfrac{7}{4}$ (16) 16   (17) -125 (18) -423

13. (1) -108 (2) -166 (3) -168 (4) $-\dfrac{1}{6}$ (5) -4

(6) -11 (7) 17   (8) -10 (9) -25 (10) -12

(11) $\dfrac{4}{15}$ (12) $-\dfrac{67}{9}$ (13) $\dfrac{1275}{196}$ (14) $-\dfrac{245}{8}$

14. $-\pi < -3 < -2 < -\sqrt{3} < -1.5 < -\sqrt{2} < -1$

15. 8 or -2

16. $7\dfrac{19}{21}$

17. $\pm 2, \pm 3, \pm 4$

18. $x-y < x < x+y < y < y-x$

19. $-\dfrac{1}{3}$

20. $a = \pm b$ and $b \neq 0$

21. 0

22. (1) $0 \neq a \in \mathbb{R}$; (2) $a = 0$

23. -17

24. 0

25. 5

Chapter 3

Section 3.1

1. (1) F   (2) F   (3) F   (4) F   (5) F   (6) T   (7) F   (8) T
   (9) F   (10) F   (11) F   (12) T

2. (1) D   (2) B   (3) C   (4) A   (5) D   (6) D   (7) B   (8) D
   (9) D   (10) B   (11) D   (12) A

3. (1) $5x + \dfrac{y}{7}$ (2) $3a + 2b - \dfrac{3}{5}$ (3) $-t - 2$

   (4) $\dfrac{a}{2} - 2b$ (5) $-\dfrac{(-a)^2}{3}$ (6) $\dfrac{1}{a^3 + b^3}$

   (7) $|a - b| - 3$ (8) $2x^2 + b^2$ (9) $\dfrac{1}{-|a + b|}$

   (10) $m + \dfrac{15}{m}$ (11) $n(18 - n)$ (12) $\dfrac{1}{3}(2a - b^2)$

4. Monomials: $0, -\dfrac{xy^2}{4}, m, \pi, 5a^3b^6c^6d^2, -8$

   Non-polynomials: $\dfrac{1}{3a}, x + \dfrac{1}{y}, \dfrac{3y^2}{x}, \dfrac{b + 1}{a}, 4 - \dfrac{2m}{3n}, \sqrt{x + 1}$

5. Format: degree, coefficient

   (1) $5, \dfrac{2}{3}$ (2) $2, 15$ (3) $-\infty, 0$ (4) $16, 1$ (5) $3, -\dfrac{3}{4}$ (6) $0, -5$

6. Format: degree, constant term

   (1) $3, 0$ (2) $4, 1$ (3) $1, 0$ (4) $1, 5$ (5) $0, -4$ (6) $4, -7$
   (7) $7, -1$ (8) $4, 0$ (9) $-\infty, 0$

7. (1) $2a + 2b, ab$ (2) $4x, x^2$ (3) $\pi D, \dfrac{\pi D^2}{4}$ (4) $\left(\dfrac{s}{4}\right)^2$

   (5) $\pi r^2 + \pi(10 - r)^2$ (6) $\dfrac{1}{2}(a + 3a)(3a - 2) = 2a(3a - 2)$

8. (1) -1 (2) 0 (3) -20 (4) -18 (5) $\dfrac{1}{4}$ (6) 122 (7) 0.01 (8) $-\dfrac{11}{48}$

   (9) 10 (10) 13 (11) 4 (12) $\dfrac{4}{13}$ (13) $\dfrac{7}{19}$ (14) $-\dfrac{35}{36}$ (15) -1

9. (1) 17 (2) 8 (3) -1 (4) -1 (5) 2 (6) $\dfrac{7}{2}$

Section 3.2

1. (1) D   (2) A   (3) D   (4) C   (5) C   (6) C   (7) B   (8) D

2. (1) $5xy$ (2) $2x^3 - y^3$ (3) $5a + b$
   (4) $3m^2 - 3n^2 + 3$ (5) $-3x^2 + 7x - 1$ (6) $-x + y$

(7) $-4a + 7b$

(8) $x^2 + 2xy + y^2$

(9) $-4x^2y + 6xy^2$

(10) $-19b^2$

(11) $2a^3b - 2a^2b$

(12) $-5a$

(13) $3a + b$

(14) $-3x^4 - x^3 + 5x^2 + 2x - 2$

(15) $-\dfrac{a}{6} + 10b$

(16) $16a^2 - 7b^2 - 12c^2$

(17) $\dfrac{5}{2}a^2 - 9ab$

(18) $-\dfrac{4}{3}x^2y^2 - xy^2 - 3$

(19) $b^2 + b + 1$

(20) $-8x^2y - 3xy - 4x + 2x^2y^2$

(21) $16x^2 - 24xy + 25y^2$

(22) 2

(23) $-2x^3 + 3x^2 + x - 4$

(24) $4x^4 - x^3 - 2x^2 - 2x + 9$

(25) $16a^2 - 10ab - 12a + 6b$

3.   (1) 31   (2) 0   (3) -16 or 0 (4) 27   (5) 14

  (6) $\dfrac{3}{4}$ (7) 19   (8) $-\dfrac{8}{9}$ (9) $-\dfrac{88}{3}$ (10) 60

Section 3.3

1.   (1) T   (2) F   (3) F   (4) F   (5) T   (6) F   (7) T   (8) F

  (9) T   (10) F

2.   (1) C   (2) D   (3) C   (4) D   (5) D   (6) C   (7) C   (8) C

  (9) B   (10) C   (11) D

3.   (1) 1   (2) 1   (3) -1 (4) $-\dfrac{1}{32}$ (5) $\dfrac{1}{9}$ (6) $-\dfrac{1}{8}$

  (7) $\dfrac{1}{81}$ (8) $-\dfrac{1}{64}$ (9) $\dfrac{1}{125}$ (10) $\dfrac{1}{625}$ (11) 0.01   (12) -0.001

  (13) 49   (14) -27 (15) $\dfrac{3}{8}$ (16) -48

4.   (1) $-a^{3n-5}$ (2) $(x-y)^{n+3}$ (3) $3m$ (4) $2a^{2m}b^{3n}c$

  (5) $2m$ (6) -4 (7) 9   (8) $3ab^5$

  (9) $\dfrac{a^4}{b^2}, \dfrac{a}{3b^2}, \dfrac{b}{a}$ (10) t^2, t^p

5.   (1) a^6 (2) y^{n-m} (3) a^{mn+1} (4) $(y-x)^3$

  (5) $\dfrac{1}{16}x^4y^8$ (6) x^{2mn} (7) $x^{m^2+m^2}$ (8) 1

  (9) -1 (10) $-\dfrac{1}{6561}$ (11) $\dfrac{b^6}{a^9}$ (12) a^9

  (13) x^2 (14) $\dfrac{a^5}{b^5}$ (15) 1   (16) 0

  (17) a^{n^2+2n} (18) y^{66} (19) $(-1)^p(m-n)^{8p+5}$ (20) $(x^2-x+1)^{m+2n+1}$

Section 3.4

1.   (1) A   (2) C   (3) D   (4) B   (5) C   (6) C   (7) B   (8) D

(9) C   (10) D

2. (1) $2x^6y^3$ (2) $18a^4b^3c$ (3) $-8a^3 - 12a^2 + 4a$

    (4) $6x^3y^2 - 12x^2y^2$ (5) $\frac{1}{3}a^2b^3 - a^2b^2 + \frac{2}{3}ab^2$ (6) $48x^6 + 7x^4$

    (7) $3m^2 - 7mn + 2n^2$ (8) $-a^{19m}$ (9) $x^3 + y^3$

    (10) $5x^3 + 8x^2 + 12x + 15$ (11) $\frac{x^4}{3} - 3a^4$ (12) $-3a^2 - 7a + 19$

    (13) $13x^2y^4$ (14) $12x^7 - 3x^6 - 8x^5 + 21x^4 - 13x^2 - 7x^2 + 5x - 15$

    (15) $-6a^3b - 2a^2b^2 + 5ab$ (16) $3x^2 + 24x - 35$ (17) $2x^2 + 5xy + 8x - 3y^2 - 4y$

    (18) $6yz$ (19) $-2x^2 - 4$

3. (1) 81   (2) -30 (3) 1   (4) 246

4. -14

5. $-6y^8 + 24xy^7$

6. $-\frac{1}{4}$

7. $m = -5, n = 6$

9. (1) m (2) any nonnegative integer $k \leq m$ (3) $m + n$

Section 3.5

1. (1) F   (2) F   (3) T   (4) T   (5) F   (6) F   (7) T   (8) T

    (9) F   (10) T

2. (1) B   (2) A   (3) C   (4) C   (5) A   (6) C   (7) C   (8) D

    (9) A   (10) A   (11) B   (12) B

3. (1) $-4ab$ (2) $4ab$ (3) $4b^4$

    (4) $25x^6 - 10x^3y^2 + 4y^4$ (5) y^2, y^4, y^6 (6) $\frac{81}{16}, 2x - \frac{9}{4}$

    (7) $x^4 + 4x^2a^2 + 16a^4$ (8) $a + (b + c)^2$

4. (1) $a^3 - 8$ (2) $a^3 + b^3$ (3) $a^3 + 8b^3$

    (4) $w^2 - 4v^2$ (5) $13x^2 - 25y^2$ (6) $-x^2y - xy^2$

    (7) $a^4 + a^2b^2 + b^4$ (8) $4x^2 - 12xy + 9y^2$ (9) $9a^2 + 6ab + b^2$

    (10) $2a^4 + 18a^2$ (11) 3   (12) $x^4 - 1$

    (13) $-4x^2y^2$ (14) $x^6 - \frac{1}{27^2}$ (15) $-3x^2 - 2xy + 5y^2$

    (16) $-4yz$ (17) $2a^4 - 18a^2$ (18) $x^2 - 6xy + 9y^2$

    (19) $-15x^4 - 42x^2 + 624$ (20) $a^8 + a^4 + 1$ (21) $x^4 - y^4$

    (22) 3

5. (1) $\frac{255}{128}$ (2) $\frac{49}{4}$ (3) -60 (4) -17

(5) 2 $\qquad$ (6) $16a^2 + 8a$ $\qquad$ (7) 23 $\qquad$ (8) 18

Section 3.6

1. (1) $\dfrac{a}{3}$ $\qquad$ (2) $2x^2y$ $\qquad$ (3) $3m^4$ $\qquad$ (4) x^2

(5) $18xyz^2$ $\qquad$ (6) $5xy^2$ $\qquad$ (7) y^{n+4} $\qquad$ (8) $9ab^3$

(9) $\dfrac{2}{3}ab$ $\qquad$ (10) x^4 $\qquad$ (11) $-8a^6$ $\qquad$ (12) $2a^2 - 3$

(13) x^{cm+dm} $\qquad$ (14) $-2a + 3$ $\qquad$ (15) $\dfrac{x^2}{3} + 2xy - \dfrac{y^2}{3}$ $\quad$ (16) $-\dfrac{16}{3}xz^2 + 6x^2y + \dfrac{8}{3}y^2z$

2. (1) $3x - 5x^2 + 7x^4$ $\qquad$ (2) $-10 - 5y + 5y^2 + y^3 - 2y^4$

(3) $4 - 2x - 5x^2 - x^3$ $\qquad$ (4) $-7 + 5a + 12a^2 + 15a^3 - 12a^5 + 21a^6 - 8a^7$

3. (1) $2x^2 + 11x - 3$ $\qquad$ (2) $-65z^6 - 23z^4 + 37z^2 - 9$

(3) $5b^9 + b^8 - 4b^4 - 3b^2 + 2$ $\qquad$ (4) $-35y^5 + 22y^4 + 76y^3 - 62y^2 - 51y - 100$

4. (1) F $\quad$ (2) F $\quad$ (3) T $\quad$ (4) F $\quad$ (5) T $\quad$ (6) T $\quad$ (7) F $\quad$ (8) T

(9) F $\quad$ (10) T

5. (1) A $\quad$ (2) A $\quad$ (3) B $\quad$ (4) D $\quad$ (5) D $\quad$ (6) C $\quad$ (7) C $\quad$ (8) C

6. (1) -35 $\quad$ (2) -12 $\quad$ (3) 39 $\quad$ (4) 0 $\quad$ (5) -15 $\quad$ (6) -0.5

7. (1) $x - y$ $\qquad$ (2) $2x^2 + 4x - 3$ $\quad$ (3) 6 $\qquad$ (4) -18 $\qquad$ (5) $-10, -3$

(6) $x^2 + 3x + 1$ $\quad$ (7) $-6, 8$ $\qquad$ (8) 2 $\qquad$ (9) -9 $\qquad$ (10) -5

8. (1) $x + 2x - 5x^2 = (x - 2)(-5x - 8) - 12$

(2) $x^5 - 4x^3 + 2x^2 - 1 = (x + 2)(x^4 - 2x^3 + 2x - 4) + 7$

(3) $4x^3 + 2x^2 - 1 = (2x - 4)(2x^2 + 5x + 10) + 39$

(4) $2x^4 - 5x^3 - 26x^2 - x + 28 = (2x^2 + x - 3)(x^2 - 3x - 10) - 2$

(5) $x^4 + 1 = (x^3 + x^2 + x)(x - 1) + x + 1$

(6) $x - 6x^3 + x^6 - 18 = (x^2 - 2)(x^4 + 2x^2 - 6x + 4) - 11x - 10$

(7) $6x^3 - 8x - 5 = (3x + 2)(2x^2 - \dfrac{4}{3}x - \dfrac{16}{9}) - \dfrac{13}{9}$

(8) $2x^3 + 3x - 3 + 9x^2 = (4x + x^2 - 3)(2x + 1) + 5x$

(9) $6x^4 - 3x^2 - 7x - 3 = (2x^2 - x - 2)(3x^2 + 3) - 4x + 3$

(10) $x^5 - 2x^4 - 4x^3 + 1 = (x^3 + 2x + 1)(x^2 - 2x - 6) + 3x^2 + 14x + 6$

(11) $2x^4 + 7x^3 - 12x^2 - 27x = (2x^2 + 3x)(x^2 + 2x + 9)$

(12) $6x^6 - 4x^5 + 2x^4 - x - 5 = (2x^4 - x - 3)(3x^2 - 2x + 1) + 3x^3 + 7x^2 - 6x - 2$

9. (1) $27x^6 + x$ $\qquad$ (2) $2x^4 + x^3 + x$ $\quad$ (3) $2x^2 + 5x + 10$ $\quad$ (4) 7 $\qquad$ (5) $3x^2 - 4x + 5$

Chapter 4

Section 4.1

1.   (1) F   (2) T   (3) T   (4) T   (5) F   (6) T   (7) F   (8) T
     (9) F   (10) T

2.   (1) A   (2) D   (3) B   (4) A   (5) C   (6) D

3.   (1) $3(x-2)$ (2) $5a(a+3)$ (3) $2\pi(R-r)$
     (4) $3x^2y(2+3xy)$ (5) $-m(m^2+5m-15)$ (6) $-2xy(6xy-3x^2y^2+2)$
     (7) $3a^2b(2-a-4a^2b)$ (8) $x^2(x^2-x+2)$ (9) $a^m(ab+c^2)$
     (10) $-8m^2(m^2+3m-6)$ (11) $-3xy(1+5z-9ab)$ (12) $-3x^2yz^2(1-4xy-3z)$
     (13) $9a^2bc(2a-5bc+3b^2)$ (14) $7ab(x+3by-7axy)$

4.   (1) $0.25(a-b)(2x+1)$ (2) $(2x-3y)(a^2+b^2)$ (3) $(x-y)(a-2b)$
     (4) $(a-b)^2(3+a-b)$ (5) $(x-y)(a-b-c)$ (6) $4(a-b)^2(5x-4ay+4by)$
     (7) $(a-3)(a+b-c)$ (8) $xy(a-2b)(x+2)$ (9) $a^2(b+c)$
     (10) $-4mn(m+2n)$ (11) $y(2x+1)(2x+y+1)$ (12) $(m-n)(x-y-z)$
     (13) $(a-m)(a-n)(m+n)$ (14) $2x^2(x+2y)$ (15) $a(a-x)(b-x)(1+b)$
     (16) $(2m-n)^2$ (17) $2(a+b)(x+z)$ (18) $(a-2b)(7a-2b)$

Section 4.2

1.   (1) $(m+1)(m-1)$ (2) $(x+2y)(x-2y)$ (3) $(a-4)(a+4)$
     (4) $(5-2b)(5+2b)$ (5) $(6p-7q)(6p+7q)$ (6) $-(2x+3y)(4x+3y)$
     (7) $3y(2x+y)$ (8) $8a^3(a-3)(a+3)$ (9) $m^2(m-n)(m+n)$
     (10) $(5a^2x^5+3b^3y^4)(5a^2x^5-3b^3y^4)$ (11) $(-7a+11b-16c)(11a-7b+20c)$
     (12) $16(3a+4b)(3a-4b)$ (13) $(2x^4-yz)(2x^4+yz)(4x^8+y^2z^2)$
     (14) $(5ab^2c^8-1)(5ab^2c^8+1)$ (15) $(x-y)(x+y)(x^2+y^2)$
     (16) $(a-4b)(a+2b)(a^2-2ab+10b^2)$ (17) $(3a+2b)(3a-2b)(9a^2+4b^2)$
     (18) $(a+c)(2b-a-c)$ (19) $a^{m-1}(1-a)(1+a)$
     (20) $4(2m-n)(2n-m)$

2.   (1) $(a-1)^2$ (2) $(t+2)^2$ (3) $(x-2y)^2$
     (4) $(3m-1)^2$ (5) $(4x-5y)^2$ (6) $(1-3ab^3)^2$
     (7) $(x^2+1)^2$ (8) $(2a-1)^2$ (9) $2x(x-y)^2$
     (10) $(xy-7)^2$ (11) $xm(mn+1)^2$ (12) $(x+y+2)^2$
     (13) $-(a-b)^2$ (14) $-(x-y-5)^2$ (15) $(2x-5y)^2$
     (16) $(x+y)^2(x-y)^2$ (17) $(2x-3y-1)^2$ (18) $(x^2+x+2)^2$

(19) $(a - 7b)^2$ (20) $(a + b)^2 c^2$

3.   (1) $(x + 2)(x^2 - 2x + 4)$ (2) $(1 - y)(1 + y + y^2)$ (3) $(4m - 1)(16m^2 + 4m + 1)$

(4) $(2x + 3)(4x^2 - 6x + 9)$ (5) $(a - b^2)(a^2 + ab^2 + b^4)$ (6) $(3x - 2y)(9x^2 + 6xy + 4y^2)$

(7) $9(x + y)(x^2 + xy + y^2)$ (8) $(1 - 2a + 2b)(1 + 2a - 2b + 4a^2 - 8ab + 4b^2)$

(9) $(a + 1)(7a^2 + 5a + 1)$ (10) $2(3x^2 + 1)$

(11) $n(3m^2 - 9mn + 7n^2)$ (12) $(\frac{x}{2} - \frac{y}{3})(\frac{x^2}{4} + \frac{xy}{6} + \frac{y^2}{9})$

(13) $(2a - 5b^2 x)(4a^2 + 10ab^2 x + 25b^4 x^2)$ (14) $(x + y + 2)(x^2 + 2xy + y^2 - 2x - 2y + 4)$

(15) $(z - y - 2x)(x^2 + y^2 + z^2 + xy + yz - xz)$

(16) $(x + y + 5)(x^2 + 2xy + y^2 - 5x - 5y + 25)$

(17) $(2x + 2y - 1)(4x^2 + 8xy + 4y^2 + 2x + 2y + 1)$

(18) $(a - b)(a^2 + ab + b^2)(a^6 + a^3 b^3 + b^6)$

4.   (1) $(a - 2b)(x + y)(x - y)$ (2) $(a - b)^2 (a + b)^2$ (3) $(a^2 - 3ab + b^2)^2$

(4) $(2x - y)(4x^2 + 2xy + y^2)(3a + 4b)(9a^2 - 12ab + 16b^2)$

(5) $(x - y - 1)(x - y + 1)(x + y - 1)(x + y + 1)$ (6) $3(x - 1)^2 (x + 1)^2$

(7) $x^2 (a^2 + 2y)^2$ (8) $(x - 1)^4$ (9) $(x - y + 6z)^2$

(10) $(ax + by + ay - bx)^2$ (11) $3(x - 2y)(x + 2y)(x^2 + 4y^2)$

(12) $(a - \frac{1}{a})^2 (a + \frac{1}{a})^2$ (13) $(x + y)(x^2 - xy + y^2)(x^6 - x^3 y^3 + y^6)$

(14) $(x - y)(x + y)(x^2 + y^2)(x^2 - xy + y^2)(x^2 + xy + y^2)(x^4 - x^2 y^2 + y^4)$

(15) $(x + y)^2 (x - y)^2 (x^2 - xy + y^2)(x^2 + xy + y^2)$ (16) $\frac{m}{9}(m + 3)(m - 3)$

5.   (1) $(x + 1)^3$ (2) $(2 - b)^3$ (3) $y^2 (y - 1)^3$

(4) $(a + b)^3 (a - b)^3$ (5) $6x^2 y(xy + 1)^3$ (6) $8(a^2 + b^2)^3$

Section 4.3

1.   (1) $(x + 6)(x - 1)$ (2) $(x - 6)(x + 1)$ (3) $(x + 2)(x + 3)$

(4) $(x - 6)(x + 5)$ (5) $(x + 24)(x - 6)$ (6) $(x + 24)(x + 6)$

(7) $(x - 5)(2x + 3)$ (8) $(2x - 1)(9x - 5)$ (9) $(2x + 3)(3x - 2)$

(10) $(2x - 3)(3x + 2)$ (11) $(3x - 2)(2x - 3)$ (12) $(3 - 5a)(2 + 7a)$

(13) $(x - 11)(x + 7)$ (14) $(2x - 3)(3x + 1)$ (15) $(x - 3)(5x - 2)$

(16) $(2x + 3)(3x - 5)$ (17) $(4 - x)(3 + 2x)$ (18) $(x - 25)(x + 3)$

(19) $(3x - 7)(2x + 5)$ (20) $(x - 5)(6x - 7)$

2.   (1) $(x + 4y)(x - 7y)$ (2) $(x - 3y)(x - y)$ (3) $(4x - 7y)(5x - 2y)$

(4) $-(3x - y)(x - 3y)$ (5) $(x + 15y)(x - 2y)$ (6) $(2a + 3b)(3a - 5b)$

(7) $(2a - 7b)(3a + 5b)$ (8) $(2a - 5b)(3a - 7b)$ (9) $(2a - 7b)(3a - 5b)$

(10) $-(7mn+1)(5mn-6)$ (11) $(xy-2)(xy+8)$ (12) $(6m^2-5n^2)(3m^2-n^2)$

3. (1) $(x^2+4)(x^2-2)$ (2) $3(x^2+3)(x+1)(x-1)$

(3) $x(x-2)(x+2)(x^2+2)$ (4) $(x-2)(x+2)(x-3)(x+3)$

(5) $(x^2+2y^2)(x-y)(x+y)$ (6) $(x+1)(x+2)(x-1)(x-2)$

(7) $(m-n-2)(m-n-1)$ (8) $(x-3)(x-4)(x^2-7x-2)$

(9) $(x+4)(x-3)(x+2)(x-1)$ (10) $(m+4)^2(m-1)^2$

(11) $(x-a)(x-b)$ (12) $(ax-b)(bx-a)$

(13) $(x+1)(ax-bx+a+b)$ (14) $(x+m)(x-\dfrac{1}{m})$

(15) $(ax+bx-1)(x-a+b)$ (16) $(x+y+z)(x-y-z)(x+y-z)(x-y+z)$

(17) $(x+y+5)(x-y-2)$

Section 4.4

1. (1) $(m+n)(2x-3y)$ (2) $(2a-b)(m+2n)$ (3) $(x-2)(x+1)$

(4) $(q-7)(p+3)$ (5) $(2h+f)(f-1)$ (6) $(3x+2)(y-1)$

(7) $(m+n)(m+4)$ (8) $(p-q)(2x-y)$ (9) $(x-y)(n-2)$

(10) $(b-c)(3a+c)$

2. (1) $(x-\dfrac{1}{3}y)^2$ (2) $(x+\dfrac{1}{2})(x^2-\dfrac{x}{2}+\dfrac{1}{4})$ (3) $(\dfrac{a}{2}-b)^2$

(4) $a^2(a-\dfrac{3}{2})^2$ (5) $(2x-\dfrac{1}{3})(2x+\dfrac{1}{3})$ (6) $(x-\dfrac{y}{2}+1)(x-\dfrac{y}{2}-1)$

3. (1) $3mx(xy+1)^2$ (2) $-(x-y-5)^2$ (3) $(2a+b)(8a^2-4ab+2b^2-1)$

(4) $xy(x-y)$ (5) $(x-y+2)(x-y-2)$ (6) $(a-b-1)^2$

(7) $(x+1)(x-1)(y+1)(y-1)$(8) $(a-b)(b-c)(a-c)$ (9) $(x-2y+3)(x-2y-2)$

(10) $(a+b)(x+y)[(a^2+b^2)xy-3abxy+(x^2+y^2)ab]$

4. (1) $-(xy+x-y+1)(xy-x+y+1)$ (2) $(x^2+x-1)(x^2+x+4)$

(3) $(x-2)(x+2)(x^2+x+4)$ (4) $(x-1)(x^2+x-8)$

(5) $(x-y)(x+y-2)$ (6) $(x-1)(x+2)^2$ (7) $(a^2+a+1)^2$

(8) $(x-1)(x^2+x+1)(x^6+2x^3+3)$ (9) $(x+1)(x+2)(x+3)$

(10) $-a^3(a^2+a+1)(a^2+a-1)$ (11) $b^2(a^2+a+1)(a^2-a+1)$

(12) $(2x^2+2x+1)(2x^2-2x+1)$ (13) $(x-1)(x^2+x+1)(x^2-x+1)$

(14) $(x^2+6x+18)(x^2-6x+18)$ (15) $(x+1)(x^2+2x+2)$

5. (16) $(3a+b)(3a-b-2)$ (17) $(a+b-c)^2$ (18) $(x^2+y^2+z^2)^2$

5. $a^3+b^3-3ab+1=(a+b+1)(a^2+b^2-ab-a-b+1)$

Chapter 5

Section 5.1

1. (1) T (2) F (3) T (4) F (5) T (6) T (7) T (8) T
   (9) F (10) F (11) T (12) F (13) T (14) F (15) T (16) T

2. (1) D (2) C (3) A (4) B (5) C (6) C (7) D

3. (1) $7x - 6 = -9$ (2) $5y - 7 = 3y + 8$ (3) $\dfrac{x}{3} + 4 = 2x - 9$ (4) $|4x| - (-3) = 2$

   (5) $-x - 7 \times 7 = 8$ (6) $|3x - (-4)| = 9$ (7) $(7x - 1) = \dfrac{2}{3} + 3(2x + 1)$

4. (1) multiply by 3 on both sides   (2) subtract $2x$ on both sides
   (3) add $7x + 5$ on both sides   (4) add $-x + 3$ on both sides
   (5) multiply by 7 on both sides   (6) multiply by $\dfrac{1}{10}$ on both sides
   (7) multiply by 4 on both sides

5. (1) F. Multiply by -3 on both sides: $x = -18$
   (2) F. Subtract $x - 2$ on both sides: $x - 2 = 0$
   (3) F. Multiply by 6 on both sides: $9x - 6 = 4x$
   (4) F. Multiply by $\dfrac{1}{10}$ on both sides: $\dfrac{x}{2} - \dfrac{y}{3} = \dfrac{0.4}{10} = \dfrac{1}{25}$
   (5) F. Multiply by 12 on both sides: $8a = 36 - 3(a - 1) = 36 - 3a + 3$
   (6) T.

Section 5.2

1. (1) F (2) F (3) T (4) F (5) T (6) F (7) F (8) F
   (9) T (10) F (11) F (12) F (13) F (14) F (15) T

2. (1) D (2) C (3) D (4) A (5) D (6) B (7) D (8) B
   (9) C (10) A

3. (1) $\{7\}$ (2) $\{\dfrac{13}{2}\}$ (3) $\{-5\}$ (4) $\{\dfrac{29}{9}\}$ (5) $\{-\dfrac{9}{2}\}$
   (6) $\{-1\}$ (7) $\mathbb{R}$ (8) $\{-6\}$ (9) $\{6\}$ (10) $\{4\}$
   (11) $\{-\dfrac{7}{2}\}$ (12) $\{\dfrac{143}{43}\}$ (13) $\{\dfrac{1}{2}\}$ (14) $\{4\}$ (15) $\{-\dfrac{5}{2}\}$
   (16) $\{\dfrac{126}{25}\}$ (17) $\{-1\}$ (18) $\emptyset$ (19) $\{-3\}$ (20) $\{\dfrac{30}{7}\}$

4. (1) $x = \dfrac{1}{12}$ (2) $y = 72$ (3) $x = \dfrac{2}{3}$ (4) $x = \dfrac{8}{19}$ (5) $x = -\dfrac{1}{10}$
   (6) $x = \dfrac{1}{13}$ (7) $x = 5$ (8) $y = 1$ (9) $y = \dfrac{5}{7}$ (10) $x = \dfrac{163}{11}$

(11) $x = \dfrac{17}{7}$ (12) $x = -\dfrac{9}{4}$ (13) $x = \dfrac{52}{9}$ (14) $x = -80$ (15) $x = \dfrac{75}{23}$

(16) $y = \dfrac{38}{13}$ (17) $y = -\dfrac{48}{41}$ (18) $y = -\dfrac{128}{23}$ (19) $x = \dfrac{11}{5}$ (20) $x = -\dfrac{38}{45}$

(21) $x = 2$ (22) $x = -\dfrac{14}{45}$

5. (1) $b = 2a - c$ (2) $4 = \dfrac{C}{2\pi} - R$ (3) $h = \dfrac{S - 2\pi r^2}{2\pi r}$ (4) $b = \dfrac{2S}{h} - a$

(5) $a = \dfrac{2S}{b}$ (6) $h = \dfrac{3V}{\pi r^2}$ (7) $r = \dfrac{S}{2\pi h}$ (8) $R = \dfrac{R_1 R_2}{R_1 + R_2}$

(9) $a = \dfrac{2(S - vt)}{t^2}$ (10) $r = \dfrac{mv^2}{F}$ (11) $F = \dfrac{9C}{5} + 32$ (12) $b = \dfrac{y - ax^2 - c}{2x}$

(13) $c = \dfrac{b}{ab - 1}$

6. (1) $x = -b^2$ (2) $x = \dfrac{m^2 + n^2}{m - n}$ (3) $x = \dfrac{a^2 + ab + b^2}{a - b}$

(4) $x = \dfrac{a}{b + c - 1}$ (5) $x = m$ (6) $x = \dfrac{2mn}{n - m}$

7. $11, 13, 15$

8. $\dfrac{15}{8}$

9. $26s$

10. $32; 9$

11. $\dfrac{15}{4}$ hours

12. $\dfrac{25}{2}$ days

13. 37

14. 172km

15. Lulu: 8; Nunu: 6

16. 12π

17. 2km/h

Section 5.3

1. (1) F   (2) T   (3) F   (4) T   (5) F   (6) F   (7) T   (8) F

(9) F   (10) F   (11) T

2. (1) C   (2) B   (3) D   (4) D   (5) A   (6) A

3. (1) $x = \pm 9$ (2) $x = -\dfrac{1}{2}, \dfrac{3}{2}$ (3) $x = \pm 3$ (4) no solution

(5) $x = 3, -9$ (6) $x = \pm 8$ (7) $x = 12, -40$ (8) $x = 5$

(9) $x = 6$ or $-\dfrac{1}{2}$ (10) no solution (11) $x = \dfrac{1}{2}$ (12) $x = 1, -\dfrac{7}{5}$

(13) no solution (14) $x = 2$ (15) $x = -6, 2$ (16) $x = \dfrac{9}{2}$

(17) $x = 1, -\dfrac{11}{6}$ (18) $x = -1, 6$

4. (1) ± 6 (2) $-1, 3, -8$ (3) $-2, 4$ (4) $-6, 1$

   (5) $\pm\dfrac{1}{3}$ (6) $\dfrac{3}{2}$ (7) 1 (8) -4

Section 5.4

1. (1) T (2) F (3) T (4) F (5) F (6) T (7) F (8) F
   (9) F (10) T (11) F (12) F (13) T (14) T (15) F (16) T
   (17) T (18) F

2. (1) C (2) D (3) D (4) C (5) A (6) C (7) A (8) D
   (9) C (10) A (11) D (12) A (13) B (14) D (15) D (16) C
   (17) D (18) D

3. (1) $|a| \geq 0$ (2) $<$ (3) $<$ (4) $>$ (5) $\geq$ (6) $\geq$ (7) $\geq$ (8) $<$
   (9) $>$ (10) $\leq$ (11) $>$ (12) $<$ (13) $\leq$ (14) $<$

4. (1) $(-2, 0)$ (2) $[-6, 4)$ (3) $[-1, 3]$ (4) $(-4, -2]$
   (5) $(1, +\infty)$ (6) $(-\infty, -1]$ (7) $[2, +\infty)$ (8) $(-\infty, 4)$

5. (1) $(-3, -1)$ (2) $[2, 5)$ (3) $\emptyset$ (4) $\{0\}$
   (5) $(-6, -3]$ (6) $\emptyset$ (7) $[-2, 7)$ (8) $[-2, +\infty)$
   (9) $(-7, 7]$ (10) $[2, 6)$

6. (1) multiply by 4 on both sides, then subtract 3 on both sides
   (2) subtract -3 on both sides of $a < b$: $a - 3 < b - 3$; add b on both sides of $-3 < 3$:
   $b - 3 < b + 3$; by the transitive property $a - 3 < b - 3 < b + 3$
   (3) $a \neq 0$, so $a^2 > 0$; multiply by a^2 on both sides
   (4) multiply by $-\dfrac{2}{7}$ on both sides, then add 3 on both sides
   (5) add 4 on both sides, then multiply by $-\dfrac{1}{2}$

Section 5.5

1. (1) T (2) F (3) T (4) F (5) T (6) T (7) T (8) F
   (9) F (10) F (11) T (12) T

2. (1) C (2) C (3) B (4) B (5) C (6) B (7) B (8) C

3. (1) $[-10, +\infty)$ (2) $[-15, +\infty)$ (3) $\left(-\infty, \dfrac{6}{5}\right)$ (4) $(-\infty, 6]$

(5) $[\frac{3}{4},+\infty)$ (6) $(-\infty,5)$ (7) $(-3,+\infty)$ (8) $(-\infty,2)$

(9) $(-\infty,\frac{5}{2})$ (10) $[2,+\infty)$ (11) $(-\infty,41)$ (12) $(\frac{9}{4},+\infty)$

(13) $(-5,+\infty)$ (14) $[\frac{11}{7},+\infty)$ (15) $(-\infty,\frac{3}{2})$ (16) $(-\infty,+\infty)$

(17) $(-\infty,\frac{1}{10})$ (18) $[-\frac{15}{4},+\infty)$

4. (1) $[1,3]$ (2) $[5,6)$ (3) $[-1,8]$ (4) $[13,18)$

(5) $\emptyset$ (6) $(8,+\infty)$

Section 5.6

1. (1) F   (2) T   (3) F   (4) F   (5) T   (6) F   (7) F   (8) T

(9) T   (10) F

2. (1) C   (2) D   (3) B   (4) B   (5) D   (6) D

3. (1) $(-\infty,-\frac{5}{2}]\cup[-\frac{1}{2},+\infty)$ (2) $\{\frac{7}{4}\}$ (3) $(-\infty,+\infty)$

(4) $(-\frac{1}{3},3)$ (5) $\emptyset$ (6) $[-3,10]$ (7) $(-1,1)$

(8) $[1,3]$ (9) $(-\infty,\frac{3}{4}]$ (10) $\emptyset$ (11) $(-\infty,+\infty)$

(12) $[1,+\infty)$ (13) $[\frac{1}{3},+\infty)$ (14) $[1,+\infty)$ (15) $\emptyset$

(16) $\emptyset$ (17) $(-1,9)$ (18) $[-\frac{3}{4},1]$ (19) $(-\infty,-\frac{3}{5})$

(20) $(-\frac{1}{3},+\infty)$

4. $[0,2)$

Chapter 6

Section 6.1

1. (1) F   (2) F   (3) F   (4) F   (5) T   (6) F   (7) T   (8) F
   (9) T   (10) F

2. (1) D   (2) B   (3) C   (4) C   (5) D   (6) D   (7) C   (8) C
   (9) C   (10) D

3. (1) $\{x|x \neq 0\}$ (2) $\{p|p \neq 3\}$ (3) $\{y|y \neq 0\}$ (4) $\{a|a \neq \frac{1}{2}\}$
   (5) $\{x|x \neq \frac{9}{5}\}$ (6) $\mathbb{R}$ (7) $\{r|r \neq -4, 3\}$ (8) $\{r|r \neq \pm 1\}$

4. (1) $\dfrac{3y^2}{5x}$ (2) $-\dfrac{3b^2}{7a^3c^2}$ (3) $\dfrac{2}{3xz}$ (4) $\dfrac{x+2y}{3x-y}$

   (5) $-\dfrac{1}{3}$ (6) $\dfrac{a}{a+b}$ (7) $\dfrac{a+2}{a+5}$ (8) $-\dfrac{a+3}{a+7}$

   (9) $-\dfrac{y-4}{y+5}$ (10) $\dfrac{y+2}{y+3}$ (11) $-\dfrac{b+5}{b+9}$ (12) $\dfrac{x+3}{2x}$

   (13) $\dfrac{x-3}{x+4}$ (14) $\dfrac{3x+2}{4x+3}$ (15) $\dfrac{p-3q}{5p-q}$ (16) $\dfrac{y-2}{2z+1}$

   (17) $2x - y$ (18) $\dfrac{3(x+y)}{4}$ (19) $\dfrac{x^2+x+1}{(x+1)(x^2+1)}$ (20) $\dfrac{x+1}{x-1}$

   (21) $\dfrac{a^2+2ab+4b^2}{2a+3b}$ (22) $x + y - z$

5. (1) none   (2) $x = \dfrac{3}{2}$ (3) $s = -4$ (4) $t = 3$ (5) $x = y \neq 1$ (6) $x = -2$

Section 6.2

1. (1) F   (2) F   (3) T   (4) T   (5) F   (6) F

2. (1) B   (2) B   (3) C   (4) C   (5) C   (6) C   (7) C   (8) C
   (9) C   (10) D

3. (1) $x < 1$ (2) $a = -1$ (3) $\dfrac{(-1)^n}{a-1}$ (4) $1 + x$ (5) $b = 0, a \neq 0$

4. (1) $\dfrac{6x^2}{7yz}$ (2) $\dfrac{2b^2x^2z}{3ay}$ (3) $8xy^3$ (4) $\dfrac{2y}{3}$

   (5) $x - 1$ (6) $\dfrac{2(x-3y)}{xy^2}$ (7) $-x^2y$ (8) $\dfrac{x^2+x+1}{x+1}$

   (9) $\dfrac{(x+5)(x+1)}{x+2}$ (10) $\dfrac{2(x-y)}{3}$ (11) $\dfrac{x-4y}{x+y}$ (12) $\dfrac{1}{2x+y}$

   (13) 1 (14) $-\dfrac{3x+1}{3x-5}$ (15) $\dfrac{2a+b}{3a^2b}$ (16) $\dfrac{6x+5y}{3x+4y}$

(17) $\dfrac{4(3n-2)}{3(n-2)}$ (18) $\dfrac{x^2-3x+9}{x^2+3x+9}$

5. (1) $\dfrac{1-(a+1)b^2}{(1+b)^2}$ (2) $\dfrac{2}{x-2}$ (3) $\dfrac{(x-2y)(2x-y)}{3x-y}$

(4) $\dfrac{1}{a+b}$ (5) $\dfrac{1}{(x+1)(x-2)}$ (6) $\dfrac{(x-2)(x-5)}{x+3}$

(7) 1   (8) 3   (9) $a(x-a)$ (10) $\dfrac{2}{x}$

6. (1) 3   (2) 1   (3) 1

Section 6.3

1. (1) F   (2) T   (3) F   (4) F   (5) F   (6) T   (7) F   (8) F

(9) F   (10) T

2. (1) A   (2) C   (3) C   (4) A   (5) D   (6) D   (7) C   (8) A

(9) B   (10) A   (11) C

3. (1) $-\dfrac{2a+19}{24}$ (2) $\dfrac{(3y-2x)(3y+2x)}{6xy}$ (3) $\dfrac{x(1+2x-2y)}{2(x-y)}$

(4) $\dfrac{2x^2-3x-3}{x(x+1)}$ (5) $\dfrac{4x+1}{x^2-7}$ (6) $\dfrac{n^2+2mn-m^2}{mn}$

(7) $\dfrac{15ab-a^2}{5b}$ (8) $\dfrac{x^2+3x-3}{(x-4)(x+3)}$ (9) $\dfrac{x^2+9}{x-3}$

(10) $-\dfrac{2(x+3)}{(x-2)(x+1)(x-3)}$ (11) $-\dfrac{7p-60}{(3p-5)(4p+3)}$ (12) $\dfrac{12x^2+14x-33}{(5x-6)(3x+2)(4x-3)}$

(13) $\dfrac{2x^2-9x+3}{x^2-25}$ (14) $\dfrac{18x^3-37x^2+22x+3}{(4x^2-1)(3x-5)}$ (15) $-\dfrac{3}{x^2-4}$

(16) $\dfrac{p^3+q^3}{pq}$ (17) $\dfrac{2c-b}{b^2-c^2}$ (18) $\dfrac{2a}{a-1}$

(19) $-\dfrac{17ab-5ac+4bc}{24abc}$ (20) $\dfrac{10bc-8ac+9ab}{12a^2b^2c}$ (21) $a-a^2+a^3$

(22) $-\dfrac{1}{x+1}$ (23) $\dfrac{a^2+b^2+c^2}{abc}$ (24) $-\dfrac{ab-ac+bc}{abc}$

(25) $\dfrac{ab-2a^2-3b^2}{6ab(a+b)^2(a-b)}$ (26) $\dfrac{a}{b}$ (27) 0

(28) $\dfrac{5-x}{2(x-1)}$ (29) $\dfrac{a^3}{a-1}$ (30) $\dfrac{(a^2+1)(a^4+1)}{a^3}$

4. (1) $\dfrac{5x^2-4x+6}{(x+1)(2x-3)(2x+3)}$ (2) $\dfrac{-2x^2+8x-3}{(x-6)(x+1)(x-10)}$ (3) $\dfrac{x^3-3x-4}{x(x-2)^2}$

(4) $\dfrac{(-x^3-x^2+11x+10)}{(x+1)(x^2-9)}$ (5) $\dfrac{4}{(3a-4)(a-3)}$ (6) $\dfrac{4ab}{(a-b)^2}$

(7) $-\dfrac{1}{(x-2)^2}$ (8) $\dfrac{xy}{3}$ (9) $\dfrac{4}{x^2-4}$

(10) 1 (11) $\dfrac{8x^2(x^2+1)}{(x^2-1)^3}$ (12) $\dfrac{3}{1-x^8}$

(13) $\dfrac{16z^7}{1-z^{16}}$ (14) $\dfrac{x^2+10x+1}{x^2+x-2}$ (15) 1

5. (1) -3 (2) $\dfrac{1}{3}$ (3) 1

Section 6.4

1. (1) T (2) F (3) T (4) F (5) T (6) F (7) F (8) T
(9) T (10) T

2. (1) D (2) C (3) B (4) D (5) D

3. (1) $k=1$ (2) none (3) $a=3$ (4) $h=1$ or $-\dfrac{1}{2}$ (5) $x\neq\pm1$

4. (1) $\dfrac{b}{4}$ (2) $\dfrac{(x-3)(x+5)}{x+4}$ (3) $y-1$ (4) $\dfrac{11-2x}{2}$

(5) $\dfrac{-3x+8}{5x-6}$ (6) $\dfrac{a^2}{a-1}$ (7) $-\dfrac{x+1}{2x-5}$ (8) $\dfrac{x-5}{x+1}$

(9) $\dfrac{x+b}{x+a}$ (10) $\dfrac{5x(x-2)}{2x-1}$ (11) $\dfrac{2x}{2x^2+1}$ (12) $\dfrac{a+1}{a}$

(13) $\dfrac{4x^2}{4x^4+2x^2+1}$ (14) $\dfrac{2a}{a^2-b^2}$ (15) $\dfrac{1}{x-y}$ (16) $\dfrac{4(x+y)}{x-3y}$

5. (1) $-\dfrac{15}{4}$ or $-\dfrac{3}{2}$ (2) $-\dfrac{1}{27}$ (3) -4 (4) $-\dfrac{8}{3}$ or 0 (5) $\dfrac{7}{13}$

Section 6.5

1. (1) $x=\dfrac{14}{3}$ (2) $x=2$ (3) $x=-2$ (4) no solution (5) no solution

(6) $x=-2$ (7) $x=\dfrac{10}{3}$ (8) $x=-1$ (9) $x=1$ (10) $x=\pm2$

2. (1) $x\neq1,2$ (2) $x=\dfrac{3}{2}$ (3) $x=\dfrac{7}{3}$ (4) $x=0$ (5) $x=\dfrac{11}{2}$

(6) $x=\dfrac{2}{3}$ (7) $x=1$ (8) no solution (9) $x=-\dfrac{7}{17}$ (10) no solution

3. (1) $x=-\dfrac{3}{2}$ (2) $x=\dfrac{25}{8}$ (3) $x=\dfrac{95}{3}$ (4) $x=3$ (5) $x=\dfrac{15}{11}$

(6) $x=-3$ (7) $x=-3$ (8) $x=-\dfrac{8}{5}$ (9) $x=4$ (10) $x=-\dfrac{6}{5}$

4. 10

5. 30km/h, 12km/h

Chapter 7

Section 7.1

1. (1) F   (2) T   (3) F   (4) T   (5) F   (6) T   (7) T   (8) F
   (9) F   (10) F   (11) T   (12) F

2. (1) D   (2) B   (3) B   (4) A   (5) D   (6) B   (7) C   (8) A
   (9) B   (10) A   (11) B   (12) B   (13) C   (14) C   (15) D   (16) C

3. Points on the x-axis: $(-2,0)$, $(0,0)$, $(3,0)$

   (1) Points on the y-axis: $(0,3)$, $(0,0)$, $(0, -\frac{3}{2})$

   (2) Points in Quadrant I: $(2, \frac{1}{2})$, $(\frac{1}{3}, \frac{1}{2})$

   (3) Points in Quadrant II: $(-1, 2)$, $(-\frac{7}{3}, 1)$, $(-\frac{5}{2}, \frac{8}{3})$

   (4) Points in Quadrant III: $(-3, 2)$, $(-1, -2.5)$, $(-2, -\frac{3}{2})$, $(-1, -1)$

   (5) Points in Quadrant IV: $(1, -3)$, $(2, -1)$

4. (1) $(-1, 3)$ (2) $(\frac{5}{2}, -\frac{7}{3})$ (3) $(-7, -7)$ (4) $(\frac{3}{2}, -\frac{7}{2})$ (5) $(-\frac{19}{3}, 6)$ (6) $(-\frac{49}{9}, \frac{10}{7})$

5. Format: symmetric point about the x-axis; about the y-axis; about the origin

   (1) $(-1,0); (1,0); (1,0)$ (2) $(-2, \pi); (2, -\pi); (2, \pi)$

   (3) $(-5, 17); (5, -17); (5, 17)$ (4) $(0,0); (0,0); (0,0)$

   (5) $(0, 100); (0, -100); (0, 100)$ (6) $(23, 32); (-23, -32); (-23, 32)$

   (7) $(5, -5); (-5, 5); (-5, -5)$ (8) $(-12, -12); (12, 12); (12, -12)$

6. Format: distance to the x-axis; distance to the y-axis

   (1) $17; 1.7$ (2) $\sqrt{3}; \sqrt{2}$ (3) $0; 0$ (4) $0.23; 0$ (5) $4; 5$ (6) $7 - 2\pi; \pi - 3$

7. (1) $A \cup D \cup E$ (2) $C \cup D \cup E \cup F \cup H \cup \{(0,0)\}$ (3) $A \cup C \cup E \cup F \cup G \cup H \cup \{(0,0)\}$
   (4) $B \cup D$ (5) $A \cup B \cup E \cup F \cup G \cup H \cup \{(0,0)\}$

8. (1) the point $(1, 2)$

   (2) the line parallel to the x-axis and passing through $(0, -2)$

   (3) the y-axis

   (4) the angle bisector of the second and fourth quadrants

   (5) the line through $(-1, -1)$ and $(1, 3)$

10. $D(-3, 9)$

Section 7.2

1. (1) T   (2) F   (3) T   (4) F   (5) F   (6) F   (7) F   (8) T

(9) F   (10) T   (11) F   (12) T   (13) T   (14) F   (15) F   (16) T

(17) F   (18) T

2. (1) C   (2) B   (3) A   (4) B   (5) B   (6) C   (7) D   (8) A

(9) D   (10) B   (11) C   (12) A   (13) D   (14) C   (15) A   (16) A

3. (1) $7x - 8y + 77 = 0$ (2) $7x + 10y + 14 = 0$ (3) $5x + 3y + 15 = 0$ (4) $3x - y = 0$

(5) $14x + 2y - 43 = 0$ (6) $y = 0$ (7) $x = 0$ (8) $x = \pm 3$

(9) $y = -\pi$ (10) $x + y = 0$

4 Format: slope, x-intercept, y-intercept

(1) D.N.E, -2, none   (2) $-2, 0, 0$ (3) $\dfrac{3}{2}, -2, 3$ (4) $-\dfrac{4}{5}, -\dfrac{7}{4}, -\dfrac{7}{5}$

5. $\dfrac{7}{6}$ or $-\dfrac{1}{6}$

6. $-\dfrac{3}{8}$

7. $-\dfrac{21}{5}$

8. $(5, -16)$

9. $-\dfrac{1}{10}$

10. $\dfrac{13}{6}$

11. -5

Section 7.3

1. (1) T   (2) F   (3) F   (4) T   (5) F   (6) T   (7) T   (8) F

(9) T   (10) T   (11) T   (12) F   (13) T   (14) T   (15) F   (16) T

(17) F   (18) T

2. (1) C   (2) C   (3) A   (4) C   (5) B   (6) D   (7) B   (8) B

(9) A   (10) C

3. (1) perpendicular   (2) parallel   (3) intersecting   (4) intersecting

(5) perpendicular   (6) parallel   (7) coincided   (8) perpendicular

(9) perpendicular

4. $3x + y - 15 = 0$

5. $m = \dfrac{3}{4}$

6. $2x - y - 12 = 0$

7. $m = -\dfrac{5}{9}$

8. 16

9. $-\dfrac{13}{2}$

10. 64

11. $7x + 3y + 86 = 0$

12. D.N.E

13. (1) $a_1b_2 - a_2b_1 = 0$

(2) $a_1b_2 - a_2b_1 = a_1c_2 - a_2c_1 = b_1c_2 - b_2c_1 = 0$

(3) $a_1b_2 - a_2b_1 \neq 0$

14. (x_0, y_0) should satisfy $\begin{cases} a_1x_0 + b_1y_0 = c_1 \\ a_2x_0 + b_2y_0 = c_2 \end{cases}$

Chapter 8

Section 8.1

1. (1) F   (2) T   (3) F   (4) T   (5) T   (6) F   (7) F   (8) T
    (9) F   (10) T   (11) T   (12) T   (13) T   (14) F   (15) F

2. (1) B   (2) B   (3) B   (4) A   (5) A   (6) C   (7) A   (8) D
    (9) C   (10) C   (11) B   (12) A   (13) C   (14) A   (15) C

3. (1) $(\frac{5}{13}, \frac{7}{13})$ (2) $(\frac{7}{17}, -\frac{16}{17})$ (3) $(-\frac{5}{17}, \frac{35}{17})$ (4) $(\frac{39}{19}, \frac{22}{19})$

    (5) no solution   (6) $(-\frac{19}{5}, -\frac{17}{5})$

4. (1) $(\frac{19}{43}, \frac{17}{43})$ (2) $(-6, -6)$ (3) $(4, 1)$ (4) $(\frac{24}{31}, \frac{22}{31})$

    (5) $(\frac{19}{4}, 4)$ (6) $(-26, 46)$

5. (1) $(\frac{3}{2}, \frac{7}{4})$ (2) $(\frac{13}{11}, -\frac{25}{11})$ (3) $(76, 105)$ (4) $(51, -9)$

    (5) $(\frac{17}{11}, \frac{25}{11})$ (6) $(1, 6)$ (7) $(-\frac{19}{4}, \frac{15}{2})$ (8) $(\frac{175}{17}, \frac{205}{34})$

    (9) $(2, -1)$ (10) $(-\frac{7}{23}, \frac{5}{23})$

6. (1) $(\frac{1}{10}, 4)$ (2) $(\frac{8}{7}, \frac{16}{5})$ (3) $(-2, 5)$ (4) $(-3, -2)$

7. potato-head: 10min; dinosaur: 15min

8. bananas: 3lb; apples: 5lb

9. 20 chickens; 10 rabbits

10. (1) intersecting at $(0, 0)$ (2) intersecting at $(5, -3)$ (3) not intersecting

Section 8.2

1. (1) B   (2) B   (3) D   (4) B   (5) A   (6) C

2. (1) $(1, 2, 7)$ (2) $(4, 3, 8)$ (3) $(1, 2, -1)$ (4) $(2, 0, \frac{4}{3})$ (5) $(5, 0, -3)$
    (6) $(3, 4, 5)$ (7) $(12, 4, 9)$ (8) $(1, -1, 3)$ (9) $(2, 4, -3)$ (10) $(2, -1, -2)$
    (11) $(3, 9, 15)$ (12) $(2, -1, 2)$

3. (1) $(\frac{1}{2}, -1, \frac{1}{3})$ (2) $(-1, 2, \frac{1}{3})$ (3) $(\frac{7}{3}, \frac{4}{3}, \frac{10}{3})$ (4) $(\frac{1}{16}, -\frac{1}{2}, \frac{3}{4})$

    (5) $(\frac{1}{8}, \frac{1}{7}, 1)$ (6) $(\frac{1}{6}, \frac{1}{4}, \frac{1}{3})$

4. (B): 14kg; (W): 21kg; (Y): 14kg

5. flat: 30miles; up-hill: 42miles; down-hill: 70miles

6. Christine: 180 days; Tina: 90 days; Tommy: 60 days; together: 30 days

Section 8.3

1. (1) D   (2) D   (3) C   (4) B   (5) C   (6) D   (7) A   (8) C

3. $\begin{cases} x + 2y \geq 1 \\ x - y \geq -2 \\ 2x + y \leq 5 \end{cases}$

4.

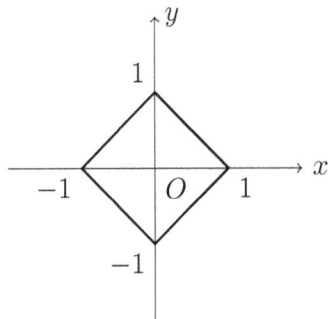 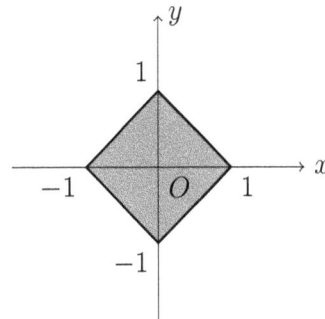

Chapter 9

Section 9.1

1. (1) T   (2) F   (3) T   (4) F   (5) T   (6) F   (7) F   (8) F
   (9) T   (10) F   (11) F   (12) T   (13) F   (14) T   (15) F   (16) F
   (17) T   (18) F   (19) T

2. (1) C   (2) A   (3) A   (4) C   (5) C   (6) D   (7) A   (8) D
   (9) A   (10) C   (11) C   (12) D   (13) C   (14) C   (15) C   (16) A

3. (1) $\pm\dfrac{5}{8}$; 1.96   (2) -7.12 (3) $0, -1$

   (4) $\pm 1, 0$; $0, 1$; $0, 1$ (5) 4   (6) arithmetic square root

   (7) -0.778; 0.190   (8) ≤ 3; $\in \mathbb{R}$ (9) 1; 1

   (10) -1 (11) -2 (12) $\dfrac{3}{2}$

   (13) $\sqrt{13}$; $\sqrt{3}$ (14) 3; -2; D.N.E   (15) ± 7

   (16) ± 10 (17) 2; ± 3 (18) -5; $\pm\sqrt{10}$

4. $2.131131113\cdots, \dfrac{\pi}{4}, \sqrt{5}, \pi^2, -\sqrt{2}, \sqrt{\pi}, \dfrac{\pi}{3}, 2 + \sqrt{3}$

5. Format: square roots; arithmetic square root

   (1) ± 3; 3   (2) $\pm\dfrac{9}{4}$; $\dfrac{9}{4}$ (3) 0; 0   (4) D.N.E; D.N.E

   (5) $\pm\sqrt{15}$; $\sqrt{15}$ (6) $\pm\sqrt{\pi}$; $\sqrt{\pi}$

6. (1) -2 (2) 0.1   (3) $\sqrt[3]{7}$ (4) -4

7. (1) 6   (2) 9   (3) 0.1   (4) 41   (5) 4   (6) D.N.E   (7) 0.7   (8) 1.2
   (9) -4 (10) 5   (11) 0.6   (12) 2   (13) $\dfrac{3}{2}$ (14) $\dfrac{12}{7}$ (15) 2.2   (16) $\dfrac{7}{2}$
   (17) $-\dfrac{5}{3}$ (18) $\dfrac{14}{9}$ (19) $\dfrac{3}{2}$ (20) $-\dfrac{5}{3}$ (21) -6 (22) $\dfrac{5}{8}$ (23) 9   (24) $\dfrac{5}{12}$
   (25) $-\dfrac{1}{2}$ (26) $\dfrac{3}{4}$ (27) -4 (28) 3

8. (1) $\sqrt[3]{0.25} < \sqrt[3]{0.26}$ (2) $\sqrt[3]{18} > \sqrt[3]{\dfrac{1}{18}}$

   (3) $\sqrt[3]{-3} < \sqrt[3]{-2} < \sqrt[3]{0.1} < \sqrt[3]{5}$ (4) $-\sqrt[4]{3.1} < -\sqrt[4]{3} < -\sqrt[4]{2.9}$

   (5) $2\sqrt{2} < 3 < 2\sqrt{3}$ (6) $\sqrt{15} < 3\sqrt{2} < 2\sqrt{5} < 5$

   (7) $\sqrt[4]{2} < \sqrt[3]{2} < \sqrt{2}$ (8) $\sqrt{\dfrac{1}{2}} < \sqrt[3]{\dfrac{1}{2}} < \sqrt[4]{\dfrac{1}{2}}$

   (9) $\sqrt[4]{x} < \sqrt[3]{x} < \sqrt{x}$ (10) $\sqrt{x} < \sqrt[3]{x} < \sqrt[4]{x}$

   (11) $\sqrt[3]{5} < \sqrt{3} < \sqrt[4]{10}$ (12) $\sqrt{5} < \sqrt[3]{12} < \sqrt{5.5} < \sqrt[3]{13} < \sqrt[3]{14} < \sqrt{6}$

   (13) $\sqrt{50} < 7\dfrac{1}{2}$ (14) $-\dfrac{22}{7} < -\pi < \pi < \sqrt{10}$

(15) $-\dfrac{1}{3} < -\dfrac{1}{\pi} < -\dfrac{1}{\sqrt{10}}$ (16) $-\dfrac{\sqrt{2}}{2} < -\dfrac{7}{10}$

9. (1) $a \geq \dfrac{5}{2}$ (2) $x \in \mathbb{R}$ (3) $[0, \dfrac{1}{2})$ (4) $b \in \mathbb{R}$

(5) $x \in \mathbb{R}$ (6) $a > \dfrac{3}{2}$ (7) $x \geq \dfrac{3}{2}, x \neq 3$ (8) $x > 4$

(9) $x \geq \dfrac{7}{8}, x \neq \dfrac{4}{3}$ (10) $x \leq \dfrac{3}{7}, x \neq -\dfrac{1}{5}$

10. (1) $\pm\dfrac{10}{13}$ (2) ± 5 (3) ± 10 (4) $7, -3$ (5) $\dfrac{3}{5}, -\dfrac{13}{5}$ (6) $\pm\dfrac{7}{5}$

(7) -1 (8) $\dfrac{5}{2}$ (9) $4, -3$ (10) -7 (11) 2 (12) -3

11. (1) $1.\overline{4} - \sqrt{2}$ (2) $\sqrt{3} - 1.732$ (3) 2 (4) 1

12. (1) $\pi - 1$ (2) $\sqrt{15} - 2$

Section 9.2

1. (1) F   (2) T   (3) F   (4) F   (5) F   (6) T   (7) T   (8) T
(9) F   (10) F   (11) T   (12) F

2. (1) A   (2) B   (3) A   (4) B   (5) C   (6) B   (7) A   (8) C
(9) C   (10) C   (11) B   (12) D   (13) B   (14) C   (15) C   (16) C
(17) D   (18) D

3. (1) $2\sqrt{6}$ (2) $3\sqrt{3}$ (3) $3\sqrt{5}$ (4) $4\sqrt{3}$ (5) $-16\sqrt{2}$

(6) $15\sqrt{2}$ (7) $-\sqrt{5}$ (8) $\dfrac{\sqrt{2}}{2}$ (9) $\dfrac{3\sqrt{2}}{5}$ (10) $\dfrac{\sqrt{65}}{65}$

(11) $-4\sqrt[3]{6}$ (12) $-\dfrac{5}{2}\sqrt[3]{4}$ (13) $\dfrac{5}{3}\sqrt{3}$ (14) $\dfrac{\sqrt{6}}{2}$ (15) $\dfrac{2}{51}\sqrt{102}$

(16) $\dfrac{\sqrt{30}}{20}$ (17) $\dfrac{\sqrt{30}}{6}$ (18) $\sqrt{3}$ (19) $-\dfrac{2}{15}\sqrt{105}$ (20) $\dfrac{\sqrt{77}}{98}$

(21) $\dfrac{2}{3}\sqrt[3]{3}$ (22) $-\dfrac{3}{2}\sqrt[3]{4}$ (23) $\dfrac{\sqrt[3]{12}}{4}$ (24) $-\dfrac{2}{5}\sqrt[3]{50}$ (25) $\dfrac{\sqrt[3]{450}}{6}$

(26) $\dfrac{2}{3}\sqrt{6}$ (27) $\dfrac{3}{2}\sqrt{3}$ (28) $\dfrac{2}{15}\sqrt{35}$ (29) $\dfrac{2}{15}\sqrt{33}$ (30) $\dfrac{5}{8}\sqrt{10}$

(31) $\dfrac{\sqrt[3]{18}}{4}$ (32) $\dfrac{3}{10}\sqrt[3]{300}$

4. (1) $3\sqrt{10}$ (2) $4\sqrt{3}$ (3) $-20\sqrt{6}$ (4) $15\sqrt{15}$

(5) $2\sqrt[3]{6}$ (6) $18\sqrt[3]{2}$ (7) $2\sqrt{11}$ (8) $5\sqrt{3} - \sqrt{35}$

(9) $24\sqrt{3} - 60\sqrt{2}$ (10) $16\sqrt{6} - 40\sqrt{3}$ (11) -2 (12) $14 - 4\sqrt{6}$

(13) $\sqrt{3} + 2\sqrt{2}$ (14) $4\sqrt{35} - 2\sqrt{15} - 6\sqrt{14} + 3\sqrt{6}$ (15) $-6 - 2\sqrt{6}$

(16) $\dfrac{26\sqrt{3}}{3}$ (17) $-4 + \sqrt{3}$ (18) $-46 + 7\sqrt{14}$ (19) $4\sqrt{6} - 6\sqrt{7}$

(20) $8\sqrt{6}$ 　　　　(21) $2\sqrt{3}$ 　　　　(22) $3\sqrt{2}$ 　　　　(23) $\dfrac{15}{2}\sqrt{2}$

(24) $-\dfrac{8}{3}\sqrt{3}$ 　　(25) $\dfrac{13}{4}\sqrt{2}$ 　　(26) $5\sqrt{3}-3\sqrt{2}$ 　　(27) $7\sqrt{6}-\dfrac{5}{3}\sqrt{3}+\dfrac{\sqrt{2}}{2}$

(28) $\sqrt{6}+2\sqrt{2}$ 　　(29) $6-2\sqrt{6}$ 　　(30) $4+\sqrt{11}$ 　　(31) $-4-2\sqrt{3}$

(32) $-\dfrac{3\sqrt{5}+2}{41}$ 　　(33) $2\sqrt{2}+5$ 　　(34) $\dfrac{5\sqrt{5}+\sqrt{35}}{18}$ 　　(35) $2\sqrt{7}-4$

(36) $8+2\sqrt{11}$ 　　(37) $2-\sqrt{3}$ 　　(38) $2\sqrt{3}+\sqrt{10}$ 　　(39) $\dfrac{3\sqrt{6}-7}{5}$

(40) $2\sqrt{15}-3\sqrt{3}$ 　　(41) $\dfrac{5\sqrt{2}+\sqrt{6}}{4}$ 　　(42) $-\dfrac{3\sqrt{22}+\sqrt{66}+3\sqrt{14}+\sqrt{42}}{3}$

5. (1) $2\sqrt{15}>3\sqrt{6}$ 　　(2) $2+\sqrt{7}>\sqrt{3}+2\sqrt{2}$ 　　(3) $\sqrt{3}+\sqrt{4}>\sqrt{5}+\sqrt{2}$

(4) $\sqrt{2}-1>\sqrt{3}-\sqrt{2}>2-\sqrt{3}$ 　　　　　　(5) $-5\sqrt{\dfrac{1}{5}}<-\dfrac{3}{2}\sqrt{2}$

(6) $\dfrac{1}{2\sqrt{5}}>\dfrac{1}{\sqrt{21}}$ 　　(7) $\sqrt{7}<\dfrac{2}{\sqrt{3}-1}$ 　　(8) $3-\sqrt{5}>2\sqrt{5}-4$

(9) $2\sqrt{5}>\sqrt{11}+1$ 　　(10) $\sqrt[5]{5}<\sqrt{2}=\sqrt[4]{4}<\sqrt[3]{3}$

6. (1) $\pi-3$ 　　(2) $\sqrt{35}-4\sqrt{2}$ 　　(3) $3\sqrt{2}-4$ 　　(4) 1

(5) $\dfrac{5}{2}\sqrt{2}-8\sqrt{3}$ 　　(6) 8 　　(7) $216-36\sqrt{6}$ 　　(8) $\dfrac{\sqrt{7}-\sqrt{3}}{2}$

(9) $\sqrt{3}-\sqrt{2}$ 　　(10) $\dfrac{\sqrt{5}+\sqrt{6}+\sqrt{7}}{2}$ 　　(11) $\dfrac{3\sqrt{2}+\sqrt{6}-2\sqrt{3}-2}{2}$

(12) $-9-6\sqrt{2}$ 　　(13) 1 　　(14) $\dfrac{17}{2}$ 　　(15) $\dfrac{3-\sqrt{5}}{2}$

(16) $12\sqrt{3}-48\sqrt{2}-6\sqrt{6}+93$ 　　(17) 1 　　(18) $-60\sqrt{15}$

7. (1) $\sqrt{3}-\sqrt{2}$ 　　(2) $\sqrt{5}-\sqrt{3}$ 　　(3) $\sqrt{7}-\sqrt{5}$ 　　(4) 1

(5) $2\sqrt{3}-\sqrt{2}$ 　　(6) $\sqrt{2}-\sqrt{3}$ 　　(7) $\dfrac{\sqrt{10}+\sqrt{2}}{2}$ 　　(8) $\dfrac{\sqrt{6}-\sqrt{2}}{2}$

(9) $\dfrac{\sqrt{5}-1}{3}$ 　　(10) $2\sqrt{3}+1$ 　　(11) $\dfrac{\sqrt{5}-\sqrt{3}}{2}$ 　　(12) 16

(13) $\sqrt{3}+1$ 　　(14) $\sqrt{6}-\sqrt{2}$

8. (1) $2\sqrt{y}$ 　　(2) $\sqrt{x}+\sqrt{y}$ 　　(3) $-xm$ 　　(4) 1

(5) $-15\sqrt{m}$ 　　(6) $xy+2y-x+1$ 　　(7) $2x\sqrt{y}$ 　　(8) $\dfrac{\sqrt{2xy}}{2xy}$

(9) $2x+1+2\sqrt{x^2+x}$ (10) 0 　　(11) $-x\sqrt{x}$ 　　(12) 0

(13) $\sqrt{1+x^2}$ 　　(14) $2a$ 　　(15) $\dfrac{x^2y^2-2xy+1}{y^2}$ 　　(16) $\sqrt{x}-\sqrt{y}$

(17) 0 　　(18) $\dfrac{2(x+y)}{x-y}$ 　　(19) 0 　　(20) $\dfrac{\sqrt{x}+\sqrt{y}}{x-y}$

(21) $\dfrac{x^2 + xy + 1}{x^2 y^2}$

9. (1) $|x - 2|$ (2) $(x - 2)^2$ (3) $|x - 2|$ (4) $-x\sqrt{x^2 + y^2}$

(5) $\dfrac{\sqrt{2y - 2x}}{y - x}$ (6) $\dfrac{\sqrt{x^3 + y^3}}{xy}$ (7) $-\dfrac{\sqrt{3(x^2 - y^2)}}{2}$ (8) $-\dfrac{\sqrt{a^2 + b^2 x^2}}{a}$

(9) $-\sqrt{2x(x - 1)}$ (10) 2 (11) $-x\sqrt{x^4 - y^4}$ (12) $3x - 3$

(13) $x + 1$ (14) ± 2 (15) $(x - y)\sqrt{x + y}$ (16) $-x\sqrt{x}$

(17) -2 (18) $-a - \dfrac{1}{a}$ (19) $\dfrac{1}{4} - a$ (20) 4

(21) $\begin{cases} 2x - 7 \text{ when } x > 4 \\ 1 \text{ when } 3 \le x \le 4 \\ 7 - 2x \text{ when } x < 3 \end{cases}$ (22) $\begin{cases} 2 \text{ when } x > 2 \\ 0 \text{ when } 1 \le x \le 2 \\ -2 \text{ when } x < 1 \end{cases}$

10. (1) $x \le 0$ (2) $x \le 2$ (3) $|x| \ge 3$ (4) $x \ge 2$

11. (1) $-\dfrac{\sqrt{6}}{6}$ (2) 14 (3) $4\sqrt{2}$ (4) $4\sqrt{3}$

(5) 4 (6) $2\sqrt{2}$ (7) $\dfrac{6}{49}$ (8) $\dfrac{\sqrt{5}}{5}$

(9) $\sqrt{2}$ (10) -4 or 0 (11) $\dfrac{3}{17}$ (12) 3

(13) 4 (14) 3 (15) $\sqrt{6} + \sqrt{2} - 2 - \sqrt{3}$

(16) $2\sqrt{7} - 2\sqrt{2} - 1$

12. (1) $-5 - 2\sqrt{6}$ (2) $\sqrt{2} + \sqrt{3}$ (3) $-\dfrac{3}{8}\sqrt{2} - \dfrac{1}{2}$ (4) $\sqrt{2} + 1$

13. (1) $(x - \sqrt{5})(x + \sqrt{5})$ (2) $(x - 1)(x + 1)(x - \sqrt{5})(x + \sqrt{5})$

(3) $(x - \sqrt{3})^2$ (4) $(x^2 + 7)(x - \sqrt{6})(x + \sqrt{6})$

14. (1) $7 - 4\sqrt{3}$ (2) $1; 2\sqrt{11} - 7; 7$

Section 9.3

1. (1) B   (2) C   (3) B   (4) C   (5) D   (6) C   (7) C   (8) A

2. (1) $\dfrac{\sqrt{a}}{a}$ (2) $\dfrac{\sqrt[3]{b^2}}{b}$ (3) $\sqrt[5]{a}$ (4) $\sqrt[4]{b^3}$ (5) $\dfrac{\sqrt[5]{c^2}}{c}$

(6) $\sqrt[3]{d^2}$ (7) $\dfrac{\sqrt[6]{2^5}}{2}$ (8) $\dfrac{\sqrt[4]{3}}{3}$ (9) $\sqrt[9]{x^4}$ (10) $\dfrac{1}{y^8}$

3. (1) $(3y)^{\frac{1}{5}}$ (2) $-4(2ab)^{\frac{1}{2}}$ (3) $x^{\frac{2}{3}}$ (4) $(a + b)^{\frac{3}{4}}$

(5) $(m - n)^{\frac{2}{3}}$ (6) $(m - n)^2$ (7) $a^{-\frac{1}{4}}$ (8) $-a^{\frac{7}{3}}$

(9) $(3xy)^{\frac{1}{2}}$ (10) $(5a - 4b)^2$ (11) $a^{\frac{3}{2}}$ (12) $p^3 q^{\frac{5}{2}}$

(13) $m^{\frac{5}{2}}$ (14) $m^2 n^{-3}$ (15) $-(x + y)^{\frac{1}{3}}$ (16) $5(xy)^{\frac{5}{2}}$

4.   (1) 4   (2) $\dfrac{1}{10}$ (3) 64   (4) $\dfrac{27}{8}$ (5) 125   (6) $\dfrac{1}{16}$

     (7) $\dfrac{343}{216}$ (8) $\dfrac{8}{125}$ (9) 625   (10) $\dfrac{1}{27}$ (11) 2   (12) $\dfrac{2}{3}$

     (13) 0.1   (14) $\dfrac{7^5}{3^5}$ (15) 8   (16) 2   (17) $-4\sqrt[4]{5}$ (18) 6

     (19) $-\dfrac{33}{8}$ (20) $\dfrac{1}{4}$ (21) $\sqrt[12]{2^{11}3^{10}}$ (22) $60\sqrt[12]{3}$ (23) $\sqrt{2}$ (24) $\sqrt[35]{3}$

     (25) $\sqrt[6]{24}$ (26) $\dfrac{\sqrt[6]{12}}{2}$

5.   (1) $\sqrt[6]{a^5}$ (2) $x^2\sqrt[6]{x}$ (3) $\dfrac{\sqrt{a}}{a}$ (4) $\dfrac{x\sqrt{x}}{y^6}$ (5) $\dfrac{mn\sqrt[3]{n}}{2}$

     (6) $\dfrac{\sqrt[6]{x^3y^4}}{4x}$ (7) $\sqrt[4]{x^3}$ (8) $\dfrac{\sqrt[6]{x}}{x}$ (9) $\sqrt{x}$ (10) $b^2\sqrt[4]{a^3b^3}$

     (11) $\sqrt[3]{x}$ (12) $\dfrac{\sqrt[4]{x^3y^3}}{y}$ (13) $a-b$ (14) $\sqrt[4]{x}-\sqrt[4]{y}$ (15) $-10x\sqrt[12]{x^5}$

     (16) $4a$ (17) $\sqrt[6]{a^5}$ (18) $\dfrac{z^8}{y^3}$ (19) $\dfrac{3ab^2}{2}$ (20) $1-\dfrac{4}{x}$

     (21) $\dfrac{-4\sqrt[10]{a^9}\sqrt[6]{b^5}}{b^3}$ (22) $\dfrac{7}{6}\sqrt[14]{m^{13}}$ (23) $\dfrac{3}{4}\sqrt[15]{x^4}\sqrt[6]{y^5}$ (24) $\dfrac{2^7x^2\sqrt[5]{x^4}}{3^7y^2}$

     (25) $\dfrac{3\sqrt[5]{48}\sqrt[10]{x^9}\sqrt[15]{y^{13}}}{4y}$ (26) $-\dfrac{2\sqrt[3]{xy}}{xy}$

6.   (1) $\sqrt[6]{32}$ (2) $4\sqrt{6}$

Section 9.4

1.   (1) C   (2) D   (3) B   (4) C   (5) C   (6) A

2.   (1) $x=\dfrac{3}{2}$ (2) no solutions   (3) $x=11$ (4) $x=\dfrac{13}{5}$

     (5) no solutions   (6) $s=21$

3.   (1) $x=16$ (2) $x=1$ (3) no solutions (4) $y=\dfrac{5}{4}$ (5) $x=4$

     (6) no solutions (7) $a=\dfrac{3}{2}$ (8) no solutions (9) $x=-1$ (10) $b=\dfrac{126}{17}$

4.   (1) $x=\pm4\sqrt{2}$ (2) $x=\pm\sqrt{5}$ (3) $x=3$ (4) $x=\pm\dfrac{\sqrt{106}}{2}$

     (5) no solutions   (6) $x=\pm\dfrac{\sqrt{15}}{5}$

5.   (1) no solutions   (2) $x=\dfrac{5}{12}$ (3) no solutions   (4) $x=\dfrac{29}{8}$

     (5) $x=\dfrac{4}{5}$ (6) $x=\dfrac{45}{8}$ (7) $x=\dfrac{13}{6}$ (8) no solutions

6. (1) $x = -\dfrac{5}{2}$ (2) $x = \pm\dfrac{5}{2}$ (3) $x = -\dfrac{4}{5}$ (4) no solutions

     (5) $x = \dfrac{19}{9}$ (6) $x = \dfrac{9}{5}$ (7) $x = \dfrac{7}{5}$ (8) $x = -3$

7. (1) $x = \dfrac{3}{2}, y = -2$ (2) $x = -\sqrt{2}$ (3) $x = \pm\sqrt[3]{2}$ (4) $x = 4, y = 16$

Chapter 10

Section 10.1

1. (1) D (2) A (3) A (4) C (5) D (6) B (7) D (8) B
(9) B (10) D (11) C (12) C (13) B (14) A (15) D (16) C
(17) C (18) A

2. (1) $11x^2 = 0$ (2) $\neq 1; = 1$ (3) $x = 1$ or $-\dfrac{2}{3}$ (4) $p^2 + 4$ (5) 4

(6) $\leq \dfrac{9}{4}$ (7) 2 or -4 (8) $\dfrac{-1 \pm \sqrt{5}}{2}$ (9) $3, 4, 5$ (10) $3, -5$

(11) $1; -2$ (12) 1; linear (13) -3 (14) $\geq -\dfrac{25}{8}; > -\dfrac{25}{8}$ and $m \neq 0$

3. (1) $1; x - 1$ (2) $9; x - 3$ (3) $\dfrac{9}{4}; x - \dfrac{3}{2}$ (4) $\dfrac{49}{4}; x + \dfrac{7}{2}$

(5) $\dfrac{1}{9}; x - \dfrac{1}{3}$ (6) $\dfrac{1}{16}; x + \dfrac{1}{4}$ (7) $\dfrac{9}{5}; x + \dfrac{3}{5}$ (8) $\dfrac{4}{3}; x + \dfrac{2}{3}$

(9) $\dfrac{9}{16}; x + \dfrac{3}{8}$ (10) $\dfrac{9}{7}; x - \dfrac{3}{7}$ (11) $\dfrac{25}{12}; x - \dfrac{5}{6}$ (12) $4; 3x + 2$

(13) $1; 2x - 1$ (14) $\dfrac{9}{4}; 7x + \dfrac{3}{2}$ (15) $\dfrac{3}{4}; x - \dfrac{\sqrt{3}}{2}$ (16) $\dfrac{5}{3}; x + \dfrac{\sqrt{5}}{3}$

(17) $x + 3; 12$ (18) $x - \dfrac{\sqrt{5}}{4}; \dfrac{11}{16}$ (19) $2x + \dfrac{3}{2}; -\dfrac{17}{4}$ (20) $x + \dfrac{1}{3}; \dfrac{8}{3}$

4. (1) $x = \pm 3$ (2) no solution (3) $x = \pm\dfrac{\sqrt{6}}{2}$ (4) $x = \pm\dfrac{3\sqrt{42}}{7}$

(5) no solution (6) $y = \pm\dfrac{\sqrt{15}}{3}$ (7) $t = 2, -\dfrac{2}{3}$ (8) $x = \dfrac{-1 \pm 3\sqrt{2}}{7}$

(9) $x = \dfrac{-3 \pm 2\sqrt{3}}{5}$ (10) $z = \dfrac{20 \pm 2\sqrt{15}}{25}$ (11) $x = 0, -\dfrac{3}{2}$ (12) $x = 0, \dfrac{15}{4}$

(13) $x = 6, -1$ (14) $x = 2, 3$ (15) $x = 2, 6$ (16) $x = -12, 3$

(17) $x = -\dfrac{3}{2}, \dfrac{7}{3}$ (18) $x = \dfrac{3}{4}$ (19) $x = -6, -\dfrac{5}{4}$ (20) $x = \dfrac{7}{3}, -\dfrac{6}{5}$

(21) $x = -\dfrac{2}{5}$ (22) $x = -\dfrac{9}{4}, \dfrac{8}{9}$ (23) $x = \dfrac{7}{15}, -2$ (24) $x = -1, 0$

(25) $n = -\dfrac{4}{7}, \dfrac{1}{7}$ (26) $x = 5, \dfrac{3}{2}$

5. (1) $x = -1 \pm \sqrt{2}$ (2) $x = \pm\sqrt{6} + 3$ (3) no solution (4) $x = -2 \pm 2\sqrt{2}$

(5) no solution (6) $x = -\dfrac{5}{2} \pm \dfrac{\sqrt{33}}{2}$ (7) $x = 3\sqrt{3}, \sqrt{3}$ (8) $x = \dfrac{-3\sqrt{5} \pm 7}{2}$

(9) $x = \dfrac{-\sqrt{2} \pm \sqrt{10}}{2}$ (10) $x = \dfrac{-4 \pm \sqrt{6}}{3}$ (11) $x = \dfrac{-1 \pm \sqrt{6}}{4}$ (12) $x = \dfrac{2 \pm \sqrt{7}}{5}$

(13) $q = 1 \pm \dfrac{\sqrt{6}}{3}$ (14) $t = \dfrac{1}{2}, -\dfrac{3}{2}$ (15) $y = -2, \dfrac{2}{3}$ (16) no solution

(17) $z = \dfrac{1 \pm \sqrt{5}}{2}$ (18) $n = 5, 2$ (19) $x = \dfrac{5 \pm \sqrt{37}}{6}$ (20) $x = \dfrac{-9 \pm \sqrt{17}}{4}$

6. (1) $x = \dfrac{-1 \pm \sqrt{5}}{2}$ (2) $x = \dfrac{-3 \pm 2\sqrt{3}}{3}$ (3) $x = \dfrac{5 \pm \sqrt{17}}{5}$ (4) $x = -4 \pm 4\sqrt{2}$

(5) $x = \dfrac{5 \pm \sqrt{31}}{2}$ (6) $x = 1 \pm \sqrt{10}$ (7) $x = \dfrac{5 \pm \sqrt{7}}{3}$ (8) $x = \dfrac{9 \pm \sqrt{21}}{2}$

(9) $x = \dfrac{3 \pm \sqrt{5}}{2}$ (10) $-\sqrt{3} \pm \sqrt{6}$ (11) $x = -\sqrt{2} \pm \sqrt{6}$ (12) $x = \dfrac{-1 \pm \sqrt{6}}{2}$

(13) no solution   (14) $x = \dfrac{1 \pm \sqrt{73}}{12}$ (15) $x = \dfrac{2\sqrt{6}}{3}, -\dfrac{\sqrt{6}}{2}$ (16) no solution

7. Format: Δ, nature of solutions

(1) 33, 2 different real solutions (2) -19, no real solution   (3) 0, 2 equal real solutions

(4) -15, no real solution   (5) 0, 2 equal real solutions   (6) 17, 2 different real solutions

(7) 16, 2 different real solutions (8) 5, 2 different real solutions (9) 12, 2 different real solutions

(10) -4, no real solution

Section 10.2

1. (1) B   (2) D   (3) B   (4) C   (5) B   (6) D   (7) A   (8) B

(9) C   (10) C

2. (1) $1; -2$ (2) $4; -\dfrac{\pi}{3}$ (3) D.N.E   (4) $2\sqrt{2}; -3 - \sqrt{2}$

(5) $-\sqrt{2} - 1; \sqrt{2}$ (6) $x^2 - 25x + 12 = 0$ (7) $< \dfrac{1}{4}, m \neq 0$ (8) 8

(9) -2 (10) 0 or 3

3. (1) $(x + 2\sqrt{3})(x - 2\sqrt{3})$ (2) $(x - 2 - \sqrt{10})(x - 2 + \sqrt{10})$

(3) $(x - 4 - 2\sqrt{5})(x - 4 + 2\sqrt{5})$ (4) $(a - 3\sqrt{3})(a + \sqrt{3})$

(5) $6(x - \dfrac{3 + \sqrt{3}}{6})(x - \dfrac{3 - \sqrt{3}}{6})$ (6) $(t - \dfrac{\sqrt{3} + 1}{2})(t - \dfrac{\sqrt{3} - 1}{2})$

(7) $3(x - \dfrac{1 + \sqrt{10}}{3})(x - \dfrac{1 - \sqrt{10}}{3})$ (8) not factorable

(9) $2(x - \dfrac{5\sqrt{2} - 10}{2})(x + \dfrac{5\sqrt{2} + 10}{2})$ (10) $5(x + \dfrac{3 + \sqrt{19}}{5})(x + \dfrac{3 - \sqrt{19}}{5})$

(11) $4(x - \dfrac{\sqrt{17} - 7}{8})(x + \dfrac{\sqrt{17} + 7}{8})$ (12) $(x - \sqrt{3})(x + \sqrt{3})(x^2 + 1)$

4. (1) $x = 4, -\dfrac{2}{3}$ (2) $x = k, \dfrac{3}{2}k$ (3) $x = \dfrac{-1 \pm 2\sqrt{2}}{4}$ (4) $x = 4y, 3y$

(5) $x = \pm(1 + \sqrt{2})$ (6) $x = 5, 0$ (7) $x = -3\sqrt{2}, \sqrt{2}$ (8) $x = 14, \dfrac{10}{7}$

(9) $x = 7, 20$ (10) $x = 3, 1$ (11) $x = \pm(3a - 2b)$ (12) no solution

(13) $a = a + b, a - b$ (14) $2 \pm \sqrt{6}$ (15) $x = \pm\sqrt{6}, \pm 2\sqrt{3}$ (16) $x = \pm\dfrac{2}{3}\sqrt{3}$

(17) no solution   (18) $x = 1$

5.   (1) $\dfrac{7 + \sqrt{21}}{2}, \dfrac{7 - \sqrt{21}}{2}$ (2) $\dfrac{-\sqrt{6} + \sqrt{2}}{4}, \dfrac{-\sqrt{6} - \sqrt{2}}{4}$ (3) D.N.E

   (4) $\dfrac{-\sqrt{15} + \sqrt{3}}{2}, \dfrac{-\sqrt{15} - \sqrt{3}}{2}$ or $\dfrac{\sqrt{15} + \sqrt{3}}{2}, \dfrac{\sqrt{15} - \sqrt{3}}{2}$

   (5) $36, -4$ or $4, -36$ (6) $-2, 3$ or $2, -3$ (7) $7, -3$

6.   (1) $4\sqrt{2} - 5$ (2) $-\dfrac{2}{3}\sqrt{2}$ (3) 44 (4) $2\sqrt{14}$

   (5) $-\dfrac{22}{3}$ (6) $200\sqrt{2}$

8.   (1) $m = \dfrac{13}{4}, x = \dfrac{3}{2}$ (2) $m < \dfrac{13}{4}$

10. $k = 5 - 2\sqrt{10}$

12. $m = -1$

13. -2

14.   (1) $m = -2$ or 1(2) $m = 0$ (3) $m = \dfrac{2}{3}$ (4) D.N.E

15. $p = -3; q = 2; x = 1, 2$ or $p = 3; q = 2; x = -1, -2$

17. $k = -1$ or $\dfrac{9}{4}$

18.   5cm

Section 10.3

1.   (1) C   (2) C   (3) B   (4) C   (5) B   (6) A   (7) A   (8) C

   (9) D   (10) C

2.   (1) substitution; $\sqrt{x - 2} \geq 0$ when existing; checking solutions

   (2) $x = 4$ (3) $x = 7$ (4) $\neq 0, 1$ (5) -2 (6) $1, 3$

3.   (1) $x = -1$ (2) $x = -1, -21$ (3) $y = 0, 1$ (4) $x = 3, -5$

   (5) $x = \dfrac{1 \pm \sqrt{5}}{2}$ (6) $x = -1 \pm \sqrt{13}$ (7) $x = 3$ (8) $x = 5, -1$

   (9) $x = \dfrac{1}{3}$ (10) $x = \dfrac{5 \pm \sqrt{13}}{2}$ (11) $x = 3$ (12) $x = 1$

   (13) $x \neq \pm 3$ (14) $x = \dfrac{-3 \pm \sqrt{17}}{2}$

4.   (1) $x = -1$ (2) $x = 6$ (3) $x = \dfrac{-3 + \sqrt{5}}{2}$ (4) $x = 5$

   (5) $x = 20$ (6) $x = 4$ (7) $x = 10, 16$ (8) $x = 1$

   (9) $x = 9$ (10) no solution   (11) $x = 7$ (12) $x = 11$

5. (1) $x = 1 \pm \sqrt{2}, \dfrac{3 \pm \sqrt{17}}{4}$ (2) $x = \pm\sqrt{2}$ (3) $x = 1, 2, -3, -4$

(4) $x = -1, -3$ (5) $x = -1$ (6) $x = \pm\sqrt{3}, \pm\dfrac{\sqrt{7}}{7}$

(7) $x = 9$ (8) $x = 1, -4$ (9) $x = \pm\sqrt{2}$

(10) $x = 26$ (11) $x = 3, -1$ (12) $x = -3, 2$

(13) $x = -3, -\dfrac{1}{3}, 2, \dfrac{1}{2}$ (14) $x = \dfrac{81}{16}$ (15) $x = -1 \pm \sqrt{2}$

(16) $x = 1 \pm \sqrt{2}, \dfrac{-3 \pm \sqrt{73}}{8}$ (17) $x = 2$ (18) $x = \pm\dfrac{5}{8}\sqrt{15}$

6. -2

7. (1) $x = \pm\dfrac{2}{3}\sqrt{6}a$ (2) $x = \pm\dfrac{4}{5}a$ (3) $x = \dfrac{1}{2}$ (4) $x = \dfrac{a-b}{2}, \dfrac{a-4b}{5}$

8. (1) $a = -6, -2, 0, \dfrac{1}{4}$ (2) $a = -1, 3, \dfrac{7}{2}$ (3) $m \neq -3, 5$

Section 10.4

1. (1) F   (2) T   (3) F   (4) F   (5) T   (6) T   (7) T   (8) F

(9) T   (10) T   (11) F   (12) F   (13) T   (14) T   (15) T   (16) T

(17) F   (18) T   (19) F   (20) T

2. (1) D   (2) A   (3) D   (4) C   (5) B   (6) C   (7) A   (8) D

3. $\left\{ 1.\bar{1}, \sqrt{\dfrac{1}{4}}, 0, -1.234, \dfrac{9}{7}, \dfrac{37}{41}, \dfrac{6-3i}{i-2} \right\}$

$\left\{ \sqrt{2}, -\sqrt[4]{3}, 1.\bar{1}, \pi, \sqrt{\dfrac{1}{4}}, 0, -1.234, \dfrac{9}{7}, \sqrt[3]{-9}, \dfrac{37}{41}, \sqrt{-\pi}i, \dfrac{6-3i}{i-2} \right\}$

$\left\{ 5i, \sqrt{-7}, \sqrt{7} + \sqrt[3]{4}i, 2 - \sqrt{2}i, \dfrac{9}{7} - 0.2i, -2\sqrt{-\pi}, \dfrac{121}{13i}, 5 + 3i, \sqrt[4]{5}i, \sqrt{\sqrt{2}+\sqrt{3}}i, \dfrac{3-2i}{-9i} \right\}$

$\left\{ 5i, \sqrt{-7}, -2\sqrt{-\pi}, \dfrac{121}{13i}, \sqrt[4]{5}i, \sqrt{\sqrt{2}+\sqrt{3}}i \right\}$

4. (1) $7 - 11i$ (2) -4 (3) $6 + 3i$ (4) $-2 + 5i$

(5) $2 - i$ (6) $-4 - i$ (7) $20 - 12i$ (8) $-12 - 28i$

(9) $\dfrac{11}{10} - \dfrac{7}{15}i$ (10) $\dfrac{61}{21} + \dfrac{5}{21}i$ (11) $-\dfrac{67}{35} - \dfrac{51}{22}i$ (12) $-\dfrac{11}{4} + \dfrac{13}{18}i$

5. (1) $8i$ (2) $5i$ (3) $2\sqrt{7}i$ (4) $\sqrt{11}i$

(5) $3\sqrt{6}i$ (6) $6\sqrt{2}i$ (7) $-14\sqrt{2}i$ (8) $-18\sqrt{3}i$

(9) $121i$ (10) $\dfrac{3\sqrt{3}}{4}i$ (11) $\dfrac{\sqrt{14}}{3}i$ (12) $\dfrac{\sqrt{70}}{6}i$

6. (1) $-\sqrt{14}$ (2) $-\sqrt{30}$ (3) $-6\sqrt{3}$ (4) $-10\sqrt{15}$

(5) -16 (6) $-2\sqrt{10}$ (7) $-9\sqrt{2}$ (8) $9i$

(9) $\dfrac{\sqrt{6}}{3}$ (10) $\dfrac{\sqrt{35}}{5}$ (11) $\dfrac{\sqrt{21}}{6}$ (12) $\dfrac{\sqrt{21}}{7}$

(13) $\dfrac{\sqrt{14}}{2}$ (14) $2\sqrt{6}i$ (15) $\dfrac{2\sqrt{15}}{5}$ (16) 2

7. (1) -15 (2) -22 (3) $30 + 18i$ (4) $-7 + 21i$

(5) 82 (6) $5 - 12i$ (7) $-69 + 67i$ (8) $-59 + 152i$

(9) $-73 + 131i$ (10) $-25i$ (11) $48 - 44i$ (12) $-5 - 12i$

8. (1) $\dfrac{-3 - 5i}{2}$ (2) $-4 + 3i$ (3) $-1 + i$ (4) $\dfrac{-12 - 8i}{13}$

(5) $\dfrac{4 - 3i}{5}$ (6) $\dfrac{37 + 77i}{82}$ (7) $-\dfrac{23 + 47i}{74}$ (8) $-\dfrac{36 - 48i}{125}$

(9) $\dfrac{15 - 8i}{17}$ (10) $-\dfrac{8 + 11i}{15}$ (11) $\dfrac{-58 + 6i}{85}$ (12) $\dfrac{1 + i}{2}$

9. (1) $\sqrt{5}$ (2) 5 (3) $2\sqrt{2}$ (4) $2\sqrt{5}$

10. (1) $\pm(1 - i)$ (2) $\pm\sqrt{\pi}i$ (3) $\pm(1 + \sqrt{3}i)$ (4) $\pm(\dfrac{\sqrt{6} + \sqrt{2}i}{2})$

Section 10.5

1. (1) C   (2) D   (3) A   (4) B   (5) D   (6) D

2. (1) $x = \pm\sqrt{3}i$ (2) $x = \pm\dfrac{\sqrt{14}}{7}i$ (3) $x = \dfrac{-1 \pm \sqrt{7}i}{2}$ (4) $x = -3, -6$

(5) $x = 1, \dfrac{3}{2}$ (6) $t = \dfrac{-2 \pm \sqrt{2}i}{3}$ (7) $y = -2\sqrt{2} \pm 2$ (8) $x = \dfrac{\sqrt{6} \pm \sqrt{2}i}{2}$

(9) $x = \dfrac{3 \pm \sqrt{7}i}{2}$ (10) $x = \dfrac{3 \pm \sqrt{15}i}{12}$ (11) $x = (1 \pm \sqrt{2})i$ (12) $x = -2i, -i$

(13) $z = \pm(\sqrt{2} + i)$ (14) $x = \pm(2 + \sqrt{5}i)$

3. (1) $3(x + \sqrt{2}i)(x - \sqrt{2}i)$ (2) $2(t + \dfrac{\sqrt{10}}{2}i)(t - \dfrac{\sqrt{10}}{2}i)$

(3) $(m + 1 + 2i)(m + 1 - 2i)$ (4) $(n + 2 + 4i)(n + 2 - 4i)$

(5) $3(x + 2 - \dfrac{\sqrt{3}}{3}i)(x + 2 + \dfrac{\sqrt{3}}{3}i)$ (6) $(3n - 1 + 5i)(3n - 1 - 5i)$

(7) $(z - 6 + 7i)(z - 6 - 7i)$ (8) $(x + 5 + 3\sqrt{11})(x + 5 - 3\sqrt{11})$

4. Format: Δ, nature of solutions

(1) -59, two conjugate imaginary solutions (2) -44, two conjugate imaginary solutions

(3) 0, two equal real solutions   (4) 193, two different real solutions

(5) -159, two conjugate imaginary solutions (6) 1, two different real solutions

(7) 29 two different real solutions   (8) 320, two different real solutions

5. Format: $x_1^2 + x_2^2; |x_1 - x_2|$

(1) $-2; 2\sqrt{2}$　　(2) $-\dfrac{2}{9}; \dfrac{2\sqrt{2}}{3}$　　(3) $\dfrac{49}{16}; \dfrac{\sqrt{73}}{4}$　　(4) $-\dfrac{7}{4}; \dfrac{\sqrt{23}}{2}$

6. $1+\sqrt{3}i,\ 1-\sqrt{3}i$

Section 10.6

1. (1) F　　(2) T　　(3) T　　(4) F　　(5) F

2. (1) D　　(2) C　　(3) A　　(4) B　　(5) A

3. (1) $x=-1$　　　(2) $x=\pm\sqrt{2}, -2\sqrt{2}$　　(3) $x=0,1,2$

(4) $x=-\dfrac{4}{3}, \pm\sqrt{3}, 0, 1, \pm2$　(5) $x=1\pm\dfrac{\sqrt{3}}{3}, 2, \dfrac{1}{3}$　　(6) $x=-2, 1, \dfrac{1}{4}$

(7) $x=2, \dfrac{1}{2}$　　　(8) $x=-\dfrac{1}{2}, 0, \dfrac{2}{3}, -2\pm\sqrt{2}$

4. (1) $(-2,3)$　(2) $[\dfrac{1}{3},2]$　(3) $\varnothing$　(4) $(-\infty,+\infty)$　(5) $(-\infty,-\dfrac{1}{3})\cup(-\dfrac{1}{3},+\infty)$

(6) $(-\infty,-\dfrac{2}{3}]\cup[\dfrac{1}{2},+\infty)$　　(7) $\{-1\}$　　(8) $[\dfrac{1-\sqrt{5}}{2}, \dfrac{1+\sqrt{5}}{2}]$

(9) $(-\infty,1-\dfrac{\sqrt{17}}{4})\cup(1+\dfrac{\sqrt{17}}{4},+\infty)$　　(10) $(-\infty,1-\dfrac{\sqrt{3}}{3})\cup(1+\dfrac{\sqrt{3}}{3},+\infty)$

(11) $(-3,-1)\cup(2,+\infty)$　　(12) $\{-1\}\cup[0,+\infty)$

(13) $(-\infty,-1)\cap(2,3)\cap(3,+\infty)$　　(14) $(-\infty,-2)\cup(-1,\dfrac{1}{2})\cup(\dfrac{1}{2},1)$

(15) $[-\sqrt{7},-\dfrac{1+\sqrt{5}}{2}]\cup\{0\}\cup[\dfrac{\sqrt{5}-1}{2},\sqrt{7}]$　(16) $(-\infty,-\dfrac{5}{3}]\cup[\dfrac{3}{2},+\infty)$

5. $m<-1$ or $m\geq2$

6. $k\geq\dfrac{4}{9}$ or $k\leq0$

Section 10.7

1. (1) B　　(2) B　　(3) D　　(4) A　　(5) C

2. (1) $[-1,0)$　　(2) $(-1,0)\cup(1,+\infty)$　　(3) $(-\infty,-2)\cup(0,3)$

(4) $(-\infty,-1]\cup(2,+\infty)$　　(5) $(-\infty,\dfrac{1}{2}]\cup(\dfrac{2}{3},1)\cup(1,+\infty)$

(6) $(-\infty,-1]\cup(0,3]$　　(7) $[1,2)\cup(3,+\infty)$

(8) $(-\infty,-3)\cup(-2,1)\cup(3,+\infty)$　　(9) $(-\infty,-4)\cup(-1,1)\cup(1,2)$

(10) $(-8,-2]\cup(0,1)\cup(1,+\infty)$　　(11) $(-\infty,-1)\cup[2,3]\cup(4,+\infty)$

(12) $[-2,1]\cup(2,+\infty)$　　(13) $(-\infty,0)\cup(2,+\infty)$

3. (1) $\begin{cases} (1 - \dfrac{1}{a}, 1) \text{ when } a > 0 \\ (-\infty, 1) \text{ when } a = 0 \\ (-\infty, 1) \cup (1 - \dfrac{1}{a}, +\infty) \text{ when } a < 0 \end{cases}$ (2) $\begin{cases} (-a, a) \cup (a, +\infty) \text{ when } a > 0 \\ (-a, +\infty) \text{ when } a \leq 0 \end{cases}$

4. $(-\infty, \dfrac{1}{n}) \cup (\dfrac{1}{m}, +\infty)$

5. $(-\infty, -1] \cup [3, +\infty)$

6. $a \in (\dfrac{1}{2}, +\infty)$

7. $m < 2$

8. $k \in (1, 3)$

Chapter 11

Section 11.1

1. (1) C   (2) D   (3) C   (4) A   (5) B   (6) C   (7) C   (8) D
    (9) D   (10) D

2. (1) 5   (2) 13   (3) $\sqrt{13}$ (4) $3\sqrt{6}$ (5) $\sqrt{10}$ (6) $\dfrac{3}{2}$

3. (1) 4   (2) $5\sqrt{2}$ (3) $\sqrt{43}$ (4) $\sqrt{33}$

4. (1) right triangle   (2) acute triangle   (3) obtuse triangle   (4) right triangle

5. (1) $x^2 + y^2 + 4x - 10y - 20 = 0$ (2) $x^2 + y^2 - 41 = 0$

    (3) $x^2 + y^2 + 6x + 4y - 12 = 0$ (4) $x^2 + y^2 + 12x + 6y - 5 = 0$

    (5) $x^2 + y^2 - 2x - 2y - 11 = 0$

6. Format: center, length of radius

    (1) $(4, -3), \sqrt{26}$ (2) $(-\dfrac{5}{2}, \dfrac{3}{2}), 1$ (3) $(5\sqrt{2}, -3\sqrt{3}), \sqrt{77}$

    (4) $(\dfrac{2}{3}, 0), \dfrac{\sqrt{10}}{3}$ (5) $(-1, -2), 0$ (a point)

7. (1) $\dfrac{\sqrt{10}}{2}$

8. (1) not intersecting

    (2) intersecting at $(-7, 3), (-6, 4)$

    (3) intersecting at $(\dfrac{3}{5}, -\dfrac{4}{5}), (-1, 4)$

9. (1) $x^2 + y^2 - 4x - 8y - 1 = 0$

    (2) $x^2 + y^2 + 4x + 8y - 1 = 0$

    (3) $x^2 + y^2 + 4x - 8y - 1 = 0$

10. (1) A–(2)   (2) B–(4)   (3) C–(1)   (4) D–(5)   (5) E–(3)

11. (1) $y = -\sqrt{1 - x^2}$ (2) $x = \sqrt{16 - y^2}$ (3) $x^2 + y^2 \le 4$ (4) $x^2 + y^2 > 1$

    (5) $\begin{cases} x^2 + y^2 - 2x + 2y + 1 \le 1 \\ y \ge -1 \end{cases}$ or $-1 \le y \le \sqrt{2x - x^2} - 1$

12. intersecting at $(0, 3)$ and $(-2, 1)$

Section 11.2

1. (1) D   (2) A   (3) C   (4) C   (5) B

2. Format: opening direction, vertex, axis of symmetry

    (1) upward,$(-3, 2), x = -3$ (2) downward,$(-1, 1), x = -1$

    (3) downward,$(0, -7), x = 0$ (4) upward,$(2, -7), x = 2$

(5) downward,$(-\frac{1}{6}, -\frac{11}{12})$, $x = \frac{1}{6}$ (6) upward,$(-\frac{1}{4}, \frac{19}{4})$, $x = -\frac{1}{4}$

(7) upward,$(\frac{5}{2}, -\frac{9}{4})$, $x = \frac{5}{2}$ (8) downward,$(-\frac{3}{4}, 1)$, $x = -\frac{3}{4}$

3. (1) left: 1 unit, upward: 2 units   (2) right: 12 units, upward: 5 units

     (3) left: 1 unit, upward: 1 unit   (4) right: 2 units, downward: 6 units

     (5) left: 2 units, upward: 13 units   (6) right: $\frac{7}{2}$ units, upward: $\frac{21}{4}$ units

     (7) flip about the x axis   (8) flip about the x-axis, right: 1 unit

4. (1) $y = 7x^2 - 6x - \sqrt{2}$ (2) $y = -3(x-4)^2 + 2$ (3) $y = \frac{1}{2}(x+2)^2 + 4$

     (4) $y = (x-5)(x-4)$ (5) $y = 4x(x-1)$ (6) $y = -2x^2 + 4x - 3$

     (7) $y = -3x^2 + x + 2$

5. Format: x-intercept; y-intercept

     (1) $-\frac{2}{3}, -1$; 2   (2) none; -7 (3) -2; -16 (4) $\frac{-3 \pm \sqrt{5}}{2}$; 1

7. (1) intersecting at $(0,1)$ and $(-4,-7)$, not tangent   (2) tangent at $(-3,-2)$

     (3) intersecting at $(\frac{5+\sqrt{21}}{2}, 13 + 2\sqrt{21})$ and $(\frac{5-\sqrt{21}}{2}, 13 - 2\sqrt{21})$, not tangent

     (4) tangent at $(1,6)$ (5) not intersecting

Section 11.3

1.

focal length	semi-major axis	semi-minor axis	center	foci	vertices	axes of symmetry
$2\sqrt{11}$	6	5	$(0,0)$	$(\pm\sqrt{11}, 0)$	$(\pm 6, 0), (0, \pm 5)$	$x = 0, y = 0$
$2\sqrt{7}$	4	3	$(0,0)$	$(0, \pm\sqrt{7})$	$(\pm 3, 0), (0, \pm 4)$	$x = 0, y = 0$
4	$\sqrt{10}$	$\sqrt{6}$	$(-\frac{1}{2}, 1)$	$(\frac{3}{2}, 1)$ $(-\frac{5}{2}, 1)$	$(-\frac{1}{2} \pm \sqrt{10}, 1)$ $(-\frac{1}{2}, 1 \pm \sqrt{6})$	$x = -\frac{1}{2}, y = 1$
$\frac{\sqrt{6}}{3}$	$\frac{\sqrt{2}}{2}$	$\frac{\sqrt{3}}{3}$	$(3, 0)$	$(3 \pm \frac{\sqrt{6}}{6}, 0)$	$(3 \pm \frac{\sqrt{2}}{2}, 0)$ $(3, \pm\frac{\sqrt{3}}{3})$	$x = 3, y = 0$

Section 11.4

1. Format: focal length $2c$; length of semi-real axis a; length of semi-imaginary axis b; center; foci; vertices; axes of symmetry; asymptotes

     (1) $2c = 2\sqrt{13}$; $a = 2$; $b = 3$; $(0,0)$; $(\pm\sqrt{13}, 0)$; $(\pm 2, 0)$; $x = 0, y = 0$; $y = \pm\frac{3}{2}x$

     (2) $2c = 8$; $a = 2$; $b = 2\sqrt{3}$; $(-1, 2)$; $(-1, -2), (-1, 6)$; $(-1, 4), (-1, 0)$; $x = -1, y = 2$; $y = \pm\frac{\sqrt{3}}{3}(x+1) + 2$

(3) $2c = 4\sqrt{2}$; $a = \sqrt{6}$; $b = \sqrt{2}$; $(-1, -1)$; $(-1 \pm 2\sqrt{2}, -1)$; $(-1 \pm \sqrt{6}, -1)$; $x = -1, y = -1$; $y = \pm\dfrac{\sqrt{3}}{3}(x + 1) - 1$

(4) $2c = 2\sqrt{6}$; $a = \sqrt{3}$; $b = \sqrt{3}$; $(2, -3)$; $(2, -3 \pm \sqrt{6})$; $(2, -3 \pm \sqrt{3})$; $x = 2, y = -3$; $y = x - 5, y = -x - 1$

Section 11.5

1.

focus	directrix	vertex	axes of symmetry
$(-\dfrac{3}{2}, 0)$	$x = \dfrac{3}{2}$	$(0, 0)$	$y = 0$
$(\dfrac{1}{6}, 0)$	$x = \dfrac{7}{6}$	$(\dfrac{2}{3}, 0)$	$y = 0$
$(\dfrac{13}{4}, -1)$	$x = -\dfrac{3}{4}$	$(2, -1)$	$y = -1$
$(\dfrac{1}{2}, \dfrac{23}{16})$	$y = \dfrac{41}{16}$	$(\dfrac{1}{2}, 2)$	$x = \dfrac{1}{2}$

2.   (1) $y^2 = 4x$ (2) $y^2 = 10(x - \dfrac{1}{2})$ (3) $(y - 3)^2 = -6(x + \dfrac{1}{2})$

(4) $(x + 5)^2 = -20(y + 2)$ (5) $x^2 - 2xy + y^2 - 6x - 6y + 3 = 0$

Section 11.6

(1)hyperbola, $e = \dfrac{2}{5}\sqrt{10}$, foci: $(0, \pm 4)$, directrixes: $y = \pm\dfrac{5}{2}$

(2) parabola, $e = 1$, focus: $(0, \dfrac{3}{2})$, directrix: $y = \dfrac{5}{2}$

(3) ellipse, $e = \dfrac{1}{2}$, foci: $(\pm\sqrt{2}, 0)$, directrix: $x = \pm 4\sqrt{2}$

(4) parabola, $e = 1$, focus: $(4, -3)$, directrix: $x = 6$

(5) circle

(6) hyperbola, $e = \sqrt{3}$, foci: $(\pm\dfrac{3}{2}\sqrt{2}, 2)$, directrixes: $x = \pm\dfrac{\sqrt{2}}{2}$

(7) point

(8) ellipse, $e = \dfrac{\sqrt{3}}{3}$, foci: $(-\dfrac{13}{4}, 1), (-\dfrac{5}{4}, 1)$, directrixes: $x = \dfrac{3}{4}, x = -\dfrac{21}{4}$

Chapter 12

Section 12.1

1. (1) F   (2) F   (3) F   (4) F   (5) T   (6) T   (7) F   (8) T
    (9) F   (10) F   (11) F   (12) T

2. (1) C   (2) C   (3) B   (4) A   (5) B   (6) D   (7) B   (8) B
    (9) C   (10) D   (11) D

3. (1) $(-\infty, -3] \cup [3, +\infty)$ (2) $[2, 3) \cup (3, +\infty)$ (3) $[-\frac{1}{2}, \frac{4}{3}]$

    (4) $[0, +\infty)$ (5) $(-\infty, -6) \cup (-6, -3] \cup [5, +\infty)$ (6) $[-4, 0) \cup (0, 1]$

    (7) $(-\infty, 1)$ (8) $\mathbb{R} - \{2, -\frac{1}{2}\}$ (9) $[-3, 3]$ (10) $[-\frac{5}{2}, \frac{5}{2}]$

    (11) $(-\infty, +\infty)$ (12) $[-\frac{1}{2}, 1)$ (13) $(-\infty, -3] \cup [2, +\infty)$

    (14) $(-\infty, +\infty)$

4. (1) neither   (2) even   (3) even   (4) odd   (5) even
    (6) neither   (7) even   (8) odd   (9) neither   (10) neither

5. (1) $\mathbb{R}$ (2) $[-13, 17)$ (3) $(-\infty, +3]$ (4) $[-\frac{41}{8}, 5]$ (5) $[\sqrt{2}, +\infty)$

    (6) $(-\infty, 3]$ (7) $[0, 2\sqrt{2}]$ (8) $[0, 16]$ (9) $(0, 1]$ (10) $\mathbb{R} - \{1\}$

6. $1, \sqrt[3]{4}$

7. $\dfrac{4}{3}$

8. $[-2, 2), (-2\sqrt{6}, 2\sqrt{6})$

10. (i) even; (ii) odd

12. $a \in [0, \dfrac{3}{4})$

Section 12.2

1. (1) F   (2) T   (3) F   (4) T   (5) F   (6) T   (7) F   (8) T
    (9) F   (10) F   (11) F   (12) T   (13) F   (14) F

2. (1) C   (2) C   (3) C   (4) D   (5) D   (6) C   (7) C   (8) C
    (9) D   (10) C   (11) C   (12) A   (13) A   (14) A   (15) C

4. (1) increasing   (2) increasing   (3) increasing   (4) decreasing
    (5) decreasing   (6) decreasing   (7) decreasing   (8) increasing

5. (1) decreasing over $\mathbb{R}$ (2) $\begin{cases} (-\infty, 0) \text{ increasing} \\ (0, +\infty) \text{ decreasing} \end{cases}$ (3) $\begin{cases} (-\infty, -5] \text{ increasing} \\ [5, +\infty) \text{ decreasing} \end{cases}$

$(4)\begin{cases}(-\infty, -1] \text{ decreasing} \\ [-1, 1] \text{ increasing} \\ [1, 3] \text{ decreasing} \\ [3, +\infty) \text{ increasing}\end{cases}$ $(5)\begin{cases}(-\infty, -2] \text{ decreasing} \\ [1, +\infty) \text{ increasing}\end{cases}$ $(6)\begin{cases}(-\infty, -3) \text{ increasing} \\ (-3, +\infty) \text{ increasing}\end{cases}$

6. (1) decreasing (2) increasing (3) decreasing (4) increasing (5) decreasing (6) decreasing

7. (1) decreasing (2) increasing

8. $a \leq -3$

9. $[-5, -2]$

Section 12.3

1. (1) D (2) D (3) C (4) B (5) A (6) A (7) C (8) B

2. (1) $(f - g)(x) = -x^2 + 4x - 3$, $f \cdot g(x) = 3x^3 - 5x^2 + 5x - 2$, $f/g(x) = \dfrac{3x - 2}{x^2 - x - 1}$, $f \circ g(x) = 3x^2 - 3x + 1$, $g \circ f(x) = 9x^2 - 15x + 7$

(2) $(f - g)(x) = \sqrt{x + 1} - \dfrac{1}{x}$, $f \cdot g(x) = \dfrac{\sqrt{x + 1}}{x}$, $f/g(x) = x\sqrt{x + 1}$, $f \circ g(x) = \sqrt{\dfrac{x + 1}{x}}$, $g \circ f(x) = \dfrac{1}{\sqrt{x + 1}}$

(3) $(f - g)(x) = \dfrac{2x^2 - 5x - 5}{x - 3}$, $f \cdot g(x) = \dfrac{4x + 2}{x - 3}$, $f/g(x) = \dfrac{2}{2x^2 - 5x - 3}$, $f \circ g(x) = \dfrac{1}{x + 2}$, $g \circ f(x) = \dfrac{7 - x}{x - 3}$

(4) $(f - g)(x) = \dfrac{x^3 + x^2 - 1}{x^2}$, $f \cdot g(x) = \dfrac{x + 1}{x^2}$, $f/g(x) = x^3 + x$, $f \circ g(x) = \dfrac{x^2 + 1}{x^2}$, $g \circ f(x) = \dfrac{1}{(x + 1)^2}$

(5) $(f - g)(x) = x^2 - 3x + 1 - \dfrac{1}{\sqrt{1 + x}}$, $f \cdot g(x) = \dfrac{x^2 - 3x + 1}{\sqrt{1 + x}}$, $f/g(x) = (x^2 - 3x + 1)\sqrt{x + 1}$, $f \circ g(x) = \dfrac{x + 2 - 3\sqrt{1 + x}}{1 + x}$, $g \circ f(x) = \dfrac{1}{\sqrt{x^2 - 3x + 2}}$

3.

f	g	$f \pm g$	$f \cdot g$	f/g	$f \circ g$
odd	odd	odd	even	even	odd
odd	even	—	odd	odd	even
even	odd	—	odd	odd	even
even	even	even	even	even	even

4.

f	g	$f + g$	$f - g$	fg $(f, g > 0)$	fg $(f, g < 0)$	f/g $(f, g > 0)$	f/g $(f, g < 0)$	$f \circ g$
↗	↗	↗	—	↗	↘	—	—	↗
↗	↘	—	↗	—	—	↗	↘	↘
↘	↗	—	↘	—	—	↘	↗	↘
↘	↘	↘	—	↘	↗	—	—	↗

5. decreasing over $(-1, 1)$ when $a > 0$; increasing over $(-1, 1)$ when $a < 0$

6. $[0, \dfrac{1}{5}]$

Section 12.4

1. (1) T   (2) F   (3) F   (4) T   (5) F   (6) F   (7) T   (8) F
(9) F   (10) T

2. (1) C   (2) A   (3) D   (4) A   (5) A   (6) D   (7) C   (8) B
(9) A   (10) A   (11) C

3. (1) $(-1.4)^{\frac{2}{3}} < 2.5^{\frac{2}{3}} < (-3)^{\frac{2}{3}}$ (2) $0.16^{-\frac{3}{4}} > 0.5^{-\frac{3}{2}} > 6.25^{\frac{3}{8}}$

(3) $4^{-\frac{1}{2}} < (\frac{1}{4})^{\frac{2}{5}} < \sqrt[3]{\frac{1}{4}} < 4^{-\frac{1}{4}}$ (4) $\left(\dfrac{2}{5}\right)^{\frac{1}{2}} < 3^{-\frac{1}{3}} < \left(\dfrac{3}{2}\right)^{-\frac{2}{3}} < \left(\dfrac{3}{5}\right)^{\frac{1}{3}} < \left(\dfrac{2}{3}\right)^{\frac{1}{3}}$

5. $(-\infty, -1) \cup (0, 1)$

6. -8

7. $(-\infty, 1) \cup (1, 2]$

8. 11

9. increasing over $(-\infty, -2)$, decreasing over $(-2, +\infty)$

Section 12.5

1. (1) T   (2) T   (3) F   (4) F   (5) T   (6) F   (7) T   (8) F
(9) T   (10) T

2. (1) D   (2) C   (3) D   (4) D   (5) B   (6) A   (7) B   (8) B
(9) C   (10) A

3. (1) $f^{-1}(x) = \dfrac{x+5}{3x-2}$ $(x \neq \dfrac{2}{3})$ (2) $f^{-1}(x) = \dfrac{3 - \sqrt{25 + 4x}}{2}$ $(x \geq -6)$

(3) $f^{-1}(x) = \sqrt{\dfrac{x+1}{x}}$ $(x \geq -1)$ (4) $f^{-1}(x) = x^2 + 2x + 3$ $(x \geq -1, x \neq 0)$

(5) $f^{-1}(x) = \sqrt[3]{\dfrac{1 - 4x}{x}}$ $(x \neq 0)$ (6) not invertible

(7) $f^{-1}(x) = \begin{cases} x^2 & x \in [-1, 0) \\ x^2 - 1 & x \in [0, 1) \end{cases}$

4. (1) $f^{-1}(x) = x^2 + 2x + 2$ $(x \geq -1)$ (2) $f^{-1}(x) = \dfrac{\sqrt{-3x - 6}}{3}$ $(x \leq -2)$

6. $-5 - 3\sqrt{2}$

7. (2) domain: $(f(9), f(-1))$; range: $(-1, 9)$

8. (1) $ad - bc \neq 0$ and $a + d = 0$ (2) $a = -3$

Section 12.6

1. (1) $x = -\dfrac{1}{2}$ (2) $x = -5, x = 3$ (3) $x = -\dfrac{1}{2}$

     (4) $x = \dfrac{7}{2}$ (5) $x = -2, x = \dfrac{1}{7}$ (6) $x = -\dfrac{2}{3}$

2. (1) $y = 0$ (2) $y = -\dfrac{\sqrt{2}}{3}$ (3) none

     (4) $y = 0$ (5) none   (6) $y = -\dfrac{6}{5}$

3. (1) $y = x - 2$ (2) $y = -\dfrac{1}{2}x$ (3) $y = -3x$

     (4) $y = \dfrac{1}{3}x - \dfrac{2}{3}$ (5) $y = x$ (6) none

4. (1) $y = \dfrac{x+3}{2}, x = 1$ (2) $y = 0, x = -1, x = \pm\sqrt{5}, x = \dfrac{2}{3}$

Chapter 13

Section 13.1

1.   (1) F   (2) T   (3) F   (4) T   (5) F   (6) F   (7) T   (8) T
    (9) F   (10) T

2.   (1) B   (2) C   (3) C   (4) B   (5) D   (6) C   (7) D   (8) B
    (9) A   (10) C   (11) C   (12) D   (13) A   (14) B   (15) D

3.   (1) $x = -2$ (2) $x = 1$ (3) $x = 10$ (4) $x = 1$ (5) $x = 0, 1$ (6) $x = 1$

4.   (1) domain: $\mathbb{R} - \{2\}$; range: $(0, 1) \cup (1, +\infty)$ (2) domain: $\mathbb{R}$; range: $(1, +\infty)$
    (3) domain: $(-\infty, 2]$; range: $[0, 3)$ (4) domain: $\mathbb{R} - \{0\}$; range: $(-\infty, -2) \cup (0, +\infty)$

5.   (1) $0.6^{-\frac{4}{5}} > (\frac{5}{3})^{\frac{1}{2}}$ (2) $4.5^{4.1} > 3.7^{3.6}$

    (3) $\sqrt[3]{2} < \sqrt[9]{16} < \sqrt[8]{8} < \sqrt[5]{4} < \sqrt{2}$ (4) $\begin{cases} \sqrt[n-1]{a^n} > \sqrt[n]{a^{n+1}} \text{ if } a > 1 \\ \sqrt[n-1]{a^n} < \sqrt[n]{a^{n+1}} \text{ if } 0 < a < 1 \end{cases}$

6.   (1) $a \in (1, +\infty)$ (2) $a \in (0, 1)$ (3) $a \in (1, +\infty)$ (4) $\dfrac{2}{5}$

    (5) $a = \dfrac{1}{2}$ (6) $a = 1$

7.   (1) $\dfrac{4}{3}$ (2) -1

8.   (1) $-\dfrac{b^4}{a - b}\sqrt{a^2 - b^2}$ (2) a

10. $\text{dom}(f) = (-\infty, +\infty)$, $\text{ran}(f) = (0, \sqrt[4]{\dfrac{4}{3}})$;

    increasing over $(-\infty, \dfrac{5}{2}]$, decreasing over $[\dfrac{5}{2}, +\infty)$

11.   (1) odd   (2) $(-1, 1)$

12.   (1) even

Section 13.2

1.   (1) T   (2) F   (3) T   (4) F   (5) T   (6) T   (7) F   (8) F
    (9) F   (10) T

2.   (1) B   (2) D   (3) B   (4) B   (5) C

3.   (1) 9   (2) $\sqrt{3}$ (3) $\dfrac{2}{a}$ (4) $-4; -2$ (5) 10 or $\dfrac{\sqrt{10}}{10}$

4.   (1) $\log_5 125 = 3$ (2) $\log_2 \dfrac{1}{64} = -6$ (3) $\ln 1 = 0$ (4) $\log_{\frac{3}{2}} \dfrac{27}{8} = 3$

    (5) $\log_{0.1} 0.01 = -2$ (6) $\log_{\frac{1}{8}} \dfrac{1}{512} = 3$

5. (1) -3 (2) -2 (3) $\dfrac{13}{12}$ (4) $\dfrac{2}{3}$ (5) 10

(6) $\dfrac{9}{8}$ (7) 2 (8) $\dfrac{8}{3}$ (9) 51 (10) 1

(11) 4 (12) 1 (13) 0 (14) 4 (15) $-\dfrac{9}{8}$

(16) $\dfrac{3}{2}$

6. (1) $1+3\log_2 3$ (2) $\dfrac{3}{2}$ (3) 2 (4) $2-\log_5 2$ (5) $2-\log 99$

(6) $2\ln 2+3\ln x+2\ln(2x-1)$ (7) $\dfrac{1}{2}(\log_3 x+\log_3 y-\log_3 z)$

(8) $\dfrac{1}{2}[\log_3(3x+5)+\log_3(x-3)]$ (9) $5\ln(x^2-1)-5\ln x^2$

(10) $\log_2(3x-1)$ (11) 0

7. (1) $x=\dfrac{2+\log_2 5}{6}$ (2) $x=\dfrac{\sqrt[4]{3}}{3}$ (3) $x=2\sqrt[7]{2}$ (4) $x=\dfrac{1}{3}-\dfrac{1}{2}\log_3 2$

(5) -2 (6) $x=1$ (7) $x=\sqrt{2}$ (8) $x=2$

(9) $x=1+\sqrt{2}$ (10) no solution (11) $x=-1$ (12) $x=5,7$

(13) $x=\sqrt{3}$ (14) $x=100$ (15) $x=125$ (16) $x=2,16$

(17) $x=2$ (18) $x=4,5$ (19) $x=1,\log_2 7-1$ (20) $\ln 5-\ln 2$

8. $\dfrac{16}{9}$

9. $\log 2$

10. $\dfrac{3a}{2(b+1)}$

12. $a=10,b=0$

Section 13.3

1. (1) T (2) F (3) F (4) F (5) T (6) F (7) T (8) F
(9) F (10) T

2. (1) D (2) B (3) C (4) A (5) B (6) C (7) D (8) B
(9) C (10) A (11) C (12) C (13) A (14) B (15) C (16) D
(17) D

3. (1) $y=\log_5(x-1)$ (2) $\dfrac{4(3-a)}{a+3}$ (3) $-x^2-\ln(1-x)$

(4) 9 (5) 1 (6) odd

4. (1) $(\dfrac{2}{3},1]$ (2) $\begin{cases}(-a,0) \text{ if } a>1 \\ (0,+\infty) \text{ if } 0<a<1\end{cases}$

(3) $[2, \frac{8}{3}]$

(4) $[-\sqrt{e+1}, -\sqrt{2}] \cup [\sqrt{2}, \sqrt{e+1}]$

5. $f^{-1}(x) = \ln \frac{x}{1-x}$, $\text{dom}(f^{-1}) = (0, 1)$, $\text{ran}(f^{-1}) = (-\infty, +\infty)$

6. (1) $x = -1, 6$ (2) $x = \frac{1}{9}$ (3) $x = \frac{29}{8}$ (4) $x = e^4, \frac{1}{e}$ (5) $x = 0, 1$

    (6) $y = 2$ (7) $x = -9.9$ (8) $x = 3$ (9) $x = -1$ (10) $x = 2$

    (11) $x = 90$ (12) $x = 7, -1$ (13) $x = 10, \frac{1}{100}$ (14) no solution

8. (1) $\log_a \frac{a}{b} < \log_b \frac{b}{a} < \log_b a < \log_a b$ (2) $|\log_a(1-x)| > |\log_a(1+x)|$

    (3) $\log_2(\log_2 x) < (\log_2 x)^2 < \log_2 x^2$ (4) $1 < a < b$ or $b < 1 < a$ or $a < b < 1$

9. odd

10. (1) $\text{dom}(f) = (0, 1)$, $\text{ran}(f) = (-\infty, +\infty)$, f is increasing over $(0, 1)$.

    (2) $\text{dom}(f) = (0, +\infty)$, $\text{ran}(f) = (-\infty, \frac{1}{4}]$,

    f is decreasing over $(0, 2\sqrt{2}]$ and increasing over $[2\sqrt{2}, +\infty)$

11. $a = \frac{1}{4}, x = \frac{1}{8}$

12. $a \in (0, 1)$

13. (1) $x \in (0, +\infty)$ (3) $a = \frac{3}{2}, b = \frac{1}{2}$

14. $[0, \frac{4}{5}]$

15. (1) $\log_a \frac{m^2 + 4m}{(m+2)^2}$ (2) decreasing   (3) $\log_a \frac{5}{9}$

www.ingramcontent.com/pod-product-compliance
Lightning Source LLC
Chambersburg PA
CBHW080659220326
41598CB00033B/5263